Effects of Strength and Stiffness Degradation on Seismic Response

FEMA P440A

June 2009

Effects of Strength and Stiffness Degradation on Seismic Response

Prepared by

APPLIED TECHNOLOGY COUNCIL
201 Redwood Shores Parkway, Suite 240
Redwood City, California 94065
www.ATCouncil.org

Prepared for

FEDERAL EMERGENCY MANAGEMENT AGENCY
Department of Homeland Security (DHS)

Michael Mahoney, Project Officer
Robert D. Hanson, Technical Monitor
Washington, D.C.

ATC MANAGEMENT AND OVERSIGHT
Christopher Rojahn (Project Executive)
Jon A. Heintz (Project Quality Control Monitor)
William T. Holmes (Project Tech. Monitor)

PROJECT MANAGEMENT COMMITTEE
Craig Comartin (Project Technical Director)
Eduardo Miranda
Michael Valley

PROJECT REVIEW PANEL
Kenneth Elwood
Subhash Goel
Farzad Naeim

CONSULTANT
Dimitrios Vamvatsikos

Notice

Foreword

One of the primary goals of the Federal Emergency Management Agency (FEMA) and the National Earthquake Hazards Reduction Program (NEHRP) is to encourage design and construction practices that address the earthquake hazard and minimize the potential damage resulting from that hazard. This document, *Effects of Strength and Stiffness on Degradation on Seismic Response* (FEMA P440A), is a follow-on publication to *Improvement of Nonlinear Static Seismic Analysis Procedures* (FEMA 440). It builds on another FEMA publication addressing the seismic retrofit of existing buildings, the *Prestandard and Commentary for Seismic Rehabilitation of Buildings* (FEMA 356) and the subsequent publication, ASCE/SEI Standard 41-06 *Seismic Rehabilitation of Existing Buildings* (ASCE 41).

The goal of FEMA 440 was improvement of nonlinear static analysis procedures, as depicted in FEMA 356 and ASCE 41, and development of guidance on when and how such procedures should be used. It was a resource guide for capturing the current state of the art in improved understanding of nonlinear static procedures, and for generating future improvements to those products. One of the recommendations to come out of that work was to fund additional studies of cyclic and in-cycle strength and stiffness degradation, and their impact on response and response stability.

This publication provides information that will improve nonlinear analysis for cyclic response, considering cyclic and in-cycle degradation of strength and stiffness. Recent work has demonstrated that it is important to be able to differentiate between cyclic and in-cycle degradation in order to more accurately model degrading behavior, while current practice only recognizes cyclic degradation, or does not distinguish between the two. The material contained within this publication is expected to improve nonlinear modeling of structural systems, and ultimately make the seismic retrofit of existing hazardous buildings more cost-effective.

This publication reaffirms FEMA's ongoing efforts to improve the seismic safety of new and existing buildings nationwide. This project is an excellent example of the interagency cooperation that is made possible through the NEHRP. FEMA is proud to have sponsored the development of this resource document through the Applied Technology Council (ATC), and is grateful for work done by the Project Technical Director, Craig Comartin, the Project Management Committee, the Project Review Panel, the Project Working Group, and all other contributors who made this publication possible. All those who participated are listed at the end of this document, and FEMA appreciates their involvement.

Federal Emergency Management Agency

In September 2004 the Applied Technology Council (ATC) was awarded a "Seismic and Multi-Hazard Technical Guidance Development and Support" contract (HSFEHQ-04-D-0641) by the Federal Emergency Management Agency (FEMA) to conduct a variety of tasks, including one entitled "Advanced Seismic Analysis Methods – Resolution of Issues" (ATC-62 Project). The purpose of this project was to resolve a series of difficult technical issues that were identified during the preparation of the FEMA 440 report, *Improvement of Nonlinear Static Seismic Analysis Procedures* (FEMA, 2005).

FEMA 440 was funded by FEMA to develop improvements to nonlinear static analysis procedures contained in the FEMA 356 *Prestandard and Commentary for the Seismic Rehabilitation of Buildings* (FEMA, 2000), and the ATC-40 Report, *Seismic Evaluation and Retrofit of Concrete Buildings* (ATC, 1996). Unresolved technical issues identified in FEMA 440 included the need for additional guidance and direction on: (1) component and global modeling to consider nonlinear degrading response; (2) soil and foundation-structure interaction modeling; and (3) simplified nonlinear multiple-degree-of-freedom modeling.

Of these issues, this project has investigated nonlinear degrading response and conducted limited initial studies on multiple-degree-of-freedom effects. Work has included an extensive literature search and review of past studies on nonlinear strength and stiffness degradation, and review of available hysteretic models for capturing degrading strength and stiffness behavior. To supplement the existing body of knowledge, focused analytical studies were performed to explore the effects of nonlinear degradation on structural response. This report presents the findings and recommendations resulting from these efforts.

ATC is indebted to the members of the ATC-62 Project Team who participated in the preparation of this report. Direction of technical activities, review, and development of detailed recommendations were performed by the Project Management Committee, consisting of Craig Comartin (Project Technical Director), Eduardo Miranda, and Michael Valley. Literature reviews and focused analytical studies were conducted by Dimitrios Vamvatsikos. Technical review and comment at critical developmental stages were provided by the Project Review Panel, consisting of Kenneth Elwood, Subhash Goel, and Farzad Naeim. A workshop of invited experts was convened to obtain feedback on preliminary findings and recommendations, and input from this group was instrumental in shaping the final product. The names and affiliations individuals who contributed to this work are included in the list of Project Participants provided at the end of this report.

ATC also gratefully acknowledges Michael Mahoney (FEMA Project Officer), Robert Hanson (FEMA Technical Monitor), and William Holmes (ATC Project Technical Monitor) for their input and guidance in the preparation of this report, Peter N. Mork for ATC report production services, and David Hutchinson as ATC Board Contact.

Jon A. Heintz   Christopher Rojahn
ATC Director of Projects   ATC Executive Director

Executive Summary

Much of the nation's work regarding performance-based seismic design has been funded by the Federal Emergency Management Agency (FEMA), under its role in the National Earthquake Hazards Reduction Program (NEHRP). Prevailing practice for performance-based seismic design is based on FEMA 273, *NEHRP Guidelines for the Seismic Rehabilitation of Buildings* (FEMA, 1997) and its successor documents, FEMA 356, *Prestandard and Commentary for the Seismic Rehabilitation of Buildings* (FEMA, 2000), and ASCE/SEI Standard 41-06, *Seismic Rehabilitation of Existing Buildings* (ASCE, 2006b). This series of documents has been under development for over twenty years, and has been increasingly absorbed into engineering practice over that period.

The FEMA 440 report, *Improvement of Nonlinear Static Seismic Analysis Procedures* (FEMA, 2005), was commissioned to evaluate and develop improvements to nonlinear static analysis procedures used in prevailing practice. Recommendations contained within FEMA 440 resulted in immediate improvement in nonlinear static analysis procedures, and were incorporated in the development of ASCE/SEI 41-06. However, several difficult technical issues remained unresolved.

1.   Project Objectives

The Applied Technology Council (ATC) was commissioned by FEMA under the ATC-62 Project to further investigate the issue of component and global response to degradation of strength and stiffness. Using FEMA 440 as a starting point, the objectives of the project were to advance the understanding of degradation and dynamic instability by:

- Investigating and documenting currently available empirical and theoretical knowledge on nonlinear cyclic and in-cycle strength and stiffness degradation, and their affects on the stability of structural systems

- Supplementing and refining the existing knowledge base with focused analytical studies

- Developing practical suggestions, where possible, to account for nonlinear degrading response in the context of current seismic analysis procedures.

This report presents the findings and conclusions resulting from the literature search and focused analytical studies, and provides recommendations that can be used to improve both nonlinear static and nonlinear response history analysis modeling of strength and stiffness degradation for use in performance-based seismic design.

2. Literature Review

Past research has shown that in-cycle strength and stiffness degradation are real phenomena, and recent investigations confirm that the effects of in-cycle strength and stiffness degradation are critical in determining the possibility of lateral dynamic instability.

The body of knowledge is dominated by studies conducted within the last 20 years; however, relevant data on this topic extends as far back as the 1940s. A summary of background information taken from the literature is provided in Chapter 2. A comprehensive collection technical references on this subject is provided in Appendix A.

3. Focused Analytical Studies

To supplement the existing body of knowledge, focused analytical studies were performed using a set of eight nonlinear springs representing different types of inelastic hysteretic behavior. These basic spring types were used to develop 160 single-spring systems and 600 multi-spring systems with differing characteristics. Each system was subjected to incremental dynamic analysis with 56 ground motion records scaled to multiple levels of increasing intensity. The result is an extensive collection of data on nonlinear degrading response from over 2.6 million nonlinear response history analyses on single- and multi-spring systems.

Development of single- and multi-spring models is described in Chapter 3, analytical results are summarized in Chapter 4, and sets of analytical data are provided in the appendices. A Microsoft Excel visualization tool that was developed to view all available data from multi-spring studies is included on the CD accompanying this report.

4. Comparison with FEMA 440 Limitations on Strength for Lateral Dynamic Instability

In FEMA 440, a minimum strength requirement (R_{max}) was developed as an approximate measure of the need to further investigate the potential for lateral dynamic instability caused by in-cycle strength degradation and P-delta effects. To further investigate correlation between R_{max} and lateral dynamic instability, the results of this equation were compared to quantile incremental dynamic analysis (IDA) curves for selected multi-spring systems included in this investigation. Results indicate that values predicted by the FEMA 440 equation for R_{max} are variable, but generally plot between the median and 84th percentile results for lateral dynamic instability of the systems investigated. Observed trends indicate that an improved equation, in a form similar to R_{max}, could be developed as a more accurate (less variable) predictor of lateral dynamic instability for use in current nonlinear static analysis procedures.

5. Findings, Conclusions, and Recommendations

Findings, conclusions, and recommendations resulting from the literature review and focused analytical studies of this investigation are collected and summarized in Chapter 5, grouped into the following categories:

- Findings related to improved understanding of nonlinear degrading response and judgment in implementation of nonlinear analysis results in engineering practice.

- Recommended improvements to current nonlinear analysis procedures

- Suggestions for further study

6. Findings Related to Improved Understanding and Judgment

Results from focused analytical studies were used to identify predominant characteristics of median incremental dynamic analysis (IDA) curves and determine the effects of different degrading behaviors on the dynamic stability of structural systems. Observed practical ramifications from these studies are summarized below:

- Behavior of real structures can include loss of vertical-load-carrying capacity at lateral displacements that are significantly smaller than those associated with sidesway collapse. Use of the findings of this investigation with regard to lateral dynamic instability (sidesway collapse) in engineering practice should include consideration of possible vertical collapse modes that could be present in the structure under consideration.

- Historically, the term "backbone curve" has referred to many different things. For this reason, two new terms have been introduced to distinguish between different aspects of hysteretic behavior. These are the *force-displacement capacity boundary*, and *cyclic envelope*.

- Nonlinear component parameters should be based on a force-displacement capacity boundary, rather than a cyclic envelope. Determining the force-displacement capacity boundary from test results using a single cyclic loading protocol can result in overly conservative predictions of maximum displacement.

- Observed relationships between selected features of the force-displacement capacity boundary and the resulting characteristics of median IDA curves support the conclusion that the nonlinear dynamic response of a system can be correlated to the parameters of the force-displacement capacity boundary of that system. Of particular interest is the relationship between global deformation demand and the intensity of the ground motion at lateral dynamic instability (collapse). Results indicate that it is possible to use nonlinear static procedures to estimate the potential for lateral dynamic instability of systems exhibiting in-cycle degradation.

- It is important to consider the dependence on period of vibration in conjunction with the effects of other parameters identified in this investigation. The generalized effect of any one single parameter can be misleading.

- It is important to recognize the level of uncertainty that is inherent in nonlinear analysis, particularly regarding variability in response due to ground motion uncertainty.

- In most cases the effects of in-cycle strength degradation dominate the nonlinear dynamic behavior of a system. This suggests that in many cases the effects of cyclic degradation can be neglected.

- Two situations in which the effects of cyclic degradation were observed to be important include: (1) short period systems; and (2) systems with very strong in-cycle strength degradation effects (very steep negative slopes and very large drops in lateral strength).

7. Improved Equation for Evaluating Lateral Dynamic Instability

An improved estimate for the strength ratio at which lateral dynamic instability might occur (R_{di}) was developed based on nonlinear regression of the extensive volume of data generated during this investigation. In performing this regression, results were calibrated to the median response of the SDOF spring systems studied in this investigation. Since the proposed equation for R_{di} has been calibrated to median response, use of this equation could eliminate some of the conservatism inherent in the current R_{max} limitation on use of nonlinear static procedures. Calibrated using the extensive volume of data generated during this investigation, use of this equation could improve the reliability of current nonlinear static procedures with regard to cyclic and in-cycle degradation.

Median response, however, implies a fifty percent chance of being above or below the specified value. Use of R_{di} in engineering practice should consider whether or not a median predictor represents an appropriate level of safety against the potential for lateral dynamic instability. If needed, a reduction factor could be applied to the proposed equation for R_{di} to achieve a higher level of safety on the prediction of lateral dynamic instability.

8. Simplified Nonlinear Dynamic Analysis Procedure

Focused analytical studies comparing force-displacement capacity boundaries to incremental dynamic analysis results led to the concept of a simplified nonlinear dynamic analysis procedure. In this procedure, a nonlinear static analysis is used to generate an idealized force-deformation curve (i.e., static pushover curve), which is then used as a force-displacement capacity boundary to constrain the hysteretic behavior of an equivalent SDOF oscillator. This SDOF oscillator is then subjected to incremental dynamic analysis, or approximate IDA results are obtained using the open source software tool, *Static Pushover 2 Incremental Dynamic Analysis*, SPO2IDA (Vamvatsikos and Cornell 2006). A Microsoft Excel version of the SPO2IDA application is included on the CD accompanying this report.

The procedure is simplified because only a SDOF oscillator is subjected to nonlinear dynamic analysis. Further simplification is achieved through the use of SPO2IDA, which avoids the computational effort associated with incremental dynamic analysis. This simplified procedure is shown to have several advantages over nonlinear static analysis procedures. Use of the procedure is explained in more detail in the example application contained in Appendix F.

9. Application of Results to Multiple-Degree-of-Freedom Systems

Multi-story buildings are more complex dynamic systems whose seismic response is more difficult to estimate than that of SDOF systems. Recent studies have suggested that it may be possible to estimate the collapse capacity of multiple-degree-of-freedom (MDOF) systems through dynamic analysis of equivalent SDOF systems. As part of the focused analytical work, preliminary studies of MDOF systems were performed. Results indicate that many of the findings for SDOF systems in this investigation (e.g., the relationship between force-displacement capacity boundary and IDA curves; the equation for R_{di}) may be applicable to MDOF systems.

Results of MDOF investigations are summarized in Appendix G. More detailed study of the application of these results to MDOF systems is recommended, and additional investigations are planned under a project funded by the National Institute of Standards and Technology (NIST).

10. Concluding Remarks

Using FEMA 440 as a starting point, this investigation has advanced the understanding of degradation and dynamic instability by:

- Investigating and documenting currently available empirical and theoretical knowledge on nonlinear cyclic and in-cycle strength and stiffness degradation, and their affects on the stability of structural systems

- Supplementing and refining the existing knowledge base with focused analytical studies

Results from this investigation have confirmed conclusions regarding degradation and dynamic instability presented in FEMA 440, provided updated information on modeling to differentiate between cyclic and in-cycle strength and stiffness degradation, and linked nonlinear dynamic response to major characteristics of component and system degrading behavior. This information will ultimately improve nonlinear modeling of structural components, improve the characterization of lateral dynamic instability, and reduce conservatism in current analysis procedures making it more cost-effective to strengthen existing buildings for improved seismic resistance in the future.

Table of Contents

List of Figures

List of Tables

Chapter 1

Introduction

Much of the nation's work regarding performance-based seismic design has been funded by the Federal Emergency Management Agency (FEMA), under its role in the National Earthquake Hazards Reduction Program (NEHRP). Prevailing practice for performance-based seismic design is based on FEMA 273, *NEHRP Guidelines for the Seismic Rehabilitation of Buildings* (FEMA, 1997) and its successor documents, FEMA 356, *Prestandard and Commentary for the Seismic Rehabilitation of Buildings* (FEMA, 2000), and ASCE/SEI Standard 41-06, *Seismic Rehabilitation of Existing Buildings* (ASCE, 2006b). This series of documents has been under development for over twenty years, and has been increasingly absorbed into engineering practice over that period.

The FEMA 440 report, *Improvement of Nonlinear Static Seismic Analysis Procedures* (FEMA, 2005), was commissioned to evaluate and develop improvements to nonlinear static analysis procedures used in prevailing practice. In FEMA 440, deviation between nonlinear static and nonlinear response history analyses was attributed to a number of factors including: (1) inaccuracies in the "equal displacement approximation" in the short period range; (2) dynamic P-delta effects and instability; (3) static load vector assumptions; (4) strength and stiffness degradation; (5) multi-degree of freedom effects; and (6) soil-structure interaction effects.

FEMA 440 identified and defined two types of degradation in inelastic single-degree-of-freedom oscillators. These included cyclic degradation and in-cycle degradation, as illustrated in Figure 1-1. Cyclic degradation was characterized by loss of strength and stiffness occurring in subsequent cycles. In-cycle degradation was characterized by loss of strength and negative stiffness occurring within a single cycle. This distinction was necessary because the consequences of cyclic degradation and in-cycle degradation were observed to be vastly different. In general, systems with cyclic degradation were shown to have stable dynamic response, while systems with severe in-cycle degradation were prone to dynamic instability, potentially leading to collapse.

Recommendations contained within FEMA 440 resulted in immediate improvement in nonlinear static analysis procedures, and were incorporated in the development of ASCE/SEI 41-06. However, several difficult technical issues remained unresolved. These included the need for additional guidance and direction on: (1) expansion of component and global modeling to include nonlinear degradation of strength and stiffness; (2) improvement of simplified nonlinear modeling to include multi-degree of freedom effects; and (3) improvement of modeling to include soil and foundation structure interaction effects.

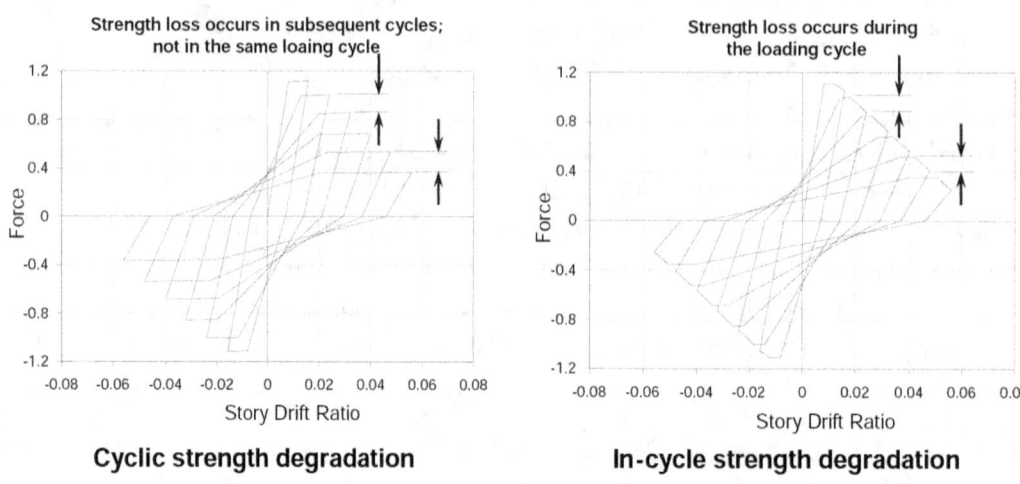

Figure 1-1   Types of degradation defined in FEMA 440.

1.1   Project Objectives

The Applied Technology Council (ATC) was commissioned by FEMA under the ATC-62 Project to further investigate the issue of component and global response to degradation of strength and stiffness. Using FEMA 440 as a starting point, the objectives of the project were to advance the understanding of degradation and dynamic instability by:

- Investigating and documenting currently available empirical and theoretical knowledge on nonlinear cyclic and in-cycle strength and stiffness degradation, and their affects on the stability of structural systems

- Supplementing and refining the existing knowledge base with focused analytical studies

- Developing practical suggestions, where possible, to account for nonlinear degrading response in the context of current seismic analysis procedures.

The result is an extensive collection of available research on component modeling of degradation, and a database of analytical results documenting the effects of a variety of parameters on the overall response of single-degree-of-freedom systems with degrading components. This report presents the findings and conclusions resulting from focused analytical studies, and provides recommendations that can be used to improve both nonlinear static and nonlinear response history analysis modeling of strength and stiffness degradation for use in performance-based seismic design.

1.2   Scope of Investigation

The scope of the investigative effort included two primary activities. The first was to assemble and review currently available research on the effects of degrading nonlinear component properties on structural system response. The second was to augment this information with supplemental analytical data, where needed.

1.2.1   Literature Review

Work included an extensive review of existing research on hysteretic models that have been developed and used for modeling nonlinear response of structures, with an emphasis on those that have incorporated degradation of stiffness and strength. The review included theoretical and empirical investigations that have studied the effect of hysteretic behavior on seismic response. Interviews with selected researchers were also conducted.

The body of knowledge is dominated by studies conducted within the last 20 years; however, relevant data on this topic extends as far back as the 1940s. In summary, past research leads to the conclusion that in-cycle strength and stiffness degradation are real phenomena, and recent investigations confirm that the effects of in-cycle strength and stiffness degradation are critical in determining the possibility of lateral dynamic instability.

Only a small number of analytical studies and experimental tests have considered the dynamic loading effects of in-cycle strength and stiffness degradation. Most experimental studies to date have only considered individual components or individual subassemblies, and have not considered larger systems of components with mixed hysteretic behavior. There are only a few studies that have considered combined effects of strength, stiffness, period of vibration together with in-cycle degradation.

A summary of background information taken from the literature is provided in Chapter 2. A comprehensive collection and summary of technical references on the development, evolution, and applicability of various hysteretic models for use in nonlinear structural analysis is provided in Appendix A.

1.2.2 Focused Analytical Studies

To supplement the current state of knowledge, a program of nonlinear dynamic focused analytical studies was developed and implemented. The purpose of this program was to investigate the response of systems comprised of degrading components, test various characteristics of degrading component behavior, and identify their effects on the dynamic stability of a system.

The basis of the focused analytical studies is a set of eight nonlinear springs representing different types of inelastic hysteretic behavior:

- Typical gravity frame (e.g., steel)

- Non-ductile moment frame (e.g., steel or concrete)

- Ductile moment frame (e.g., steel or concrete)

- Stiff non-ductile system (e.g., concentric braced frame)

- Stiff and highly pinched non-ductile system (e.g., infill wall)

- Idealized elastic-perfectly-plastic system (for comparison)

- Limited-ductility moment frame (e.g., concrete)

- Non-ductile gravity frame (e.g., concrete)

Each spring was defined with a hysteretic model based on information available in the literature. While intended to be representative of realistic degrading response that has been observed to occur in some structural components, these idealized springs are not intended to be a detailed characterization of the mechanical behavior exclusively associated with any one specific structural component or structural assembly.

Individual springs were combined to approximate the behavior of more complex systems consisting of a mixture of subassemblies having different hysteretic characteristics. Combinations included gravity frame components working with various different primary lateral-force resisting components to approximate a range of possible building types encountered in practice. For each such combined system, variations in the relative contribution of individual springs to the initial stiffness and maximum lateral strength over a range of periods were considered. Development of single-degree-of-freedom (SDOF) models used in focused analytical studies is described in Chapter 3.

Extensive parametric studies varying the strength, stiffness, period, and post-elastic properties were conducted on each component spring and combined system using Incremental Dynamic Analysis (IDA). Results of over 2.6 million nonlinear response history analyses are summarized in Chapter 4.

A limited study of multiple-degree-of-freedom (MDOF) systems was also conducted. This effort compared the results of nonlinear dynamic analyses of MDOF buildings performed by others to analytical results for SDOF representations of the same systems. The purpose was to investigate the extent to which results from nonlinear static analyses might be combined with dynamic analyses of SDOF systems to estimate the global response of MDOF systems. Preliminary MDOF investigations are described in Appendix G. Additional MDOF investigations are planned under a project funded by the National Institute of Standards and Technology (NIST).

1.3   Report Organization and Content

Chapter 1 introduces the project context, objectives, and scope of the investigation.

Chapter 2 provides background information related to modeling of component hysteretic behavior, summarizes results of past studies, and introduces new terminology.

Chapter 3 describes the development of SDOF models, and explains the analytical procedures used in the conduct of focused analytical studies.

Chapter 4 summarizes the results of focused analytical studies on single-spring and multi-spring systems, compares results to recommendations contained in FEMA 440, and explains the development of a new equation measuring the potential for lateral dynamic instability.

Chapter 5 collects and summarizes the findings, conclusions, and recommendations resulting from this investigation related to improved understanding of nonlinear degrading response and judgment in implementation of nonlinear analysis results in engineering practice, improvements to current nonlinear analysis procedures, and suggestions for further study.

Appendix A provides a comprehensive collection and summary of technical references on the development, evolution, and applicability of various hysteretic models for use in nonlinear structural analysis.

Appendix B contains plots of selected incremental dynamic analysis results for single-spring systems.

Appendix C contains normalized plots of selected incremental dynamic analysis results for multi-spring systems.

Appendix D contains non-normalized plots of selected incremental dynamic analysis results for multi-spring systems.

Appendix E explains the concepts of uncertainty and fragility, how incremental dynamic analysis results can be converted into fragilities, and how to use this information to calculate estimates of annualized probability for limit states of interest.

Appendix F provides an example application of a simplified nonlinear dynamic analysis procedure, including quantitative evaluation of alternative retrofit strategies and development of probabilistic estimates of performance using the concepts outlined in Appendix E.

Appendix G describes a set of preliminary studies of MDOF systems comparing results of MDOF analyses with results from equivalent SDOF representations of the systems, and provides recommendations for additional MDOF studies.

A compact disc (CD) accompanying this report provides electronic files of the report and appendices in PDF format, an electronic visualization tool in Microsoft Excel format that can be used to view the entire collection of multi-spring incremental dynamic analysis results, and the *Static Pushover 2 Incremental Dynamic Analysis* (SPO2IDA) software tool in Microsoft Excel format (Vamvatsikos and Cornell, 2006) that can be used to estimate the dynamic response of systems based on idealized force-displacement (static pushover) curves.

Chapter 2
Background Concepts

This chapter provides background information on modeling of component hysteretic behavior. It summarizes how various types of hysteretic behavior have been investigated in past studies, and explains how these behaviors have been observed to affect seismic response. It introduces new terminology, and explains how the new terms are related to observed differences in nonlinear dynamic response.

2.1 Effects of Hysteretic Behavior on Seismic Response

Many hysteretic models have been proposed over the years with the purpose of characterizing the mechanical nonlinear behavior of structural components (e.g., members and connections) and estimating the seismic response of structural systems (e.g., moment frames, braced frames, shear walls). Available hysteretic models range from simple elasto-plastic models to complex strength and stiffness degrading curvilinear hysteretic models. This section presents a summary of the present state of knowledge on hysteretic models, and their influence on the seismic response of structural systems. A comprehensive summary of technical references on the development, evolution, and applicability of various hysteretic models is presented in Appendix A.

2.1.1 Elasto-Plastic Behavior

In the literature, most studies that have considered nonlinear behavior have used non-degrading hysteretic models, or models in which the lateral stiffness and the lateral yield strength remain constant throughout the duration of loading. These models do not incorporate stiffness or strength degradation when subjected to repeated cyclic load reversals. The simplest and most commonly used non-deteriorating model is an elasto-plastic model in which system behavior is linear-elastic until the yield strength is reached (Figure 2-1). At yield, the stiffness switches from elastic stiffness to zero stiffness. During unloading cycles, the stiffness is equal to the loading (elastic) stiffness.

Early examples of the use of elasto-plastic models include studies by Berg and Da Deppo (1960), Penzien (1960a, 1960b), and Veletsos and Newmark (1960). The latter study was the first one to note that peak lateral displacements of moderate and long-period single-degree-of-freedom (SDOF) systems with elasto-plastic behavior were, on average, about the same as that of linear elastic systems with the same period of vibration and same damping ratio. Their observations formed the basis of what is now known as the "equal displacement approximation." This widely-used approximation implies that the peak displacement of moderate and long-period non-degrading systems is proportional to the ground motion intensity, meaning that if the ground motion intensity is doubled, the peak displacement will be on average, approximately twice as large.

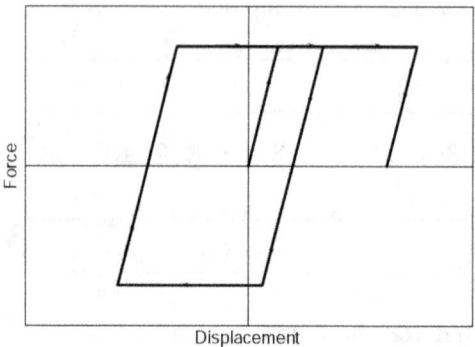

Figure 2-1   Elasto-plastic non-degrading piecewise linear hysteretic model.

Veletsos and Newmark also observed that peak lateral displacement of short-period SDOF systems with elasto-plastic behavior were, on average, larger than those of linear elastic systems, and increases in peak lateral displacements were larger than the increment in ground motion intensity. Thus, the equal displacement approximation was observed to be less applicable to short-period structures.

Using many more ground motions, recent studies have corroborated some of the early observations by Veletsos, identified some of the limitations in the equal displacement approximation, and provided information on record-to-record variability (Miranda, 1993, 2000; Ruiz-Garcia and Miranda, 2003; Chopra and Chintanapakdee, 2004). These studies have shown that, in the short-period range, peak inelastic system displacements increase with respect to elastic system displacements as the period of vibration decreases and as the lateral strength decreases. These observations formed the basis of the improved displacement modification coefficient C_1, which accounts for the effects of inelastic behavior in the coefficient method of estimating peak displacements, as documented in FEMA 440 *Improvement of Nonlinear Static Seismic Analysis Procedures* (FEMA, 2005).

2.1.2 Strength-Hardening Behavior

Another commonly used non-degrading hysteretic model is a strength-hardening model, which is similar to the elasto-plastic model, except that the post-yield stiffness is greater than zero (Figure 2-2). Early applications of bilinear strength-hardening models include investigations by Caughey (1960a, 1960b) and Iwan (1961). Positive post-yield stiffness is also referred to as "strain hardening" because many materials exhibit gains in strength (harden) when subjected to large strain levels after yield. Strength hardening in components, connections, and systems after initial yield is also caused by eventual mobilization of a full member crossection, or sequential yielding of the remaining elements in a system. This is typically the most important source of strength hardening observed in a structural system.

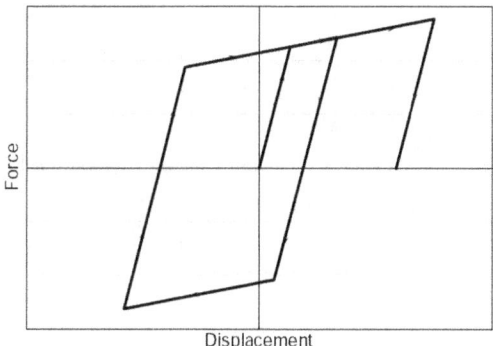

Figure 2-2   Strength-hardening non-degrading piecewise linear hysteretic model.

Although many studies have considered elasto-plastic and bilinear strength-hardening behavior, it was not until recently that comprehensive statistical studies were conducted to systematically quantify differences in peak displacements using a wide range of periods of vibration, a wide range of post-elastic stiffnesses, and large numbers of ground motions. Several recent studies have provided quantitative information on the average effects of positive post-yield stiffness on response, and on the variability in response for different ground motion records. They are in agreement that, for moderate and long-period structures, the presence of a positive post-elastic stiffness leads to relatively small (less than 5%) reductions in peak displacement (Ruiz-Garcia and Miranda, 2003; Chopra and Chintanapakdee, 2004). The magnitude of the reduction varies based on the strength of the system and period of vibration.

System strength is often characterized by a parameter, R, defined as the ratio between the strength that would be required to keep the system elastic for a given intensity of ground motion, S_{aT}, and the lateral yield strength of the system, F_y:

$$R = \frac{S_{aT}}{F_y/W} = \frac{S_{aT}g}{F_y} \tag{2-1}$$

where S_{aT} is expressed as a percentage of gravity. This R factor is related to, but not the same as, the response-modification coefficient used in code-based equivalent lateral force design procedures.

For weaker systems (systems with higher values of R), the reduction in response is greater (more beneficial). For short-period systems, the presence of a positive post-elastic stiffness can lead to significant reductions in peak lateral displacements.

Other recent studies have shown that a positive post-elastic stiffness can have a very large effect in other response parameters. In particular, MacRae and Kawashima (1997), Kawashima et al., (1998) Pampanin et al. (2002), Ruiz-Garcia and Miranda, (2006a) have shown that small increments in post-yield stiffness can lead to substantial reductions in residual drift in structures across all period ranges.

2.1.3 Stiffness-Degrading Behavior

Many structural components and systems will exhibit some level of stiffness degradation when subjected to reverse cyclic loading. This is especially true for reinforced concrete components subjected to several large cyclic load reversals. Stiffness degradation in reinforced concrete components is usually the result of cracking, loss of bond, or interaction with high shear or axial stresses. The level of stiffness degradation depends on the characteristics of the structure (e.g., material properties, geometry, level of ductile detailing, connection type), as well as on the loading history (e.g., intensity in each cycle, number of cycles, sequence of loading cycles).

Figure 2-3 shows three examples of stiffness-degrading models. In the first model, the loading and unloading stiffness is the same, and the stiffness degrades as displacement increases. In the second model the loading stiffness decreases as a function of the peak displacement, but the unloading stiffness is kept constant and equal to the initial stiffness. In the third model, both the loading and unloading stiffnesses degrade as a function of peak displacement, but they are not the same.

In order to evaluate the effects of stiffness degradation, many studies have compared the peak response of stiffness-degrading systems to that of elasto-plastic and bilinear strength-hardening systems (Clough 1966; Clough and

Johnston 1966; Chopra and Kan, 1973; Powel and Row, 1976; Mahin and Bertero, 1976; Riddell and Newmark, 1979; Newmark and Riddell, 1980; Iwan 1980; Otani, 1981; Nassar and Krawinkler 1991; Rahnama and Krawinkler, 1993; Shi and Foutch, 1997; Foutch and Shi, 1998; Gupta and Krawinkler, 1998; Gupta and Kunnath, 1998; Medina 2002; Medina and Krawinkler, 2004; Ruiz-Garcia and Miranda, 2005).

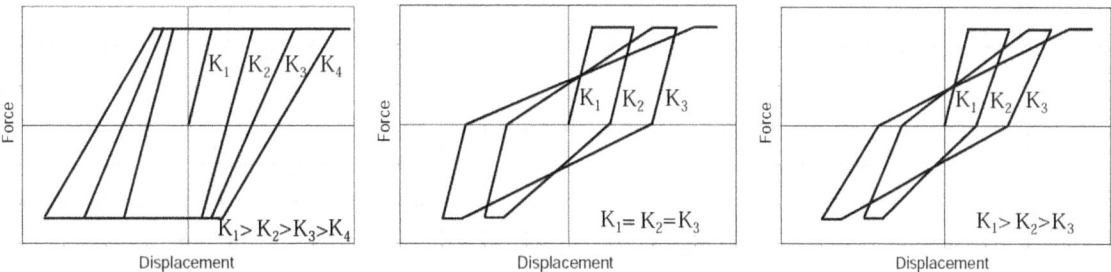

Figure 2-3   Three examples of stiffness-degrading piecewise linear hysteretic models.

These studies have concluded that, in spite of significant reductions in lateral stiffness and hysteretic energy dissipation capacity (area enclosed within hysteresis loops), moderate and long-period systems with stiffness-degrading behavior experience peak displacements that are, on average, similar to those of structures with elasto-plastic or bilinear strength-hardening hysteretic behavior. In some cases, peak displacements can even be slightly smaller. This observation suggests that it is possible to use simpler hysteretic models that do not incorporate stiffness degradation to estimate lateral displacement demands for moderate and long-period structures (systems with fundamental periods longer than 1.0s).

These same studies, however, have concluded that short-period structures with stiffness degradation experience peak displacements that are, on average, larger than those experienced by systems with elasto-plastic or bilinear strength-hardening hysteretic behavior. Differences in peak displacements between stiffness-degrading and non-degrading systems increase as the period of vibration decreases and as the lateral strength decreases.

The above studies examined the effects of stiffness degradation on structures subjected to ground motions recorded on rock or firm soil sites. Ruiz-Garcia and Miranda (2006b) examined the effects of stiffness degradation on structures subjected to ground motions recorded on soft soil sites. This study concluded that the effects of stiffness degradation are more important for structures built on soft soil, especially for structures with periods shorter than the predominant period of the ground motion.

2.1.4 Pinching Behavior

Structural components and connections may exhibit a hysteretic phenomenon called pinching when subjected to reverse cyclic loading (Figure 2-4). Pinching behavior is characterized by large reductions in stiffness during reloading after unloading, along with stiffness recovery when displacement is imposed in the opposite direction.

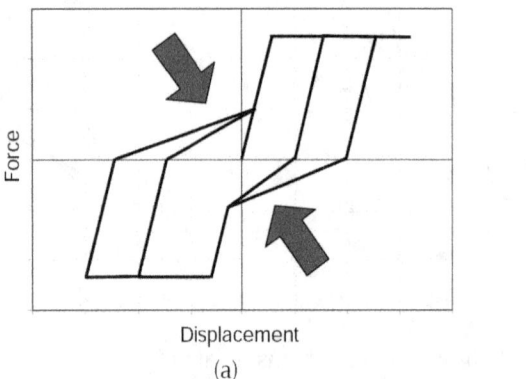 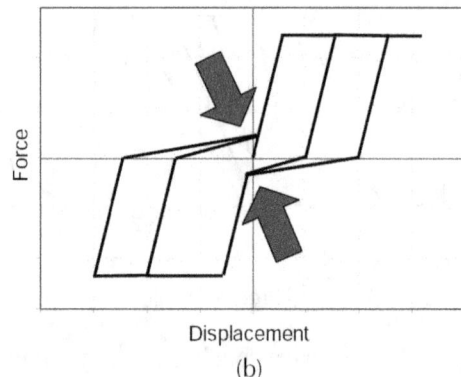

Displacement   Displacement
(a)   (b)

Figure 2-4   Examples of hysteretic models with: (a) moderate pinching behavior; and (b) severe pinching behavior.

Pinching behavior is particularly common in reinforced concrete components, wood components, certain types of masonry components, and some connections in steel structures. In reinforced concrete, pinching is typically produced by opening of cracks when displacement is imposed in one direction. Partial stiffness recovery occurs when cracks are closed during displacements imposed in the other direction. In wood structures pinching is primarily caused by opening and closing of gaps in framing elements due to nail pullout. Pinching also occurs as a result of opening and closing of flexural cracks in reinforced masonry, opening and closing of gaps between masonry infill and the surrounding structural frame, and opening and closing of gaps between plates in steel end-plate connections. The level of pinching depends on the characteristics of the structure (e.g., material properties, geometry, level of ductile detailing, and connections), as well as the loading history (e.g., intensity in each cycle, number of cycles, and sequence of loading cycles).

Several studies have shown that, for moderate and long-period systems, pinching alone or in combination with stiffness degradation has only a small affect on peak displacement demands, as long as the post-yield stiffness remains positive (Otani, 1981; Nassar and Krawinkler 1991; Rahnama and Krawinkler, 1993; Shi and Foutch, 1997; Foutch and Shi, 1998; Gupta and

Krawinkler, 1998; Gupta and Kunnath, 1998; Medina 2002; Medina and Krawinkler, 2004; Ruiz-Garcia and Miranda, 2005).

These and other studies have shown that moderate and long-period systems, with up to 50% reduction in hysteretic energy dissipation capacity due to pinching, experience peak displacements that are, on average, similar to those of structures with elasto-plastic or bilinear strength-hardening hysteretic behavior. This observation is particularly interesting because it is contrary to the widespread notion that structures with elasto-plastic or bilinear behavior exhibit better performance than structures with pinching behavior because of the presence of additional hysteretic energy dissipation capacity.

These same studies, however, have also shown that short-period structures with pinching behavior experience peak displacements that tend to be larger than those experienced by systems with elasto-plastic or bilinear strength-hardening hysteretic behavior. Differences in peak displacements increase as the period of vibration decreases and as the lateral strength decreases.

2.1.5 Cyclic Strength Degradation

Structural components and systems may experience reductions in strength generically referred to as strength degradation or strength deterioration (Figure 2-5). One of the most common types of strength degradation is cyclic strength degradation in which a structural component or system experiences a reduction in lateral strength as a result of cyclic load reversals.

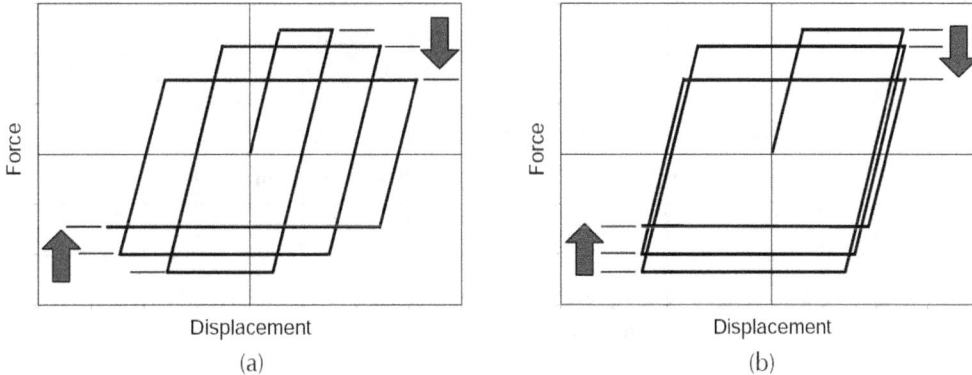

Figure 2-5   Examples of cyclic strength degradation: (a) due to increasing inelastic displacement; and (b) due to repeated cyclic displacement

In cyclic strength degradation, reductions in lateral strength occur after the loading has been reversed, or during subsequent loading cycles. Cyclic reductions in lateral strength are a function of the level of peak displacement experienced in the system (Park, Reinhorn and Kunnath, 1987; Rahnama and

Krawinkler, 1993). This is illustrated in Figure 2-5(a), which shows an elasto-plastic system experiencing strength degradation in subsequent loading cycles as the level of inelastic displacement increases. Hysteretic models that incorporate this type of strength degradation typically specify the reduction in strength as a function of the ductility ratio, which is taken as the ratio of peak deformation to yield deformation.

Cyclic strength degradation can also occur in subsequent cycles even if the level of inelastic displacement is not being increased (Park, Reinhorn and Kunnath, 1987; Rahnama and Krawinkler, 1993). This is illustrated in Figure 2-5(b), which shows an elasto-plastic system experiencing cyclic strength degradation as a result of a single level of inelastic displacement that is imposed a number of times. The reduction in lateral strength increases as the number of cycles increases. Hysteretic models that incorporate this type of strength degradation (Park, Reinhorn and Kunnath, 1987; Rahnama and Krawinkler, 1993; Mostaghel 1998, 1999; Sivaselvan and Reinhorn 1999, 2000) typically specify the reduction in strength as a function of the total hysteretic energy demand imposed on the system, taken as the area enclosed by the hysteresis loops.

Most structural systems exhibit a combination of the types of cyclic strength degradation shown in Figure 2-5. Several hysteretic models that incorporate both types of cyclic strength degradation have been developed (Park and Ang, 1985; Park, Reinhorn and Kunnath, 1987; Rahnama and Krawinkler, 1993; Valles et al., 1996; Shi and Foutch, 1997; Foutch and Shi, 1998; Gupta and Krawinkler, 1998;Gupta and Kunnath, 1998; Pincheira, Dotiwala, and D' Souza 1999; Medina 2002; Medina and Krawinkler, 2004; Mostaghel 1998, 19990; Sivaselvan and Reinhorn 1999, 2000; Chenouda, and Ayoub, 2007).

Many of these same investigators have compared the peak response of systems with cyclic strength degradation to that of elasto-plastic and bilinear strength-hardening systems. In moderate and long-periods systems, the effects of cyclic strength degradation have been shown to be very small, and in many cases can be neglected, even with reductions in strength of 50% or more. The reason for this can be explained using early observations from Veletsos and Newmark (1960), which concluded that peak displacement demands in moderate and long-period systems were not sensitive to changes in yield strength. This conclusion logically extends to moderate and long-period systems that experience cyclic changes (reductions) in lateral strength during loading.

In short-period structures, however, studies have shown that cyclic strength degradation can lead to significant increases in peak displacement demands.

This observation can also be explained by results from Veletsos and Newmark (1960), which concluded that peak displacement demands in short-period systems are very sensitive to changes in yield strength. This conclusion logically extends to short-period systems that experience cyclic changes (reductions) in lateral strength during loading.

2.1.6 Combined Stiffness Degradation and Cyclic Strength Degradation

Several recent studies have examined the effects of stiffness degradation in combination with cyclic strength degradation (Gupta and Kunnath, 1998; Song and Pincheira, 2000; Medina 2002; Medina and Krawinkler, 2004; Ruiz-Garcia and Miranda, 2005; Chenouda, and Ayoub, 2007). Examples of these behaviors are illustrated in Figure 2-6. Figure 2-6a shows a system with moderate stiffness and cyclic strength degradation (MSD), and Figure 2-6b shows a system with severe stiffness and cyclic strength degradation (SSD). In these systems, lateral strength is reduced as a function of both the peak displacement demand as well as the hysteretic energy demand on the system.

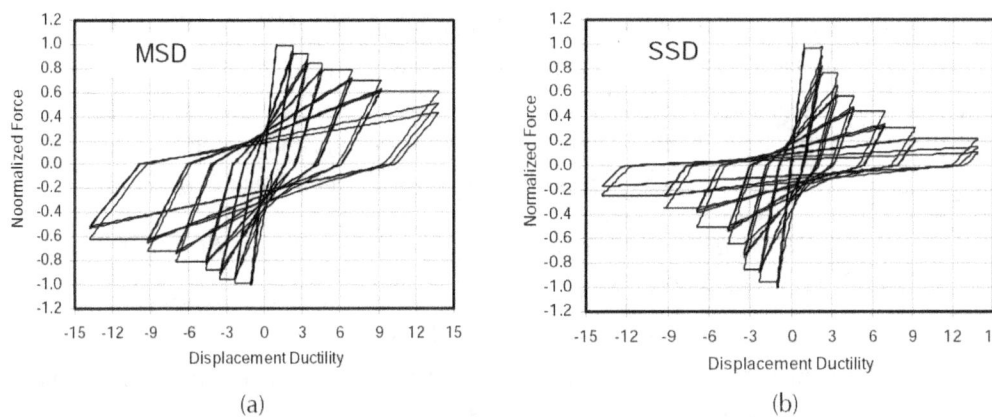

Figure 2-6   Hysteretic models combining stiffness degradation and cyclic strength degradation: (a) moderate stiffness and cyclic strength degradation; and (b) severe stiffness and cyclic strength degradation (Ruiz-Garcia and Miranda, 2005).

These studies have shown that, for moderate and long-period systems with combined stiffness and cyclic strength degradation, peak displacements are, on average, similar to those experienced by elasto-plastic or bilinear strength-hardening systems. These effects are only observed to be significant for short-period systems (systems with periods of vibration less than 1.0s).

2.1.7 In-Cycle Strength Degradation

In combination with stiffness degradation, structural components and systems may experience in-cycle strength degradation (Figure 2-7). In-cycle strength degradation is characterized by a loss of strength within the same cycle in which yielding occurs. As additional lateral displacement is imposed, a smaller lateral resistance is developed. This results in a negative post-yield stiffness within a given cycle.

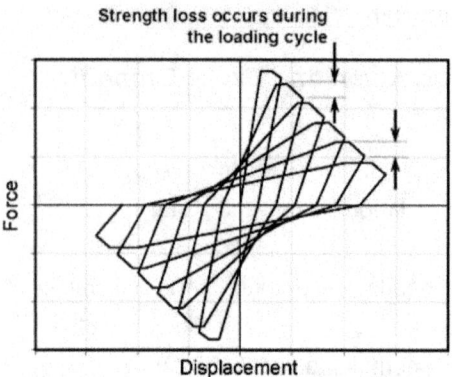

Figure 2-7   In-cycle strength degradation.

In-cycle strength degradation can occur as a result of geometric nonlinearities (P-delta effects), material nonlinearities, or a combination of these. In reinforced concrete components, material nonlinearities that can lead to in-cycle strength degradation include concrete crushing, shear failure, buckling or fracture of longitudinal reinforcement, and splice failures. In steel components, material nonlinearities that can lead to in-cycle strength degradation include buckling of bracing elements, local buckling in flanges of columns or beams, and fractures of bolts, welds, or base materials.

2.1.8   Differences Between Cyclic and In-Cycle Strength Degradation

FEMA 440 identified the distinction between cyclic and in-cycle degradation to be very important because the consequences of each were observed to be vastly different. Dynamic response of systems with cyclic strength degradation is generally stable, while in-cycle strength degradation can lead to lateral dynamic instability (i.e., collapse) of a structural system.

Figure 2-8 compares the hysteretic behavior of two systems subjected to the loading protocol shown in Figure 2-9. This loading protocol comprises six full cycles (twelve half-cycles) with a linearly increasing amplitude of 0.8% drift in each cycle. The system in Figure 2-8a has cyclic degradation and the system in Figure 2-8b has in-cycle degradation. When subjected to this loading protocol, both hysteretic models exhibit similar levels of strength and stiffness degradation, and similar overall behavior. Their behavior under different loading protocols, however, can be significantly different.

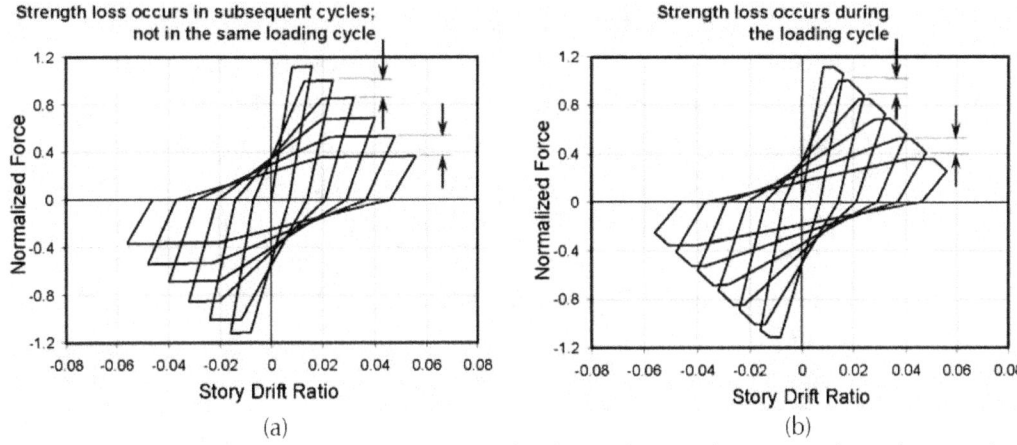

Figure 2-8   Hysteretic behavior for models subjected to Loading Protocol 1 with: (a) cyclic strength degradation; and (b) in-cycle degradation.

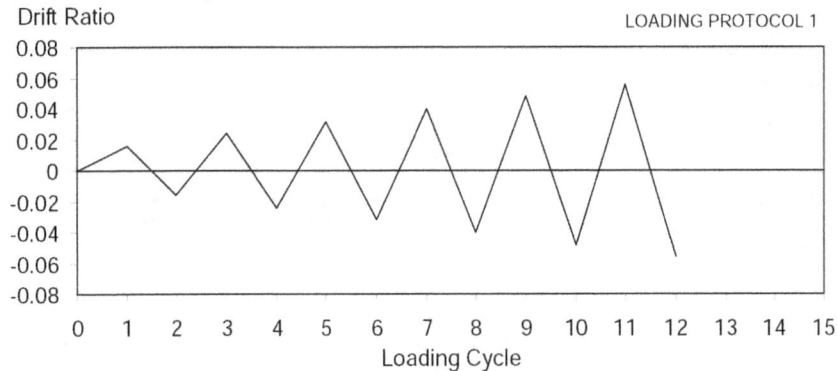

Figure 2-9   Loading Protocol 1 used to illustrate the effects of cyclic and in-cycle strength degradation.

A second loading protocol, shown in Figure 2-10, is identical to the first protocol through four half-cycles, but during the fifth half-cycle it continues to impose additional lateral displacement until a drift ratio of 7.0% is reached.

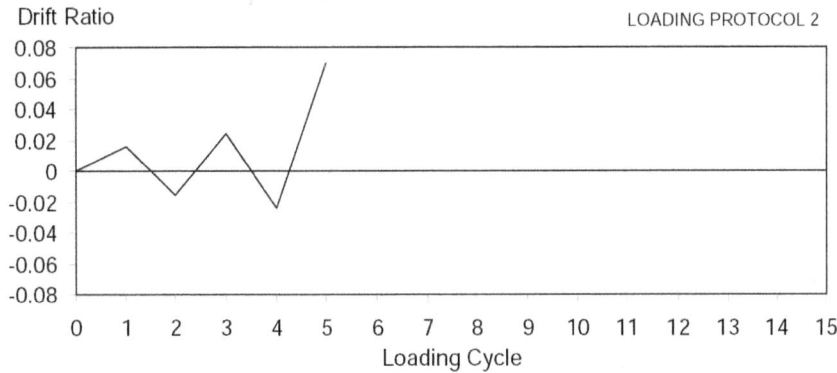

Figure 2-10   Loading Protocol 2 used to illustrate the effects of cyclic and in-cycle strength degradation.

Figure 2-11 compares the hysteretic behavior of both systems subjected to the second loading protocol. Initially, the responses are similar. During the fifth half-cycle, however, the responses diverge. The model with cyclic degradation (Figure 2-11a) is able to sustain lateral strength without loss as the drift ratio increases. In contrast, the model with in-cycle degradation (Figure 2-11b) experiences a rapid loss in strength as the drift ratio increases. While the model with cyclic strength degradation remains stable, the model with in-cycle strength degradation becomes unstable after losing lateral resistance.

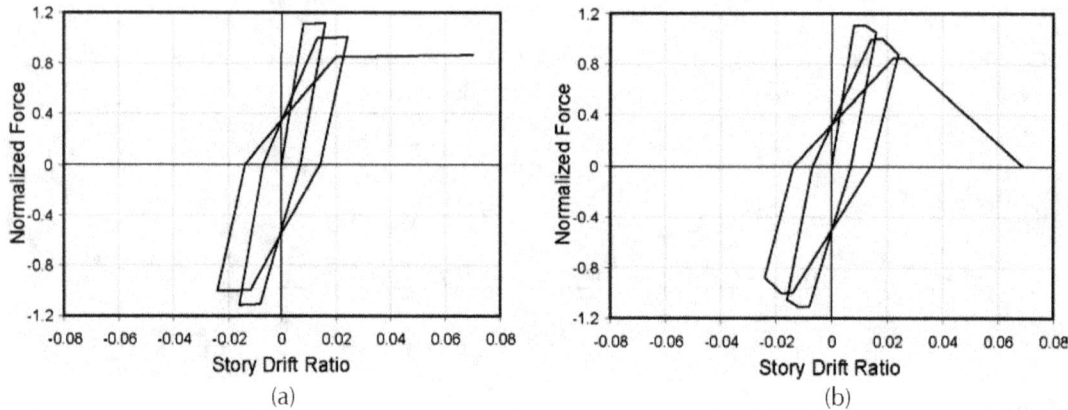

Figure 2-11   Hysteretic behavior for models subjected to Loading Protocol 2 with: (a) cyclic strength degradation; and (b) in-cycle degradation.

Figure 2-12 shows the displacement time histories for these same two systems when subjected to the north-south component of the Yermo Valley ground motion of the 1992 Landers Earthquake. The system with cyclic strength degradation (Figure 2-12a) undergoes a large peak drift ratio during the record, experiences a residual drift at the end of the record, and yet remains stable over the duration of the record. In contrast, the system with in-cycle degradation (Figure 2-12b) undergoes a similar peak drift ratio during the record, but ratchets further in one direction in subsequent yielding cycles, and eventually experiences lateral dynamic instability (collapse).

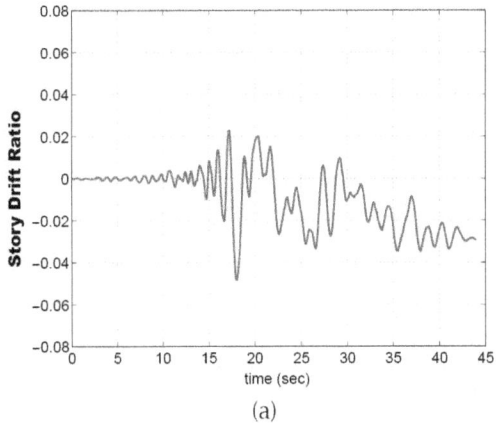 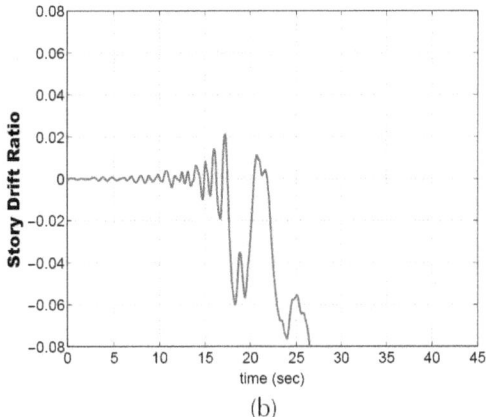

(a)   (b)

Figure 2-12   Displacement time histories for models subjected to the 1992 Landers Earthquake with: (a) cyclic strength degradation; and (b) in-cycle strength degradation.

2.2   Concepts and Terminology

Historically, the term *backbone curve* has referred to many different things. It has been used, for example, to describe limitations on the force-deformation behavior of structural components, force-displacement plots from nonlinear static pushover analyses of structural systems, curves enveloping the force-displacement response of structural components undergoing cyclic testing, and curves tracing the force-displacement response of structural components undergoing monotonic testing.

In the case of component modeling, parameters taken from one definition of a backbone curve versus another are not interchangeable, and their incorrect usage can have a significant affect on the predicted nonlinear response. For this reason, two new terms are introduced to distinguish between different aspects of hysteretic behavior. These are the *force-displacement capacity boundary*, and *cyclic envelope*.

2.2.1   Force-Displacement Capacity Boundary

Several recent models have been developed to incorporate various types of degrading phenomena (Kunnath, Reinhorn and Park, 1990; Kunnath, Mander and Fang, 1997; Mostaghel 1998, 19990; Sivaselvan and Reinhorn 1999, 2000; Ibarra, Medina, Krawinkler, 2005; Chenouda and Ayoub, 2007). A common feature in all these degrading models is that they start by defining the maximum strength that a structural member can develop at a given level of deformation. This results in an effective "boundary" for the strength of a member in force-deformation space, termed the *force-displacement capacity boundary*.

Figure 2-13 shows examples of two such boundaries commonly used in structural analysis of degrading components. These curves resemble the conceptual force-displacement relationship used to express component modeling and acceptability criteria in ASCE/SEI 41-06 *Seismic Rehabilitation of Existing Buildings* (ASCE, 2006b), commonly referred to as "backbones."

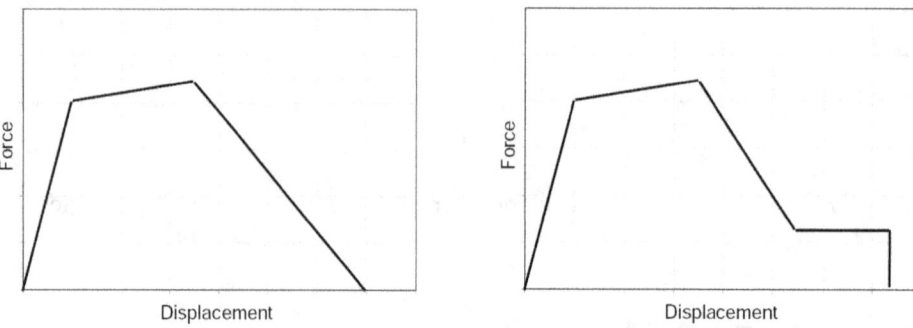

Figure 2-13   Examples of commonly used force-displacement capacity boundaries.

A cyclic load path cannot cross a force-displacement capacity boundary. If a member is subjected to increasing deformation and the boundary is reached, then the strength that can be developed in the member is limited and the response must continue along the boundary. This behavior is in-cycle strength degradation, and is shown in Figure 2-14. Note that only displacement excursions intersecting portions of the capacity boundary with a negative slope will result in in-cycle strength degradation.

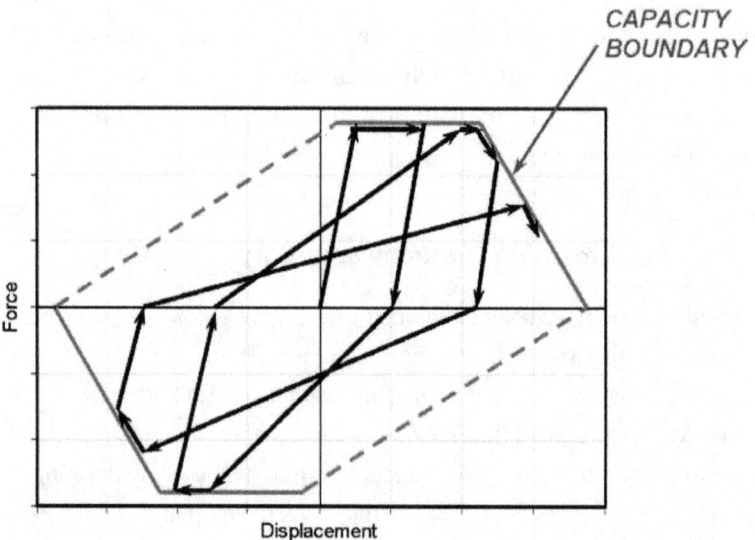

Figure 2-14   Interaction between the cyclic load path and the force-displacement capacity boundary.

In most cases, the force-displacement capacity boundary will not be static. More advanced models consider that the force-displacement capacity boundary will degrade (move inward) as a result of cyclic degradation (Figure 2-15). In some cases, it is also possible for the boundary to move outward due to cyclic strain hardening, such as in the case of structural steel elements subjected to large strains, but this behavior is not considered here.

In order to define the cyclic behavior of a component model, one must define where the force-displacement capacity boundary begins, and how it degrades under cyclic loading. In the absence of cyclic strain hardening, the initial force-displacement capacity boundary is simply the monotonic response of a component. Accordingly, the ideal source for estimating the parameters of the initial force-displacement capacity boundary comes from monotonic tests.

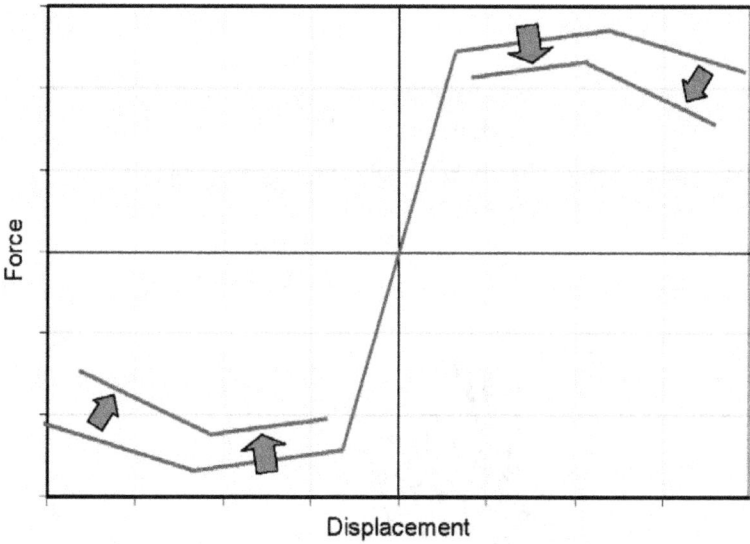

Figure 2-15   Degradation of the force-displacement capacity boundary.

Once the initial capacity boundary is defined, then cyclic degradation parameters must be estimated based on the results of cyclic tests. The use of several cyclic protocols is desirable to ensure that the calibrated component model is general enough to represent component response under any type of loading. This requires the availability of multiple identical specimens that are tested under several different loading protocols, which is a significant undertaking, and is rarely done.

When utilizing existing test data to calibrate a component model, it is uncommon to find sets of test data that include both monotonic and cyclic tests on identical specimens. It is even more uncommon to find sets of data that include monotonic tests and cyclic tests using multiple loading protocols on identical specimens. As such, there are only a small number of cases in which this kind of data exists (Tremblay et al., 1997; El-Bahy, 1999; Ingham et al., 2001; Uang et al., 2000; Uang et al., 2003).

Most existing data is based only on a single cyclic loading protocol. In such cases, cyclic degradation can be approximated directly from the test data. In the absence of monotonic test data, the initial force-displacement capacity boundary must be extrapolated from the cyclic data (since the monotonic response is unknown). Considerable judgment must be exercised in extrapolating an initial force-displacement capacity boundary because there may be several combinations of initial parameters and cyclic degradation parameters that result in good agreement with the observed cyclic test data. Such an approach has been used by Haselton et al. (2007) for reinforced concrete components and Lignos (2008) for steel components.

2.2.2 Cyclic Envelope

A *cyclic envelope* is a force-deformation curve that envelopes the hysteretic behavior of a component or assembly that is subjected to cyclic loading. Figure 2-16 shows a cyclic envelope, which is defined by connecting the peak force responses at each displacement level.

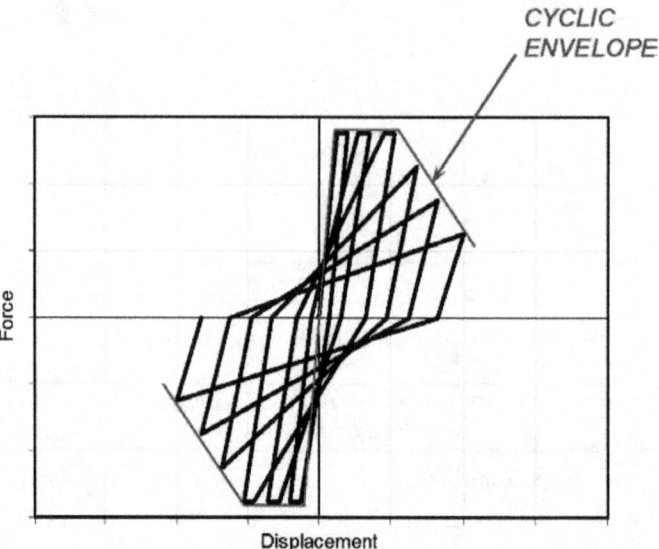

Figure 2-16   Example of a cyclic envelope.

Where loading protocols have included multiple cycles at each displacement increment, a different curve (often referred to as cyclic "backbone") has been defined based on the force at either the second or third cycle at each displacement level. Such a definition was included in FEMA 356 *Prestandard and Commentary for the Seismic Rehabilitation of Buildings*

(FEMA, 2000). In ASCE/SEI 41-06 (with Supplement No. 1) this has been changed to be more consistent with the concept of a cyclic envelope, as described above.

2.2.3 Influence of Loading Protocol on the Cyclic Envelope

The characteristics of a cyclic envelope are strongly influenced by the points at which unloading occurs in a test, and are therefore strongly influenced by the loading protocol that was used in the experimental program. Studies by Takemura and Kawashima (1997) illustrate the influence that the loading protocol can have on the characteristics of the cyclic envelope. In these studies, six nominally identical reinforced concrete bridge piers were tested using six different loading protocols, yielding six significantly different hysteretic behaviors. The loading protocols and resulting hysteretic plots are shown in Figure 2-17 through Figure 2-19.

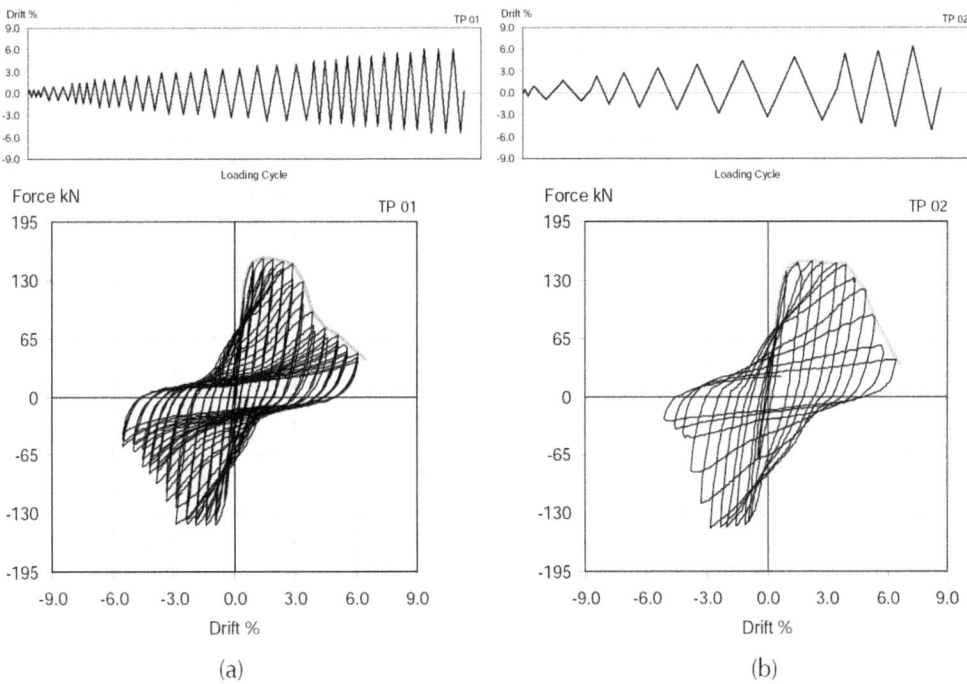

Figure 2-17   Loading protocols and resulting hysteretic plots for identical reinforced concrete bridge pier specimens: (a) Loading Protocol TP01; and (b) Loading Protocol TP02 (adapted from Takemura and Kawashima, 1997).

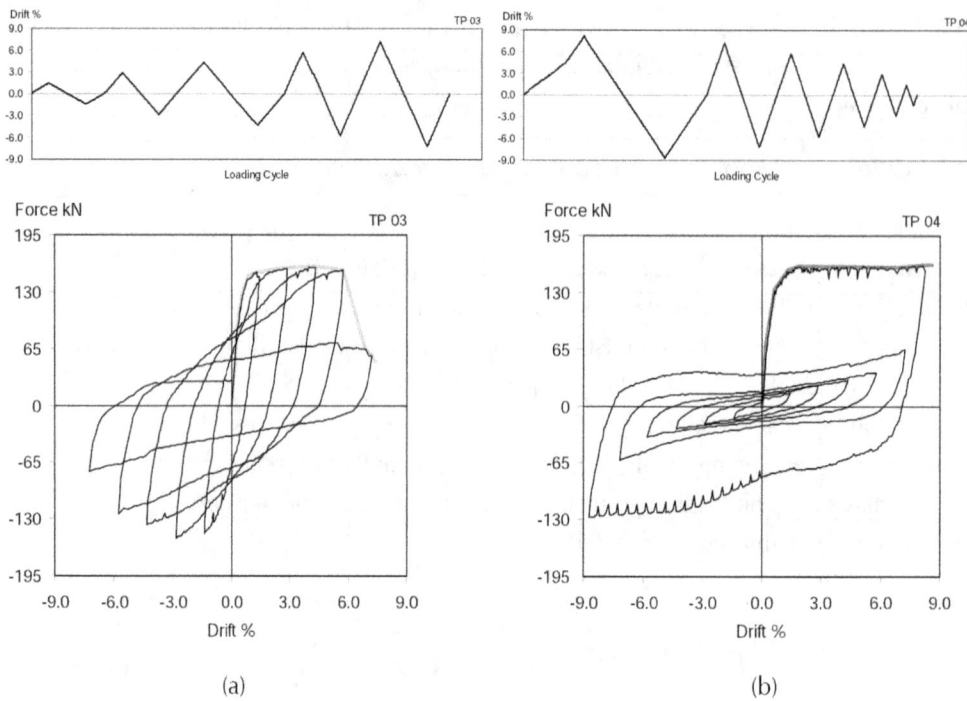

Figure 2-18   Loading protocols and resulting hysteretic plots for identical reinforced concrete bridge pier specimens: (a) Loading Protocol TP03; and (b) Loading Protocol TP04 (adapted from Takemura and Kawashima, 1997).

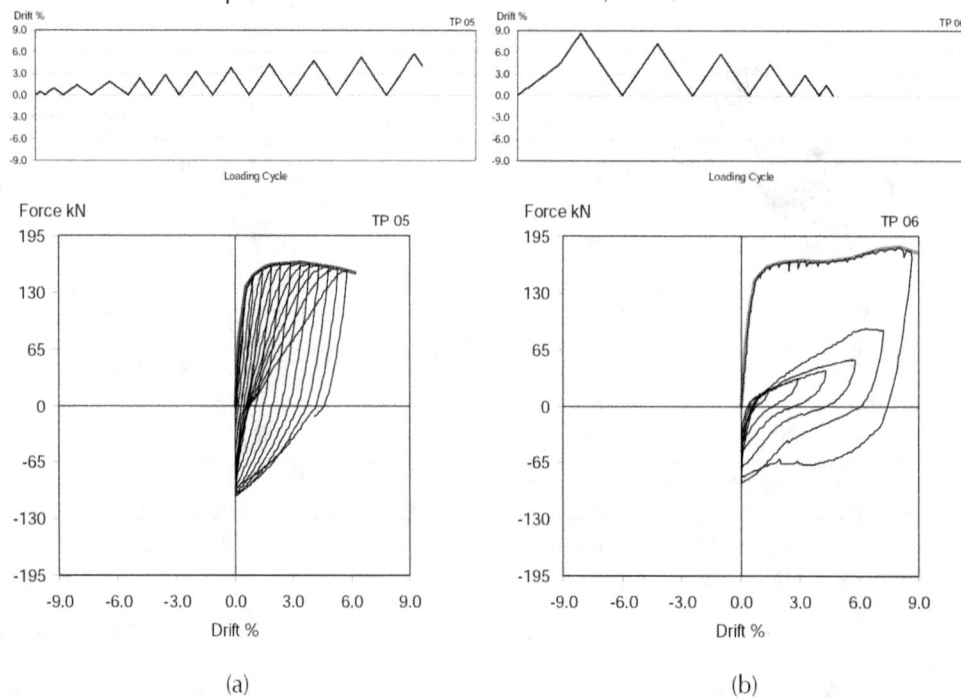

Figure 2-19   Loading protocols and resulting hysteretic plots for identical reinforced concrete bridge pier specimens: (a) Loading Protocol TP05; and (b) Loading Protocol TP06 (adapted from Takemura and Kawashima, 1997).

The resulting cyclic envelopes are plotted together in Figure 2-20 for comparison. Loading protocols with more cycles and increasing amplitudes in each cycle (e.g., TP 01, TP 02, and TP 03) resulted in smaller cyclic envelopes. Loading protocols with fewer cycles and decreasing in amplitudes in each cycle (e.g., TP 04 and TP 06) resulted in larger cyclic envelopes.

These studies show that if nominally identical specimens are loaded with different loading protocols, their cyclic envelope will change depending on the number of cycles used in the loading protocol, the amplitude of each cycle, and the sequence of the loading cycles.

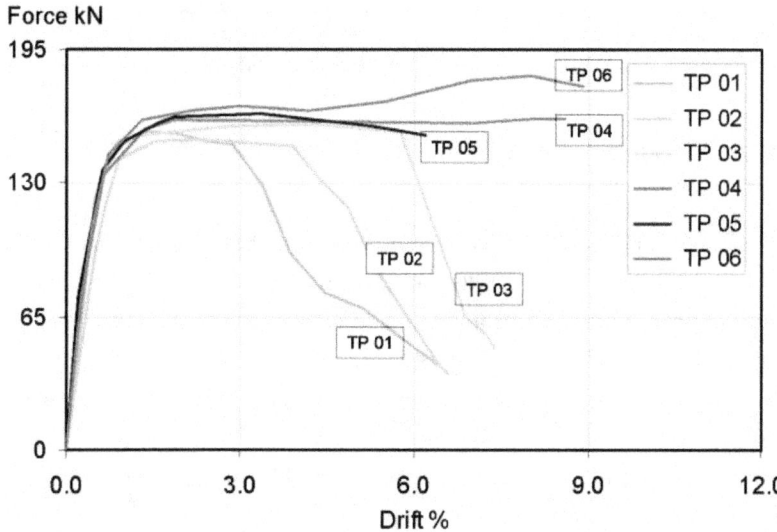

Figure 2-20   Comparison of cyclic envelopes of reinforced concrete bridge pier specimens subjected to six different loading protocols (adapted from Takemura and Kawashima, 1997).

2.2.4 Relationship between Loading Protocol, Cyclic Envelope, and Force-Displacement Capacity Boundary

For analytical purposes, a series of hysteretic rules can be specified to control the hysteretic behavior of a component within a force-displacement capacity boundary. Unless a loading protocol has forced the structural component or system to reach the force-displacement capacity boundary, the resulting cyclic envelope will be smaller, and in some cases significantly smaller, than the actual capacity boundary.

Figure 2-21 shows the cyclic envelope for a structural component subjected to a single loading protocol. In Figure 2-21a, the cyclic envelope is equal to the force-displacement capacity boundary. In Figure 2-21b, the force-displacement capacity boundary extends beyond the cyclic envelope (which would be the case if the component actually had more force-displacement capacity than indicated by a single cyclic envelope).

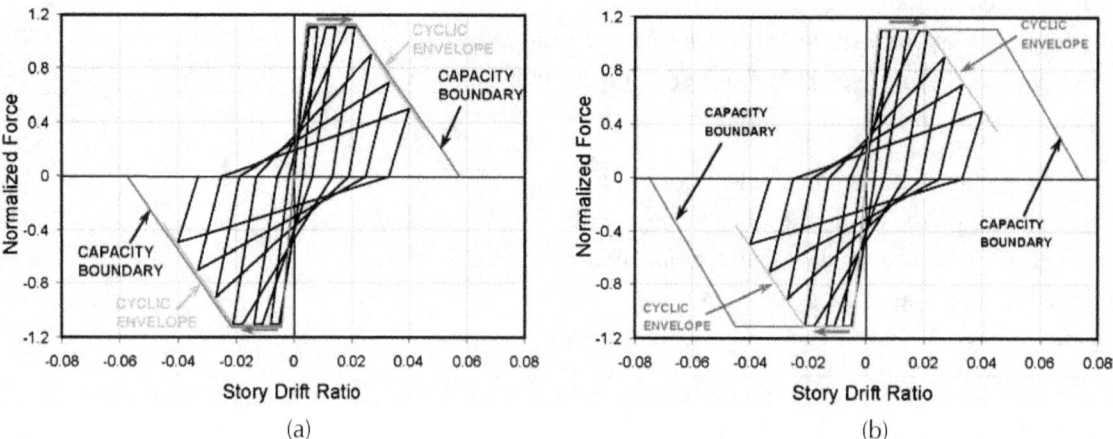

Figure 2-21   Examples of a force-displacement capacity boundary that is (a) equal to the cyclic envelope, and (b) extends beyond the cyclic envelope.

Figure 2-22 shows the hysteretic behavior of the same component subjected to a different loading protocol. In this protocol the first four cycles are the same, but in the fifth cycle additional lateral displacement is imposed up to a peak story drift ratio of 5.5%. In Figure 2-22a, the component reaches the force-displacement capacity boundary and the response is forced to follow a downward slope along the boundary (in-cycle strength degradation). Eventually, zero lateral resistance is reached before the unloading cycle can begin.

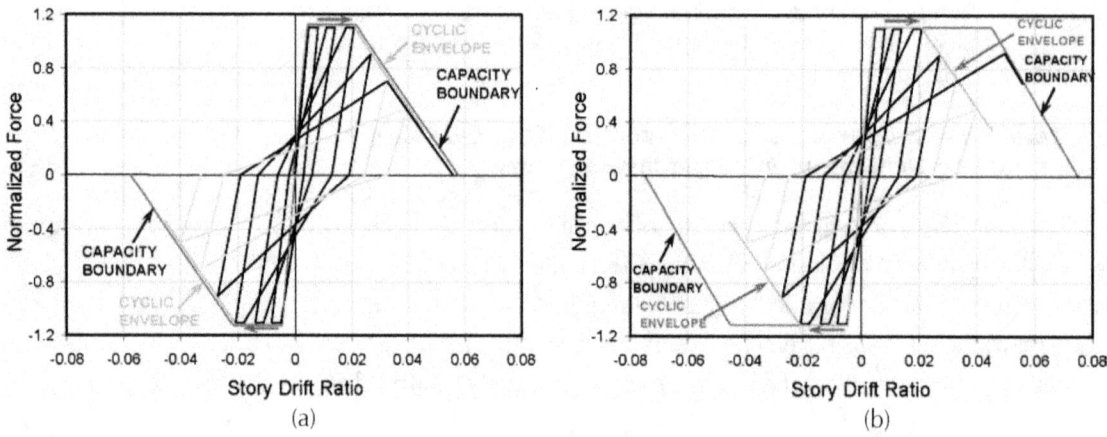

Figure 2-22   Comparison of hysteretic behavior when the force-displacement capacity boundary is: (a) equal to the cyclic envelope, and (b) extends beyond the cyclic envelope.

In Figure 2-22b, however, because the force-displacement capacity boundary extends beyond the cyclic envelope, the component has additional capacity to resist deformation. As the lateral displacement approaches 5.5%, the response continues to gain strength until the force-displacement capacity boundary is reached. The response is then forced to follow along the boundary (in-cycle strength degradation) until the unloading cycle commences at peak story drift ratio of 5.5%. In this case the component can continue to resist 70% of its peak lateral strength at a story drift ratio of 5.5%, rather than degrading to zero lateral resistance before unloading occurs.

Under lateral displacements that are less than or equal to those used to generate the cyclic envelope, differences between the cyclic envelope and the force-displacement capacity boundary are of no consequence. However, under larger lateral displacements these differences will affect the potential for in-cycle degradation to occur, which will significantly affect system behavior and response. Determining the force-displacement capacity boundary based on the results of a single cyclic loading protocol can result in overly conservative results due to significant underestimation of the actual force-displacement capacity and subsequent overestimation of lateral displacement demands.

Chapter 3
Development of Single-Degree-of-Freedom Models for Focused Analytical Studies

This chapter describes the development of single-degree-of-freedom models, and explains the analytical procedures used in the conduct of focused analytical studies.

3.1 Overview of Focused Analytical Studies

3.1.1 Purpose

From past research, it is apparent that in-cycle strength and stiffness degradation are real phenomena that have been observed and documented to cause instability in individual components. Little experimental information exists, however, on whether or not larger assemblies of components of mixed hysteretic behavior experience similar negative stiffness that could lead to dynamic instability. In order to further investigate the response of systems with degrading components, focused analytical studies were conducted. The purpose of these studies was to test and quantify the effects of different degrading behaviors on the dynamic stability of structural systems.

3.1.2 Process

Studies consisted of nonlinear dynamic analyses of single-degree-of-freedom oscillators with varying system characteristics. Characteristics under investigation included differences in hysteretic behavior, such as cyclic versus in-cycle degradation, and variations in the features of the force-displacement capacity boundary, such as the point at onset of degradation, the slope of degradation, the level of residual strength, length of the residual strength plateau, and ultimate deformation capacity (Figure 3-1).

The process used for developing, analyzing, and comparing structural system models in the focused analytical studies was as follows:

- A set of single-degree-of-freedom (SDOF) springs were developed featuring different hysteretic and force-displacement capacity boundary characteristics. While not an exact representation of the mechanical behavior of any one specific structural component, springs were intended to capture the major characteristics of force-displacement capacity boundaries for systems that would typically be encountered in practice.

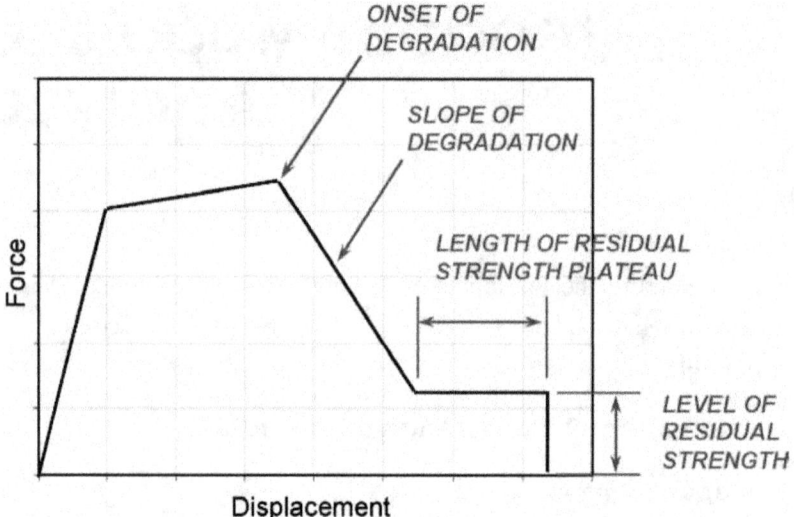

Figure 3-1   Features of the force-displacement capacity boundary varied in focused analytical studies.

- Multiple spring models were used to represent the behavior of more complex structural systems containing subsystems with different hysteretic and force-displacement capacity boundary characteristics. Multi-spring SDOF systems were developed by placing two individual springs in parallel, linked by a rigid diaphragm.

- Nonlinear response history analyses were performed using the Open System for Earthquake Engineering Simulation (OpenSEES) software (Fenves and McKenna, 2004). In OpenSEES, structural system models were subjected to the Incremental Dynamic Analysis (IDA) procedure (Vamvatsikos and Cornell, 2002) in which the nonlinear dynamic response of individual and multiple spring systems were evaluated at incrementally increasing levels of ground motion intensity.

- Results were compared in two ways: (1) among systems with different components that were tuned to have the same global yield strength and the same period of vibration; and (2) among systems composed of the same two components but having different relative contributions from each, thus exhibiting different strength and stiffness characteristics. Comparisons between systems tuned to the same yield strength and period of vibration were used to observe the influence of different hysteretic rules and force-displacement capacity boundary characteristics. Comparisons between systems composed of the same components, but with different strength and stiffness characteristics, were used to observe the relative contribution from each subsystem on overall system response.

3.1.3 Incremental Dynamic Analysis Procedure

Focused analytical studies were conducted using the Incremental Dynamic Analysis (IDA) procedure (Vamvatsikos and Cornell, 2002). Incremental dynamic analysis is a type of response history analysis in which a system is subjected to ground motion records scaled to increasing levels of intensity until lateral dynamic instability is observed. In incremental dynamic analysis, intensity is characterized by a selected intensity measure (IM), and lateral dynamic instability occurs as a rapid, nearly infinite increase in the engineering demand parameter (EDP) of interest, given a small increment in ground motion intensity.

3.1.3.1 Intensity measures

Two intensity measures were used in conducting incremental dynamic analyses. One was taken as the 5% damped spectral acceleration at the fundamental period of vibration of the oscillator, $S_a(T,5\%)$. This measure is generally appropriate for single-degree-of-freedom systems. It, however, does not allow comparison among systems having different periods of vibration. For this reason, a normalized intensity measure, $R = S_a(T,5\%)/S_{ay}(T,5\%)$ was also used, where $S_{ay}(T,5\%)$ is the intensity that causes first yield to occur in the system. This places first yield at a normalized intensity of one.

The normalized intensity measure $S_a(T,5\%)/S_{ay}(T,5\%)$ closely resembles the strength ratio, R, which is a normalized strength that is often used in studies of SDOF systems (see Chapter 2). Higher values of the normalized intensity measure $S_a(T,5\%)/S_{ay}(T,5\%)$ represent systems with lower lateral strength. Note that the R-factor discussed here is not the same as the response-modification coefficient used in code-based equivalent lateral force design procedures. Rather, it is essentially the system ductility reduction factor, R_d, as defined in the *NEHRP Recommended Provisions for Seismic Regulations for New Buildings and Other Structures, Part 2: Commentary* (FEMA, 2004b).

3.1.3.2 Engineering Demand Parameters

The engineering demand parameter of interest was taken as story drift ratio. This parameter is a normalized measure of lateral displacement that allows for non-dimensional comparison of results. Lateral dynamic instability occurs when solutions to the input ground motion fail to converge, implying infinite lateral displacements.

3.1.3.3 Collapse

Lateral dynamic instability is manifested in structural systems as sidesway collapse caused by loss of lateral-force-resisting capacity. Sidesway collapse mechanisms can be explicitly simulated in incremental dynamic analyses, and comparisons of analytical results are based on this limit state.

It should be noted, however, that behavior of real structures can include loss of vertical-load-carrying capacity at lateral displacements that are significantly smaller than those associated with sidesway collapse. Inelastic deformation of structural components can result in shear and flexural-shear failures in members, and failures in joints and connections, which can lead to an inability to support vertical loads (vertical collapse) long before sidesway collapse can be reached. Differences between sidesway and vertical collapse behaviors are shown in Figure 3-2.

(a)   (b)

Figure 3-2   Different collapse behaviors: (a) vertical collapse due to loss of vertical-load-carrying capacity; and (b) incipient sidesway collapse due to loss of lateral-force-resisting capacity (reproduced with permission of EERI).

Consideration of vertical collapse modes is beyond the scope of this investigation, however, collapse simulation and explicit consideration of both vertical and sidesway collapse modes are described in FEMA P695 *Quantification of Building Seismic Performance Factors* (FEMA, 2009).

3.1.3.4 Incremental Dynamic Analysis Curves

By plotting discrete intensity measure/engineering demand parameter pairs in an IM-EDP plane, the results of incremental dynamic analyses can be displayed as a suite of IDA curves, one curve corresponding to each ground motion record. An example of one such suite of curves is shown in Figure 3-3, where IDA curves computed from 30 different ground motions are shown. Curves in this figure are plotted with the normalized intensity measure $R = S_a/S_a^{yield}$ on the vertical axis, and normalized engineering demand parameter $\mu = \delta/\delta^{yield}$ on the horizontal axis.

The IDA curves in Figure 3-3a have a common characteristic in that they all terminate with a distinctive horizontal segment, referred to as "flatline." Horizontal segments in IDA curves mean that large displacements occur at small increments in ground motion intensity, which is indicative of lateral dynamic instability. The intensity (or normalized intensity) at which IDA curves become horizontal is taken as the sidesway collapse capacity of the system.

As shown in Figure 3-3a, the sidesway collapse capacity varies significantly from one ground motion record to another. This variability in response is known as record-to-record variability. Because of record-to-record variability, the response due to any one record is highly uncertain. For this reason, statistical information on response due to a suite of ground motions is used to quantify the central tendency (median) and variability (dispersion) of the behavior of a structural system.

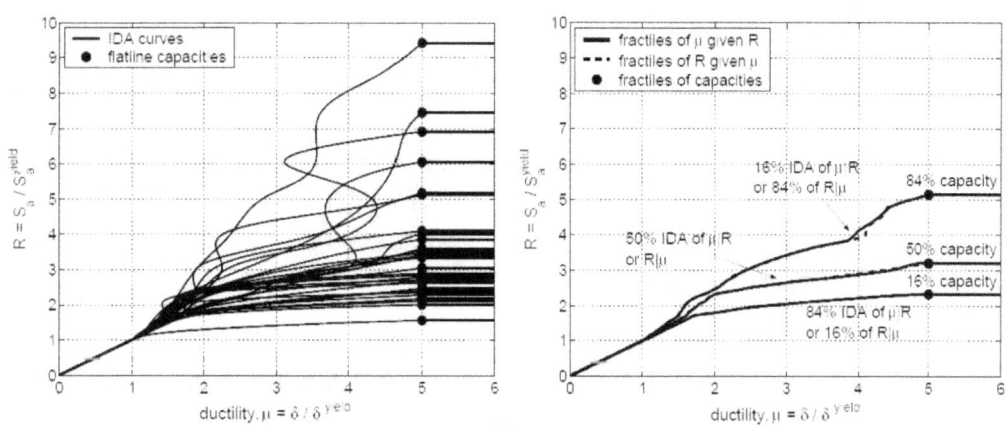

(a) Thirty IDA curves and flatline capacities   (b) Summarization into fractile IDAs, given R or μ

Figure 3-3   Examples depicting incremental dynamic analysis results; (a) suite of individual IDA curves from 30 different ground motion records; and (b) statistically derived quantile curves given μ or R (Vamvatsikos and Cornell 2006)

Figure 3-3b shows quantiles (i.e., 16th, 50th (median) and 84th percentiles) of collapse capacity derived from the results of the 30 IDA curves shown in Figure 3-3a. Also shown in Figure 3-3b, are the 16th, 50th (median) and 84th percentile curves of normalized deformation demands for given normalized ground motion intensities (μ given R), and normalized ground motion intensities for given lateral deformation demands (R given μ). In the figure, the median curve for μ given R is approximately the same as the median curve for R given μ; the 16th percentile curve for μ given R is approximately the same as the 84th percentile curve for R given μ; and the 84th percentile curve for μ given R is approximately the same as the 16th percentile curve for R given μ.

Computing normalized ground motion intensities for given lateral deformation demands (i.e., R given μ) is an iterative process (Ruiz-Garcia and Miranda, 2003). Further complicating this process is that, in certain cases, there can be multiple intensity levels corresponding to a given lateral deformation demand (Vamvatsikos and Cornell, 2002). For these reasons, results in this investigation are reported using quantiles of lateral deformation demand given ground motion intensity (i.e., μ given R).

Use of quantiles of deformation given intensity (i.e., μ given R) means that 16% of the lateral deformation demands for a given level of ground motion intensity would be to the left of the 16th percentile IDA curve, and that 84% would be to the right. Thus, the 16th percentile IDA curve for μ given R will always be above the median curve. Similarly, the 84th percentile IDA curve for μ given R will always be below the median curve.

3.1.4 Ground Motion Records

Analyses were performed using an early version of the ground motion record set selected for use in the ATC-63 Project, and provided in FEMA P695 *Quantification of Building Seismic Performance Factors* (FEMA, 2009). In general this set is intended to include far-field records from all large-magnitude events in the PEER NGA database (PEER, 2006). To avoid event bias, no more than two records were taken from any one earthquake.

In total 28 sets of two horizontal components were used (see Table 3-1). This record set is similar, but not identical, to the set ultimately selected for use in FEMA P695. All records are from firm soil sites, and none include any traces of near source directivity. Values of peak ground acceleration (PGA) and peak ground velocity (PGV) shown in the table correspond to the largest of the two horizontal components.

Table 3-1 Earthquake Records Used in Focused Analytical Studies (Both Horizontal Components)						
Event[1] Station	R[2] Km	Vs30[3] m/s	φ_1[4] Deg	φ_2[4] deg	PGA g	PGV cm/s
Northridge 1994 (M=6.7)						
1. Beverly Hills, Mullholland Dr.	9.4	356	009	279	0.52	57.2
2. Canyon Country, W Lost Canyon	11.4	309	000	270	0.48	44.8
Kern County 1952 (M=7.4)						
3. Taft Lincoln School	38.4	385	021	111	0.18	15.6
Borrego Mtn 1968 (M=6.6)						
4. El Centro Array #9	45.1	213	180	270	0.13	18.5
Duzce Turkey 1999 (M=7.1)						
5. Bolu	12	326	000	090	0.82	59.2
Hector Mine 1999 (M=7.1)						
6. Armboy	41.8	271	090	360	0.18	23.2
7. Hector	10.4	685	000	090	0.34	34.1
Imperial Valley 1979 (M=6.5)						
8. Delta	22	275	262	352	0.35	28.4
9. El centro Array #11	12.5	196	140	230	0.38	36.7
Kobe, Japan 1995 (M=6.9)						
10. Nishi-Akashi	7.1	609	000	090	0.51	36.1
11. Shin-Osaka	19.1	256	000	090	0.24	33.9
Kocaeli, Turkey 1999 (M=7.5)						
12. Duzce	13.6	276	180	270	0.36	54.1
13. Arcelik	10.6	523	000	090	0.22	27.4
Landers 1992 (M=7.3)						
14. Yermo Fire Station	23.6	354	270	360	0.24	37.7
15. Coolwater	19.7	271	long	trans	0.42	32.4
Loma Prieta 1989 (M=6.9)						
16. Capitola	8.7	289	000	090	0.53	34.2
17. Gilroy Array #3	12.2	350	000	090	0.56	42.3
Manjil Iran 1990 (M=7.4)						
18. Abbar	12.6	724	long	trans	0.51	47.3

Table 3-1 Earthquake Records Used in Focused Analytical Studies (Both Horizontal Components) (continued)

Event[1] Station	R[2] Km	Vs30[3] m/s	φ_1[4] Deg	φ_2[4] deg	PGA g	PGV cm/s
Superstition Hills 1987 (M=6.7)						
19. El Centro Imp. Co Cent	18.2	192	000	090	0.36	42.8
20. Poe Road	11.2	208	270	360	0.45	31.7
Cape Mendocino 1992 (M=7.0)						
21. Eureka – Myrtle and West	40.2	339	000	090	0.18	24.2
22. Rio Dell Overpass – FF	7.9	312	270	360	0.55	45.4
Chi-Chi, Taiwan 1999 (M=7.6)						
23. CHY101	10.0	259	090	000	0.44	90.7
24. TCU045	26.0	705	090	000	0.51	38.8
San Fernando, 1971 (M=6.6)						
25. LA Hollywood Sto FF	22.8	316	090	180	0.21	17.8
St Elias, Alaska 1979 (M=7.5)						
26. Yakutat	80.0	275	009	279	0.08	34.3
27. Icy Bay	26.5	275	090	180	0.18	26.6
Friuli, Italy 1976 (M=6.5)						
28. Tolmezzo	15.0	425	000	270	0.35	25.9

[1] Moment magnitude
[2] Closest distance to surface projection of fault rupture
[3] S-wave speed in upper 30m of soil
[4] Component

3.1.5 Analytical Models

The basis of the focused analytical studies is a set of idealized spring models representative of the hysteretic and force-displacement capacity boundary characteristics of different structural systems. The springs were modeled using the Pinching4, ElasticPP and MinMax uniaxial materials in OpenSEES. The Pinching4 material allows the definition of a complex multi-linear force-displacement capacity boundary composed of four linear segments. The ElasticPP material defines a system with an elasto-plastic force-displacement capacity boundary. The MinMax material allows the setting of an ultimate drift at which a system loses all its lateral-force-resisting capacity in both loading directions. The Pinching4 and ElasticPP materials in combination with MinMax were used to define springs with the desired force-displacement capacity boundary characteristics along with finite ultimate deformation capacities.

Parametric studies were conducted on single-degree-of-freedom (SDOF) oscillators constructed with these springs and their variants. Generic story-models were developed using single-spring systems or multi-spring systems consisting of two springs in parallel. Story models were intended to approximate the behavior of single-story systems composed of an individual subassembly or a mixture of subassemblies having complex hysteretic and force-displacement capacity boundary characteristics linked by rigid diaphragms.

3.2   Single-Spring Models

Each single-spring system model is defined by a hysteretic model confined within a force-displacement boundary (Figure 3-4) developed from information available in the literature. The single-spring systems are based on the following set of eight different hysteretic behaviors and force-displacement capacity boundary characteristics:

- Spring 1 – typical gravity frame system (e.g., steel)

- Spring 2 – non-ductile moment frame system (e.g., steel or concrete)

- Spring 3 – ductile moment frame system (e.g., steel or concrete)

- Spring 4 – stiff non-ductile system (e.g., steel concentric braced frame)

- Spring 5 – stiff, highly-pinched non-ductile system (e.g., brittle infill wall)

- Spring 6 – elastic-perfectly-plastic system (for comparison)

- Spring 7 – limited-ductility moment frame system (e.g., concrete)

- Spring 8 – non-ductile gravity frame system (e.g., concrete)

While intended to be representative of realistic degrading response that has been observed to occur in some structural components, these idealized springs are not intended to be a detailed characterization of the mechanical behavior of any one specific structural component or structural subassembly. Rather, they are used to capture the main response characteristics of components or subassemblies that are often present and combined in real structural systems. The focus was not on investigating the seismic performance of a particular structural system, but on identifying the effects of various aspects of degrading behavior on the response of one-story single-degree-of-freedom system models.

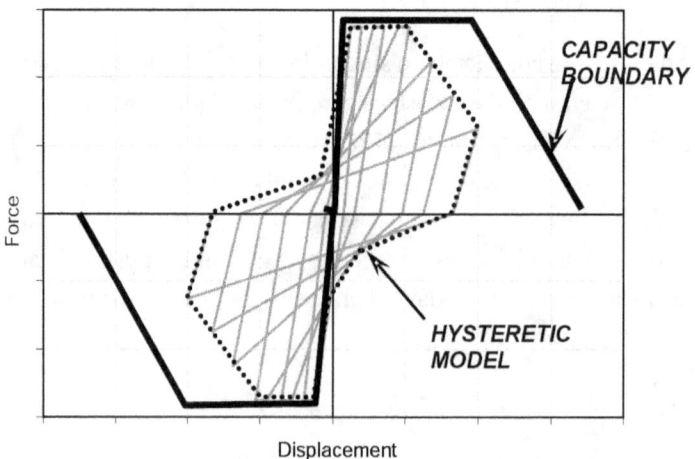

Figure 3-4   Hysteretic model confined by a force-displacement capacity boundary.

An "a" and "b" version of each spring was developed. The "a" and "b" versions differ by one or two characteristics of the force-displacement capacity boundary so that the "b" version always possesses the more favorable characteristics of the two. Sources of variation included the point at onset of degradation, the slope of degradation, the level of residual strength, and length of the residual strength plateau. To investigate period dependency, systems utilizing the "a" and "b" versions of each individual spring were tuned to periods of 0.5s, 1.0s, 1.5s, 2.0s, and 2.5s.

All springs were defined to be symmetrical, using the same force-displacement capacity boundary in both the positive and negative loading directions. All have a finite ultimate deformation capacity at which all lateral-force-resisting capacity is lost, and all, except for Spring 6 (which is elastic-perfectly-plastic), include in-cycle strength degradation.

In addition, the "a" and "b" versions of each spring (except for Spring 6) were analyzed with both a constant force-displacement capacity boundary and a degrading force-displacement capacity boundary in order to quantify the effect of cyclic degradation on system response. To do this, springs were subjected to an ATC-24 type loading protocol (ATC, 1992), consisting of two cycles at each level of drift starting at 0.5% drift, and increasing in increments of 1% drift up to a maximum of 8% drift.

The generic force-displacement capacity boundary used for all springs is shown in Figure 3-5. The values of normalized base shear, F/F_y, and story drift ratio, θ, chosen to characterize the force-displacement capacity boundary for each of the single-spring system models are listed in Table 3-2.

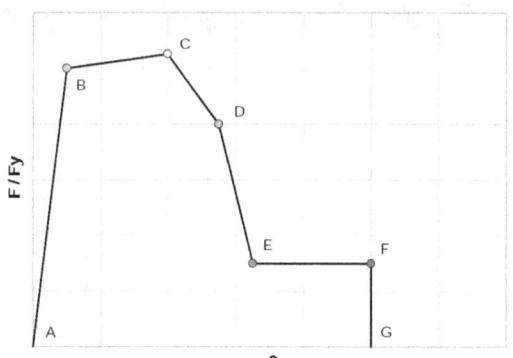

Figure 3-5   Generic force-displacement capacity boundary used for all single-spring system models.

For purposes of comparison, one version of each spring is shown in Figure 3-6. The parameters that define each spring, and the variations in each spring, are described in more detail in the sections that follow.

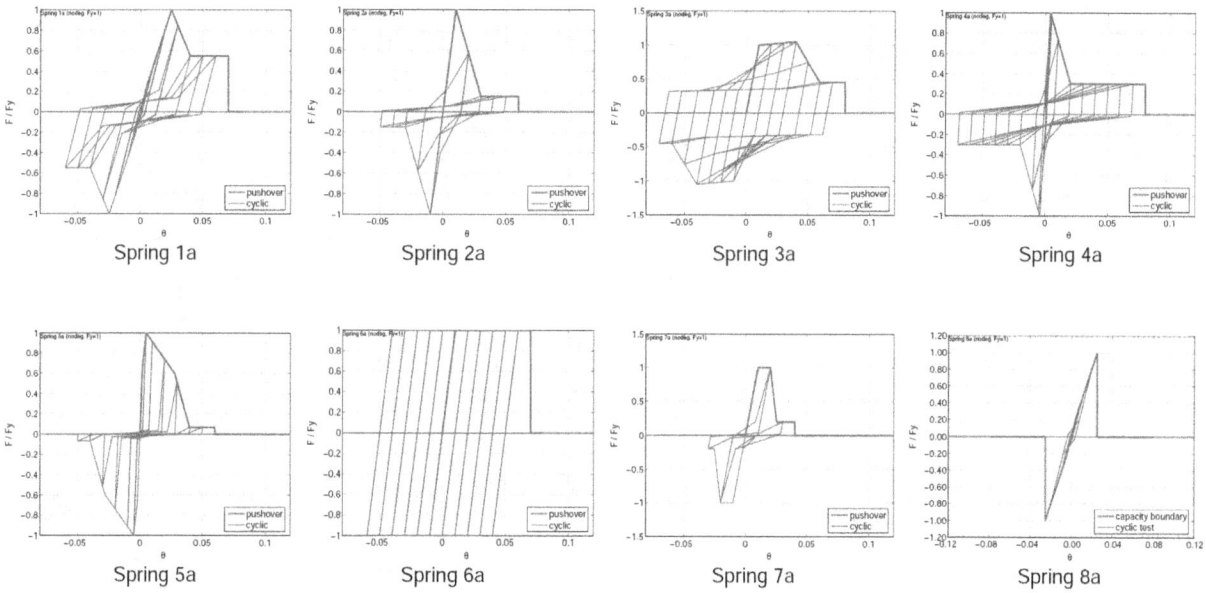

Figure 3-6   Comparison of eight basic single-spring system models.

Table 3-2   Force-Displacement Capacity Boundary Control Points for Single-Spring System Models.

Prototype	Type	Quantity	Points of the force-deformation capacity boundary						
			A	B	C	D	E	F	G
Typical gravity frame	1a	F/Fy	0	0.25	1	0.55	0.55	0.55	0
		θ	0	0.005	0.025	0.04	0.07	0.07	0.07
	1b	F/Fy	0	0.25	1	0.55	0.55	0.55	0
		θ	0	0.005	0.025	0.04	0.12	0.12	0.12
Non-ductile moment frame	2a	F/Fy	0	1	0.15	0.15	0.15	0.15	0
		θ	0	0.01	0.03	0.05	0.06	0.06	0.06
	2b	F/Fy	0	1	0.15	0.15	0.15	0.15	0
		θ	0	0.01	0.05	0.055	0.06	0.06	0.06
Ductile moment frame	3a	F/Fy	0	1	1.05	0.45	0.45	0.45	0
		θ	0	0.01	0.04	0.06	0.08	0.08	0.08
	3b	F/Fy	0	1	1.05	0.8	0.8	0.8	0
		θ	0	0.01	0.04	0.06	0.08	0.08	0.08
Stiff non-ductile system	4a	F/Fy	0	1	0.3	0.3	0.3	0.3	0
		θ	0	0.004	0.02	0.06	0.08	0.08	0.08
	4b	F/Fy	0	1	0.5	0.5	0.5	0.5	0
		θ	0	0.004	0.04	0.06	0.08	0.08	0.08
Stiff, highly pinched non-ductile system	5a	F/Fy	0	0.67	1	0.6	0.067	0.067	0
		θ	0	0.002	0.005	0.028	0.04	0.06	0.06
	5b	F/Fy	0	0.67	1	0.6	0.067	0.067	0
		θ	0	0.002	0.005	0.042	0.06	0.06	0.06
Elastic-perfectly-plastic	6a	F/Fy	0	1	1	1	1	1	0
		θ	0	0.01	0.02	0.03	0.07	0.07	0.07
	6b	F/Fy	0	1	1	1	1	1	0
		θ	0	0.01	0.02	0.03	0.12	0.12	0.12
Limited-ductile moment frame	7a	F/Fy	0	1	1	0.2	0.2	0.2	0
		θ	0	0.01	0.02	0.025	0.04	0.04	0.04
	7b	F/Fy	0	1	1	0.2	0.2	0.2	0
		θ	0	0.01	0.02	0.04	0.06	0.06	0.06
Non-ductile gravity frame	8a	F/Fy	0	1	1	0	0	0	0
		θ	0	0.025	0.025	0.025	0.025	0.025	0.025
	8b	F/Fy	0	1	1	0.55	0.55	0.55	0
		θ	0	0.025	0.025	0.03	0.04	0.04	0.04

**3: Development of Single-Degree-of-Freedom Models
for Focused Analytical Studies**

3.2.1 Springs 1a and 1b – Typical Gravity Frame Systems

Springs 1a and 1b are intended to model the behavior of typical gravity frame systems in buildings. The force-displacement capacity boundary includes a strength drop immediately after yielding that terminates on a plateau with a residual strength of 55% of the yield strength (Figure 3-7). The "a" and "b" versions of this spring differ in the length of the residual strength plateau, which extends to an ultimate deformation capacity of 7% drift in Spring 1a and 12% drift in Spring 1b. This represents the maximum ductility that is achieved by any of the spring subsystems.

The hysteretic behaviors of Spring 1a and Spring 1b, both with and without cyclic degradation, are shown in Figure 3-8 and Figure 3-9. In each figure, the initial force-displacement capacity boundary (before cyclic degradation) is overlaid onto the hysteretic plots.

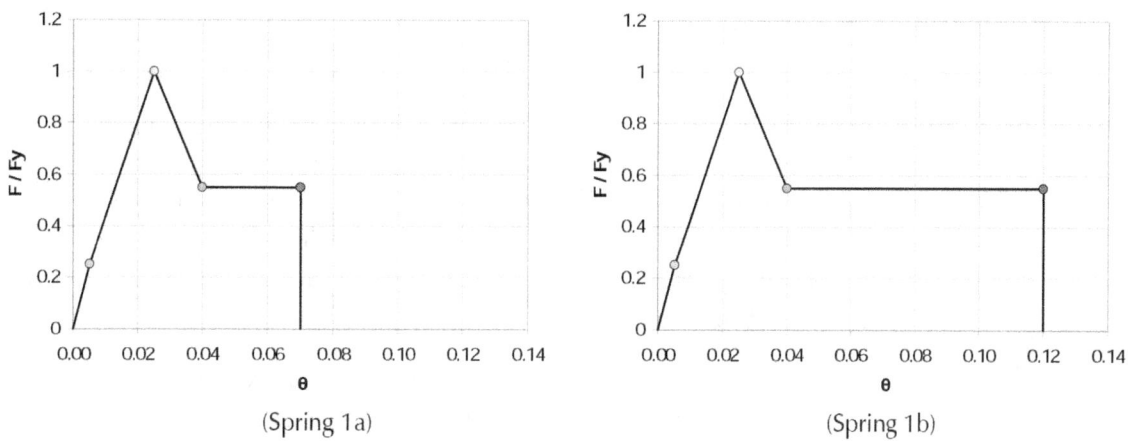

(Spring 1a)   (Spring 1b)

Figure 3-7   Force-displacement capacity boundaries for Spring 1a and Spring 1b.

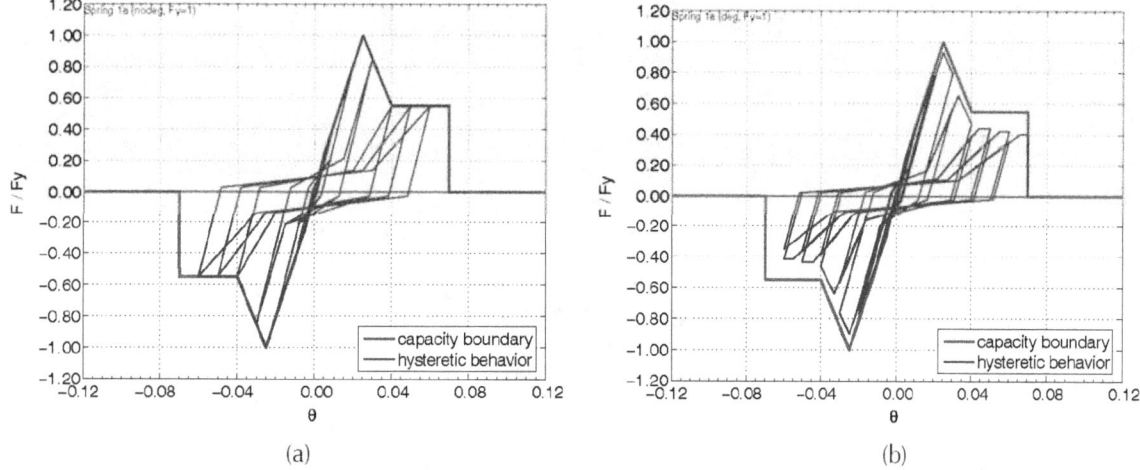

(a)   (b)

Figure 3-8   Initial force-displacement capacity boundary overlaid onto hysteretic behaviors for Spring 1a: (a) without cyclic degradation; and (b) with cyclic degradation.

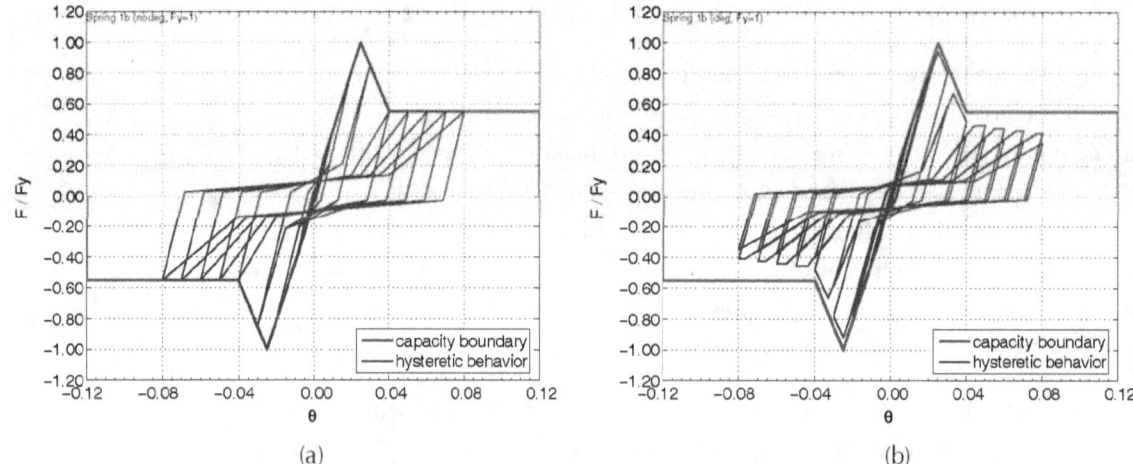

(a)   (b)

Figure 3-9   Initial force-displacement capacity boundary overlaid onto hysteretic behaviors for
                  Spring 1b: (a) without cyclic degradation; and (b) with cyclic degradation.

Springs 1a and 1b are consistent with steel gravity frame systems with classic simple shear-tab connections. Experiments have shown that the gap between the beam and column flange is a critical parameter in determining force-displacement behavior of these systems. When a joint achieves enough rotation to result in contact between the beam and column flanges, bolts in the shear tab will be subjected to bearing strength failure, and the shear connection fails (Liu and Astaneh, 2003). This limit state marks the end of the residual strength plateau.

Spring 1a is consistent with a system in which beam/column flange contact occurs relatively early (7% drift), while Spring 1b is consistent with a system in which this contact occurs later (12% drift). Results from experimental tests on steel shear tab connections (Figure 3-10) exhibit a behavior that is similar to behavior the modeled in Springs 1a and 1b.

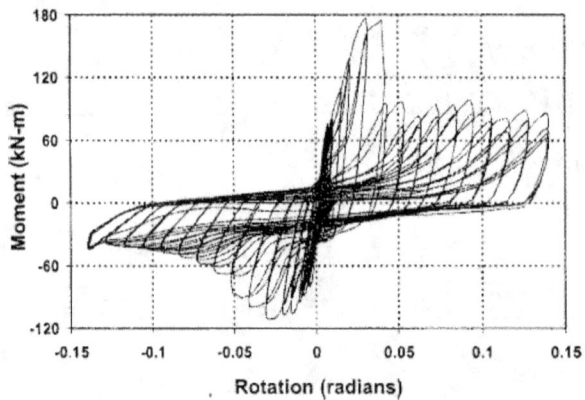

Figure 3-10   Hysteretic behavior from experimental tests on beam-to-column
                  shear tab connections (Liu and Astaneh, 2003).

3: Development of Single-Degree-of-Freedom Models   FEMA P440A
                  for Focused Analytical Studies

3.2.2 Springs 2a and 2b – Non-Ductile Moment Frame Systems

Springs 2a and 2b are intended to model the behavior of non-ductile moment-resisting frame systems in buildings. They are characterized by a force-displacement capacity boundary that includes strength degradation immediately after yielding, a low residual strength plateau at 15% of the yield strength, and an ultimate deformation capacity of 6% drift (Figure 3-11). The "a" and "b" versions of this spring differ in the negative slope of the strength-degrading segment, which is negative 43% in Spring 2a and negative 21% in Spring 2b.

The hysteretic behaviors of Spring 2a and Spring 2b, both with and without cyclic degradation, are shown in Figure 3-12 and Figure 3-13. In each figure, the initial force-displacement capacity boundary (before cyclic degradation) is overlaid onto the hysteretic plots.

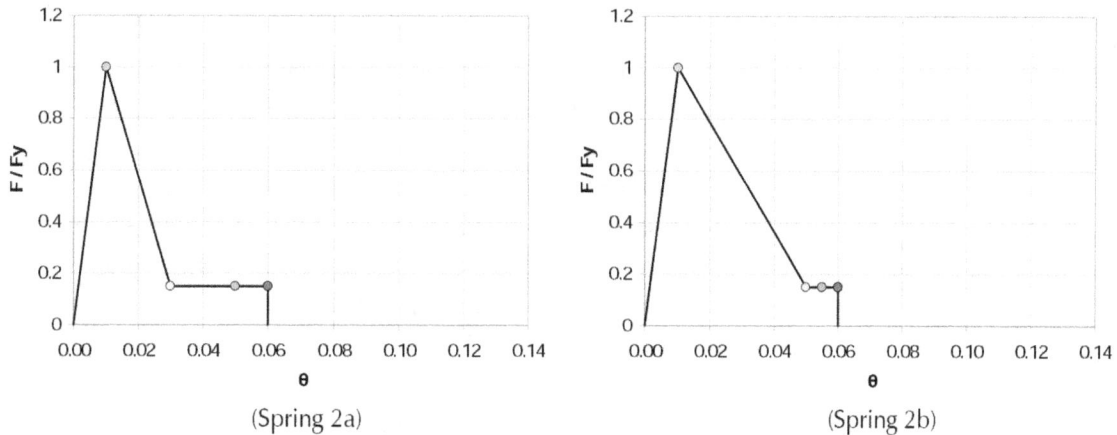

Figure 3-11 Force-displacement capacity boundaries for Spring 2a and Spring 2b.

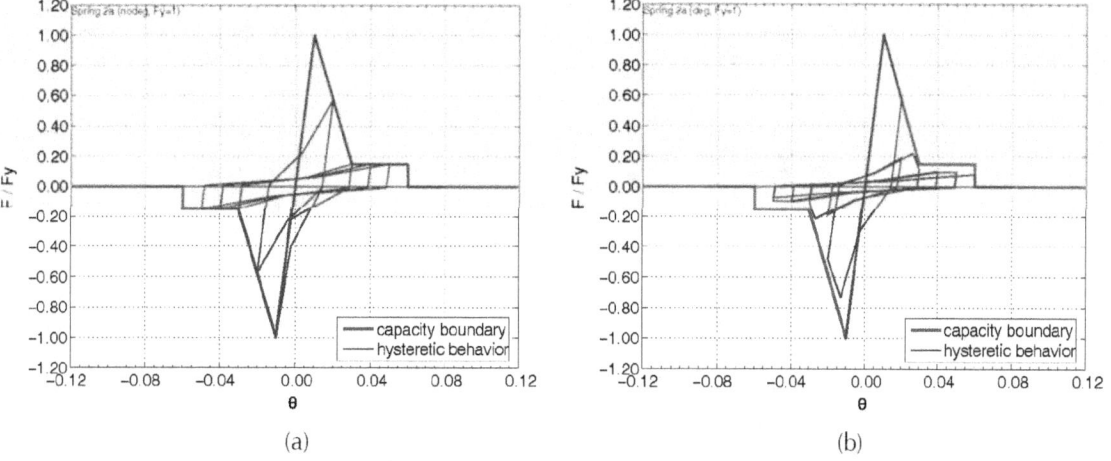

Figure 3-12   Initial force-displacement capacity boundary overlaid onto hysteretic behaviors for Spring 2a: (a) without cyclic degradation; and (b) with cyclic degradation.

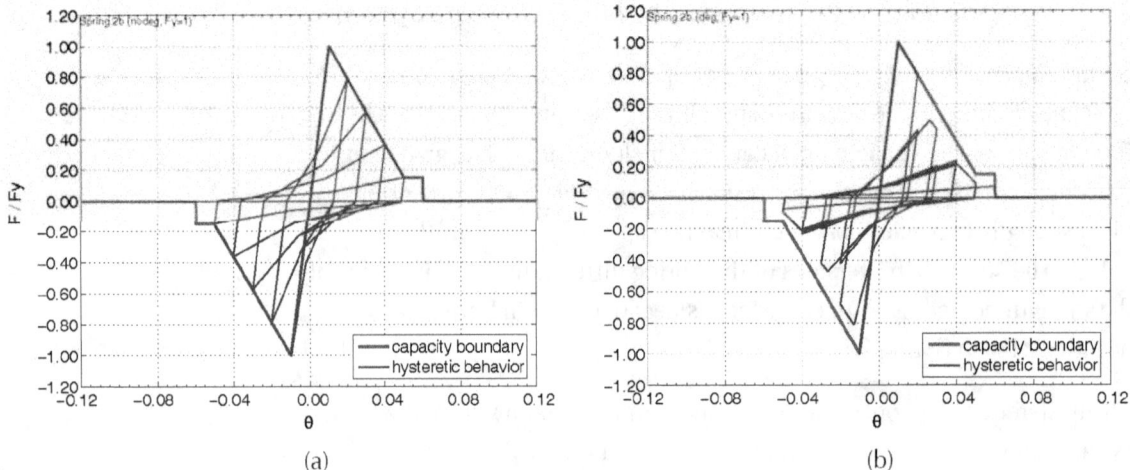

Figure 3-13   Initial force-displacement capacity boundary overlaid onto hysteretic behaviors for Spring 2b: (a) without cyclic degradation; and (b) with cyclic degradation.

Systems with this behavior could be constructed in either steel or concrete. In the case of steel, these springs would be representative of moment-resisting frames with pre-Northridge welded beam-column connections, in which connection behavior is characterized by fracture and a large reduction in lateral force resistance. In the case of concrete, these springs would be representative of older (pre-1975) concrete frames with inadequate joint reinforcement, minimal concrete confinement and other poor detailing characteristics that would be prone to shear failure. Results from experimental tests on pre-Northridge welded steel connections and shear-critical reinforced concrete columns (Figure 3-14) exhibit a behavior that is similar to the behavior modeled in Springs 2a and 2b.

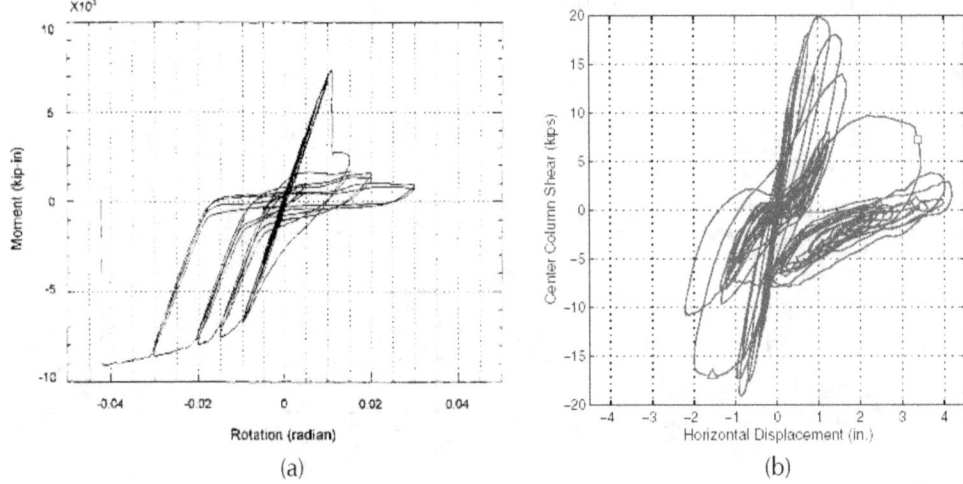

Figure 3-14   Hysteretic behavior from experimental tests on: (a) pre-Northridge welded steel beam-column connections (Goel and Stojadinovic, 1999); and (b) shear-critical reinforced concrete columns (Elwood and Moehle, 2003).

3: Development of Single-Degree-of-Freedom Models
for Focused Analytical Studies

3.2.3 Springs 3a and 3b – Ductile Moment Frame Systems

Springs 3a and 3b are intended to model the behavior of moderately-ductile moment-resisting frame systems in buildings. They are characterized by a force-displacement capacity boundary that includes a strength-hardening segment with a positive slope equal to 2% of the elastic stiffness, a strength-degrading segment that begins at 4% drift and ends at 6% drift, and a residual strength plateau with an ultimate deformation capacity of 8% drift (Figure 3-15). The "a" and "b" versions of this spring differ in the negative slope of the strength-degrading segment, which is negative 30% in Spring 3a and negative 13% in Spring 3b, and in the height of the residual strength plateau, which is 50% of yield in Spring 3a and 80% in Spring 3b.

The hysteretic behaviors of Spring 3a and Spring 3b, both with and without cyclic degradation, are shown in Figure 3-16 and Figure 3-17. In each figure, the initial force-displacement capacity boundary (before cyclic degradation) is overlaid onto the hysteretic plots.

Systems with this type of behavior could include special steel moment-resisting frames with ductile (e.g., post-Northridge) beam-column connections, or well-detailed reinforced concrete moment-resisting frames. Results from experimental tests on post-Northridge reduced beam steel moment connections (Figure 3-18) exhibit a behavior that is similar to the behavior modeled in Springs 3a and 3b.

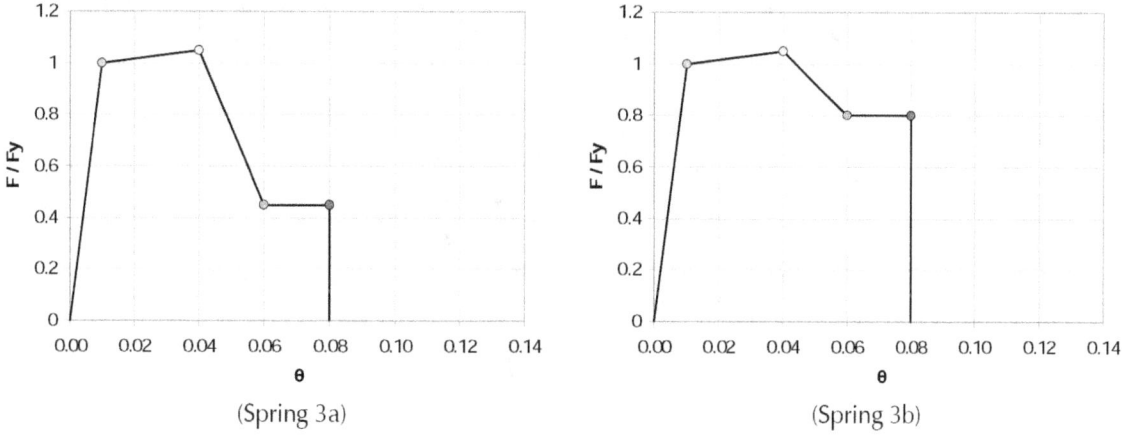

(Spring 3a)   (Spring 3b)

Figure 3-15   Force-displacement capacity boundaries for Spring 3a and Spring 3b.

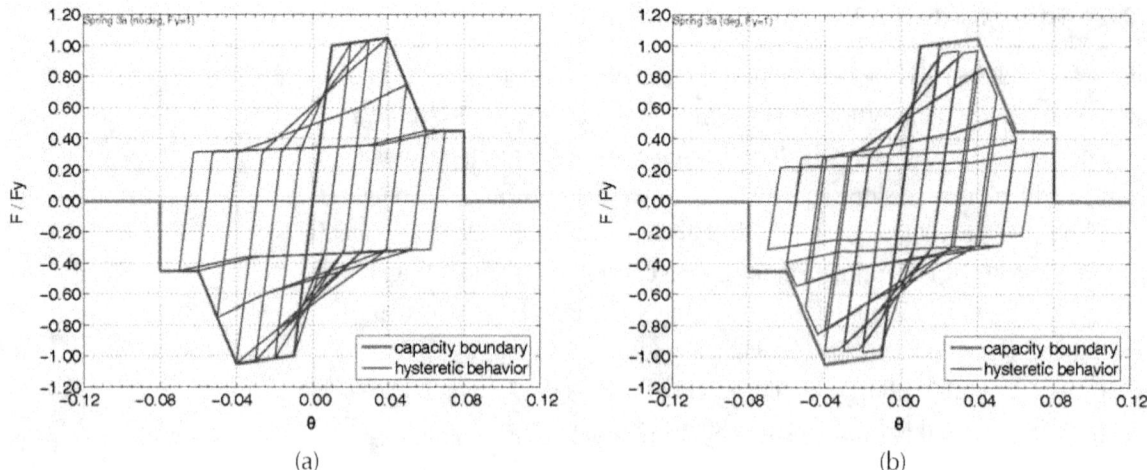

(a)   (b)

Figure 3-16   Initial force-displacement capacity boundary overlaid onto hysteretic behaviors for Spring
3a: (a) without cyclic degradation; and (b) with cyclic degradation.

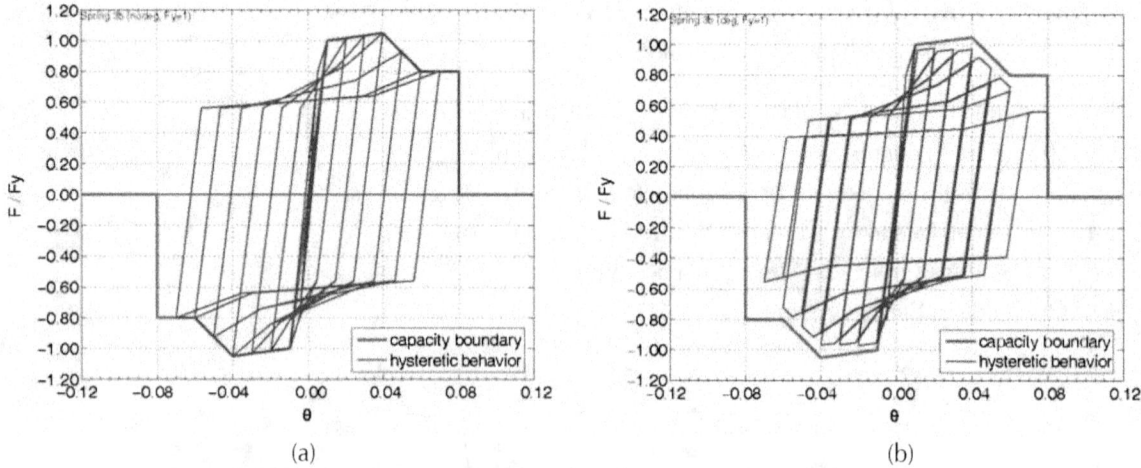

(a)   (b)

Figure 3-17   Initial force-displacement capacity boundary overlaid onto hysteretic behaviors for Spring
3b: (a) without cyclic degradation and (b) with cyclic degradation.

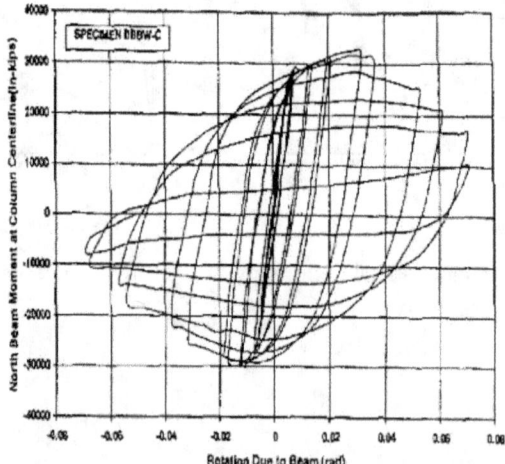

Figure 3-18   Hysteretic behavior from experimental tests on post-Northridge reduced-beam steel
moment connections (Venti and Engelhardt, 1999).

3: Development of Single-Degree-of-Freedom Models
for Focused Analytical Studies

3.2.4 Springs 4a and 4b – Stiff, Non-Ductile Systems

Springs 4a and 4b are intended to model the behavior of relatively stiff lateral-force-resisting systems that are subject to significant in-cycle strength degradation at small levels of deformation. They are characterized by a force-displacement capacity boundary that includes a strength-degrading segment beginning at 0.4% drift and terminating on a residual strength plateau with an ultimate deformation capacity of 8% drift (Figure 3-19). The "a" and "b" versions of this spring differ in the negative slope of the strength-degrading segment, which is negative 18% in Spring 4a and negative 6% in Spring 4b, and in the height of the residual strength plateau, which is 30% of yield in Spring 4a and 50% in Spring 4b.

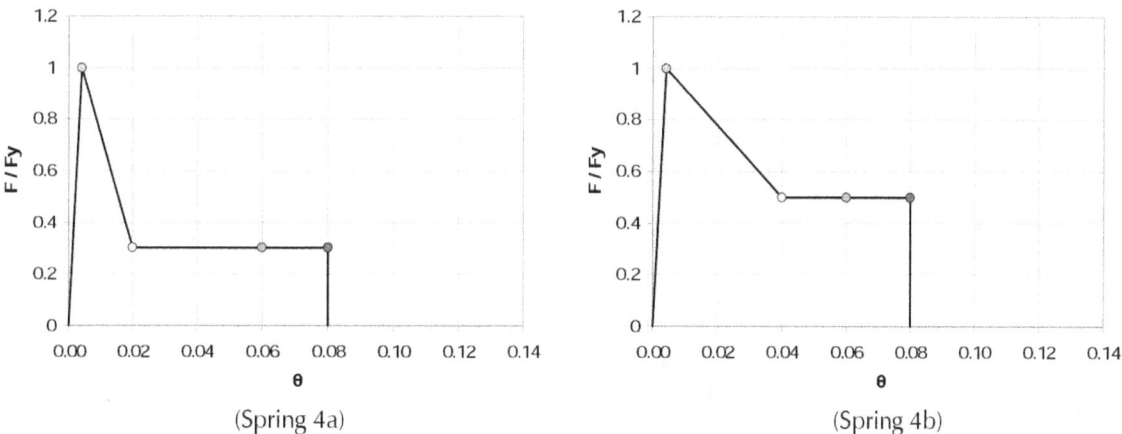

(Spring 4a)   (Spring 4b)

Figure 3-19   Force-displacement capacity boundaries for Spring 4a and Spring 4b.

The hysteretic behaviors of Spring 4a and Spring 4b, both with and without cyclic degradation, are shown in Figure 3-20 and Figure 3-21. They resemble a typical peak-oriented model with severe cyclic degradation of strength, unloading, and reloading stiffness parameters. In each figure, the initial force-displacement capacity boundary (before cyclic degradation) is overlaid onto the hysteretic plots.

Systems with this type of behavior could include steel concentric braced frames, which experience a sharp drop in strength following buckling of the braces at small levels of lateral deformation demand. Results from experimental tests on steel concentric braced frames (Figure 3-22) exhibit a behavior that is similar to the behavior modeled in Springs 4a and 4b.

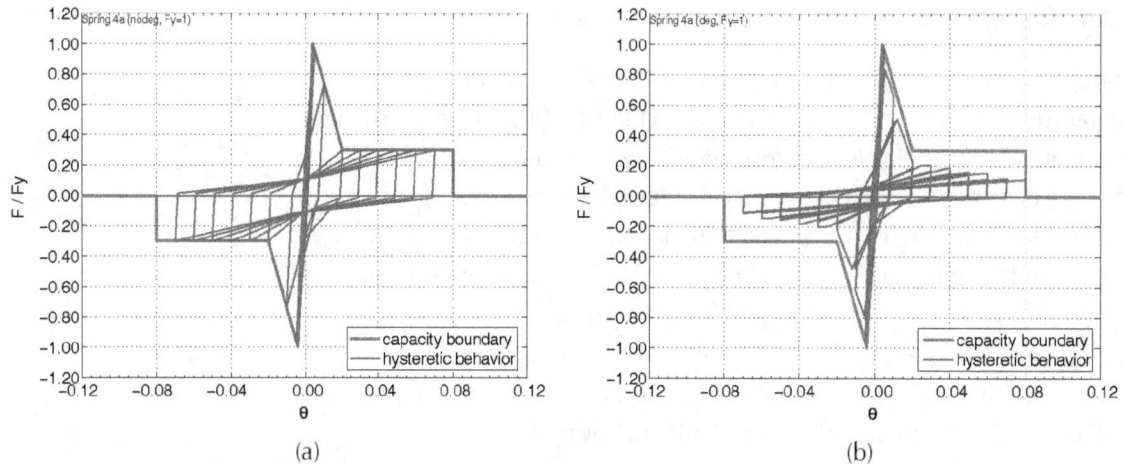

(a)   (b)

Figure 3-20   Initial force-displacement capacity boundary overlaid onto hysteretic behaviors for Spring 4a:
(a) without cyclic degradation; and (b) with cyclic degradation.

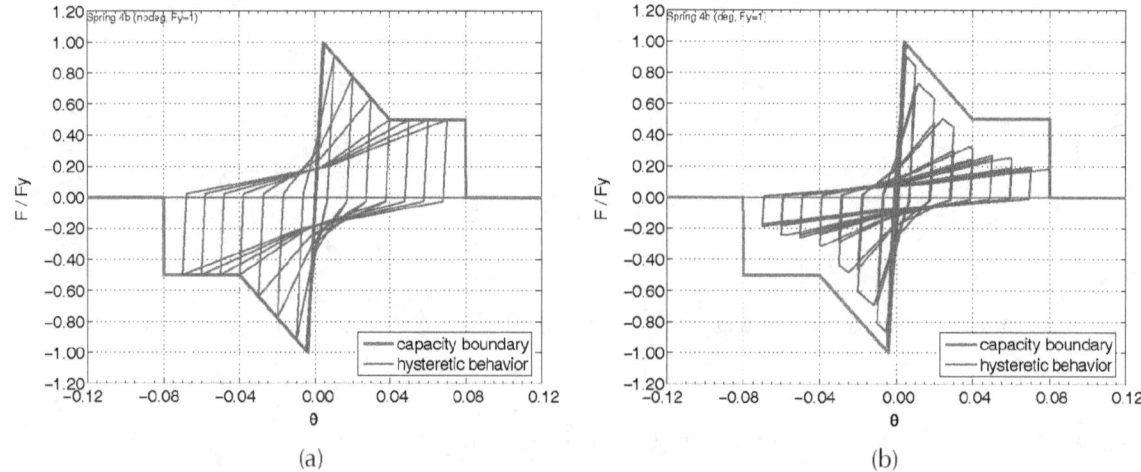

(a)   (b)

Figure 3-21   Initial force-displacement capacity boundary overlaid onto hysteretic behaviors for Spring 4b:
(a) without cyclic degradation; and (b) with cyclic degradation.

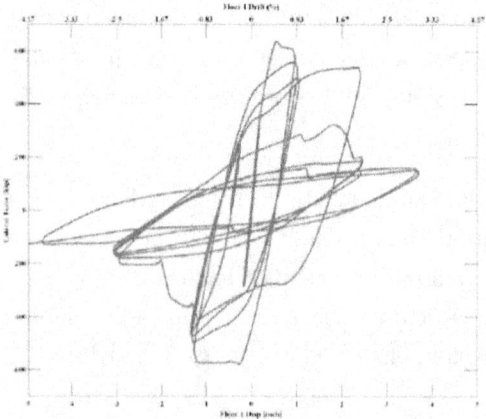

Figure 3-22   Hysteretic behavior from experimental tests on steel concentric braced frames (Uriz and
Mahin, 2004).

3.2.5 Springs 5a and 5b – Stiff, Highly-Pinched Non-Ductile Systems

Springs 5a and 5b are intended to model the behavior of stiff and highly-pinched non-ductile lateral-force-resisting systems in buildings. They are characterized by a force-displacement capacity boundary with the highest initial stiffness of any of the spring subsystems, followed by varying levels of strength degradation and an ultimate deformation capacity of 6% drift (Figure 3-23). In both the "a" and "b" versions of this spring, peak strength occurs at 0.5% drift, and initial cracking occurs at 67% of peak strength at a drift ratio of 0.2%. The "a" and "b" versions of this spring differ in the slopes of the two strength-degrading segments, which are 5% and 13% (of the initial elastic stiffness) in Spring 5a, and 3% and 9% in Spring 5b. They also differ in the presence of a residual strength plateau, which exists in Spring 5a, but not in Spring 5b.

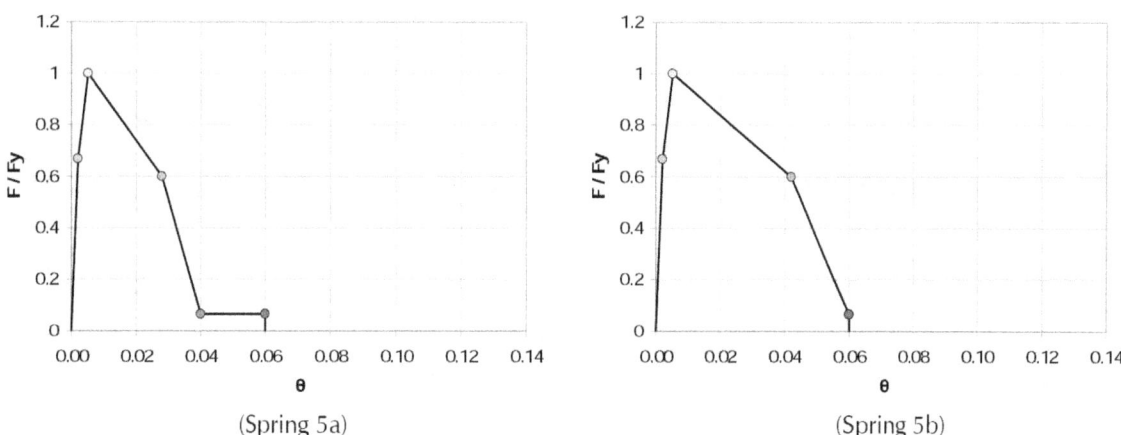

(Spring 5a) (Spring 5b)

Figure 3-23   Force-displacement capacity boundaries for Spring 5a and Spring 5b.

The hysteretic behaviors of Spring 5a and Spring 5b, both with and without cyclic degradation, are shown in Figure 3-24 and Figure 3-25. They resemble a sliding system with cyclic degradation of strength, unloading, and reloading stiffness parameters. In each figure, the initial force-displacement capacity boundary (before cyclic degradation) is overlaid onto the hysteretic plots.

Systems with this type of behavior could include masonry walls and concrete frames with masonry infill. Results from experimental tests on these systems (Figure 3-26) exhibit a behavior that is similar to the behavior modeled in Springs 5a and 5b.

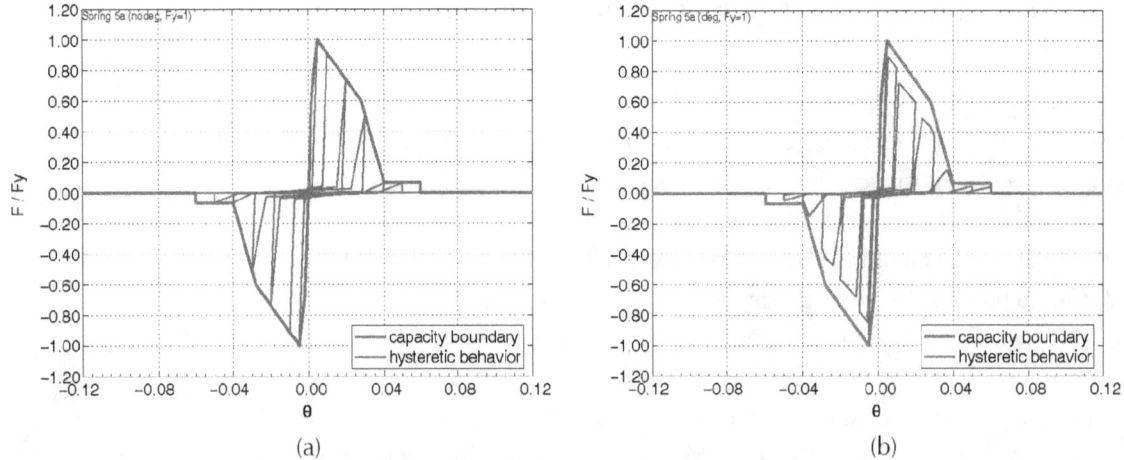

(a)   (b)

Figure 3-24   Initial force-displacement capacity boundary overlaid onto hysteretic behaviors for Spring 5a:
(a) without cyclic degradation; and (b) with cyclic degradation.

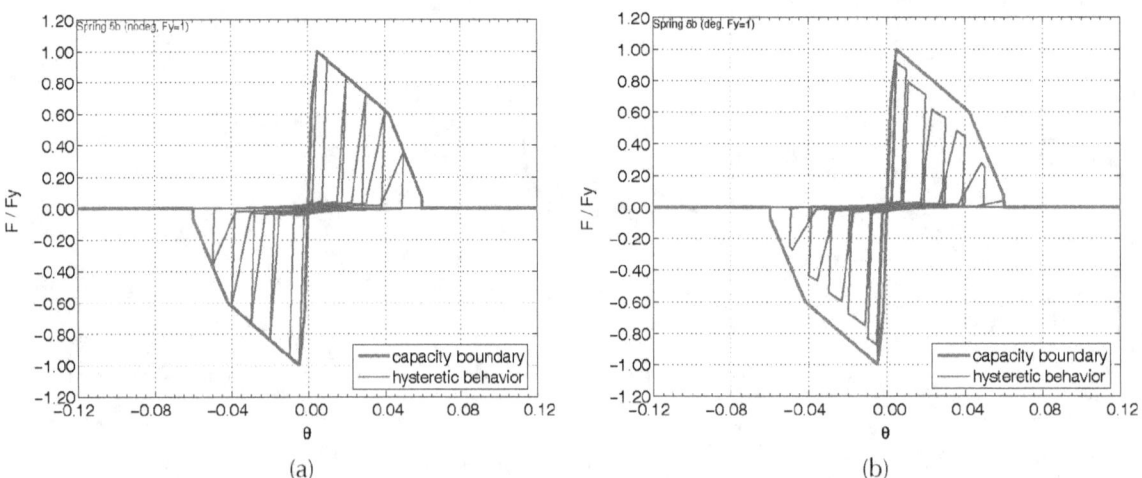

(a)   (b)

Figure 3-25   Initial force-displacement capacity boundary overlaid onto hysteretic behaviors for Spring 5b:
(a) without cyclic degradation; and (b) with cyclic degradation.

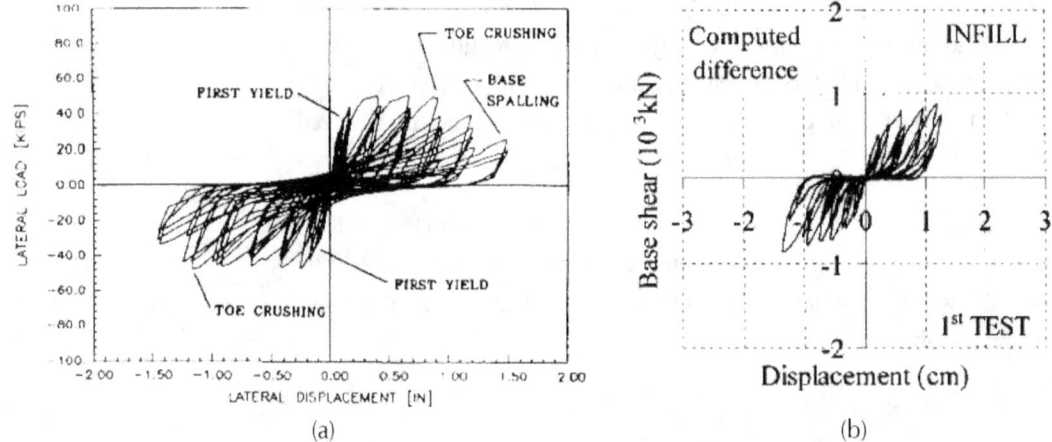

(a)   (b)

Figure 3-26   Hysteretic behavior from experimental tests on: (a) reinforced masonry walls (Shing et al.,
1991); and (b) concrete frames with masonry infill (Dolsek and Fajfar, 2005).

3.2.6   Springs 6a and 6b – Elastic-Perfectly-Plastic Systems

Springs 6a and 6b are intended to model the behavior of idealized elastic-perfectly-plastic systems with full, kinematic hysteresis loops, without any cyclic or in-cycle degradation of strength or stiffness. The force-displacement capacity boundaries are shown in Figure 3-27. The "a" and "b" versions of this spring differ in their finite ultimate deformation capacity, which is 7% drift in Spring 6a and 12% drift in Spring 6b.

Spring 6a and Spring 6b were analyzed with a constant force-displacement capacity boundary (no cyclic degradation). The resulting hysteretic behaviors are shown in Figure 3-28, with initial force-displacement capacity boundaries overlaid onto the hysteretic plots.

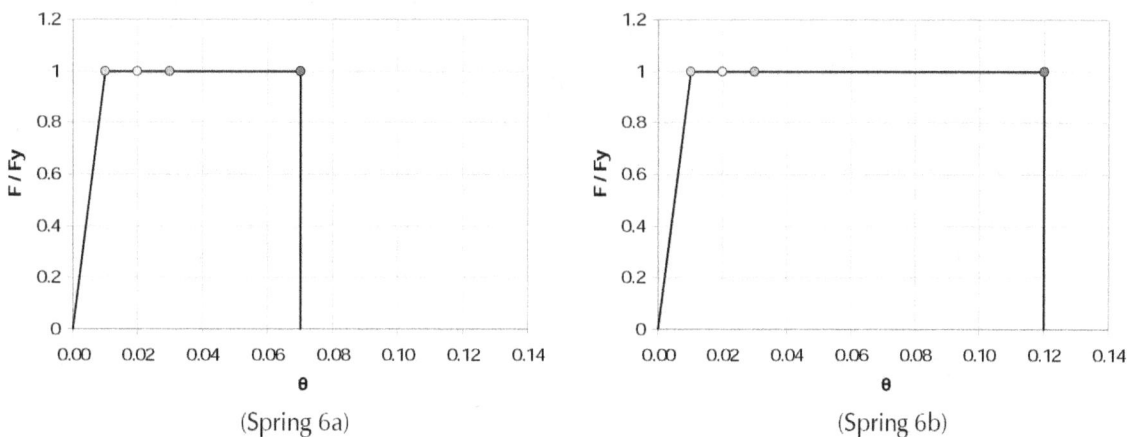

(Spring 6a)   (Spring 6b)

Figure 3-27   Force-displacement capacity boundaries for Spring 6a and Spring 6b.

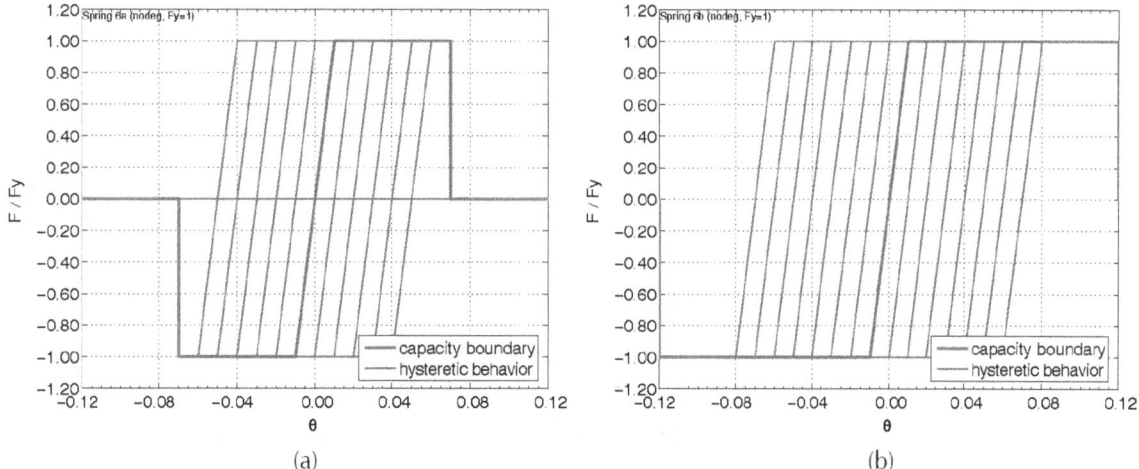

(a)   (b)

Figure 3-28   Force-displacement capacity boundary overlaid onto hysteretic behaviors for: (a) Spring 6a without cyclic degradation; and (b) Spring 6b without cyclic degradation.

This is a highly idealized system developed for comparison of results. Practically speaking, only selected buckling-restrained braces or base-isolated systems would be capable of emulating this behavior under repeated cycles of large deformation demand.

3.2.7 Springs 7a and 7b – Limited-Ductility Moment Frame Systems

Springs 7a and 7b are intended to model the behavior of limited-ductility moment-resisting frame systems in buildings. They are characterized by a force-displacement capacity boundary with a short yielding plateau that maintains strength until a drift of 2%, followed strength degradation that terminates on a short residual strength plateau set at 20% of the yield strength (Figure 3-29). The "a" and "b" versions of this spring differ in the negative slope of the strength-degrading segment, which is negative 160% in Spring 7a and negative 40% in Spring 7b, and in the ultimate deformation capacity, which is 4% drift in Spring 7a and 6% drift in Spring 7b.

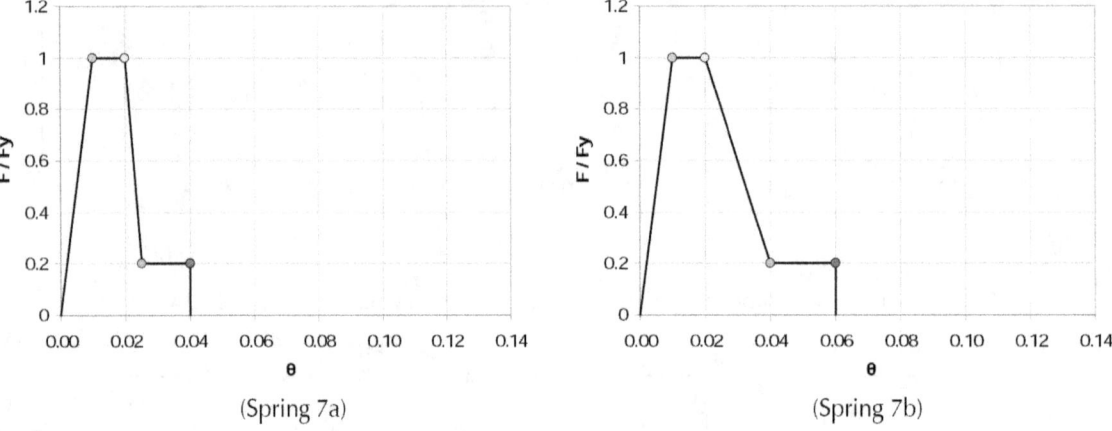

(Spring 7a)   (Spring 7b)

Figure 3-29   Force-displacement capacity boundaries for Spring 7a and Spring 7b.

The hysteretic behaviors of Spring 7a and Spring 7b, both with and without cyclic degradation, are shown in Figure 3-30 and Figure 3-31. In each figure, the initial force-displacement capacity boundary (before cyclic degradation) is overlaid onto the hysteretic plots.

Systems with this type of behavior could include older reinforced concrete frames not designed for seismic loads, which can be lightly reinforced, and may have inadequate joint reinforcement or concrete confinement. Results from experimental tests on lightly reinforced concrete columns (Figure 3-32) exhibit a behavior that is similar to the behavior modeled in Springs 7a and 7b.

3: Development of Single-Degree-of-Freedom Models for Focused Analytical Studies

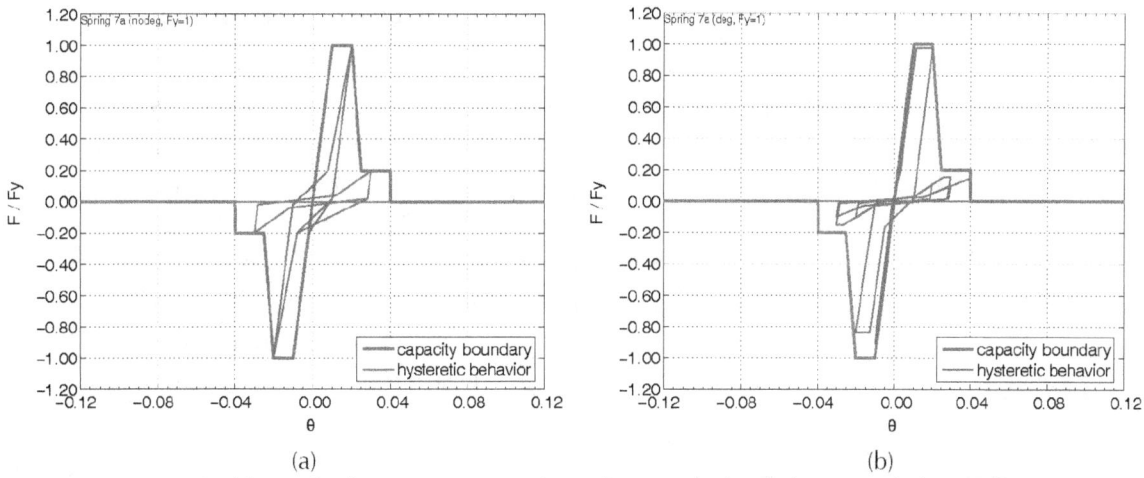

(a)

(b)

Figure 3-30   Initial force-displacement capacity boundary overlaid onto hysteretic behaviors for Spring 7a: (a) without cyclic degradation; and (b) with cyclic degradation.

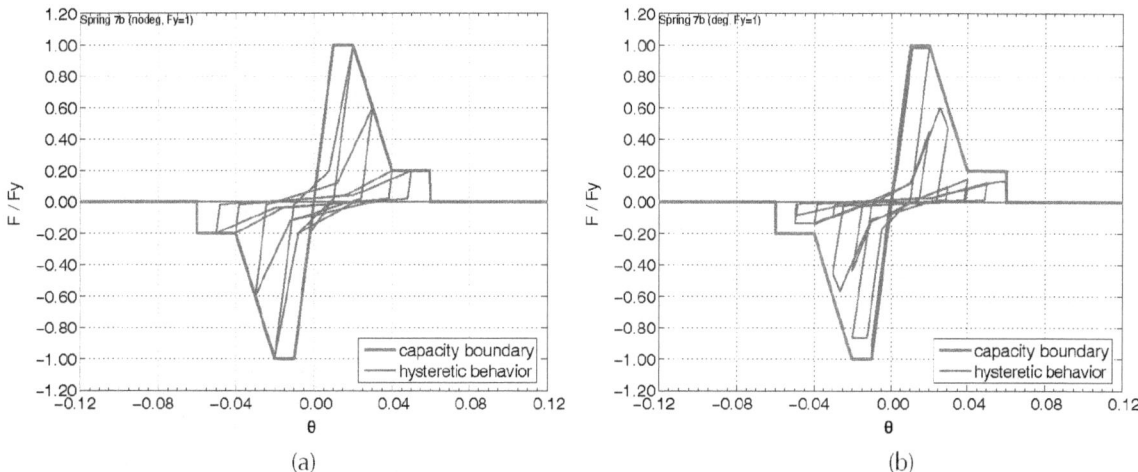

(a)

(b)

Figure 3-31   Initial force-displacement capacity boundary overlaid onto hysteretic behaviors for Spring 7b: (a) without cyclic degradation; and (b) with cyclic degradation.

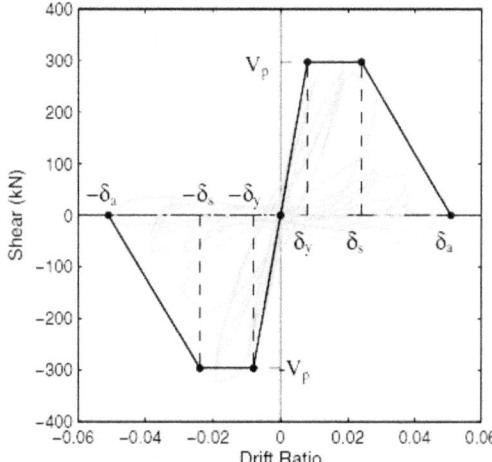

Figure 3-32   Hysteretic behavior from experimental tests on lightly reinforced concrete columns (Elwood and Moehle, 2006; Sezen, 2002).

3.2.8 Springs 8a and 8b – Non-Ductile Gravity Frame Systems

Springs 8a and 8b are intended to model the behavior of non-ductile gravity frame systems in buildings. The force-displacement capacity boundary includes significant strength degradation immediately after yielding, and limited ultimate deformation capacity (Figure 3-33). The "a" and "b" versions of this spring differ in the strength that is lost after yield, which is 100% in Spring 8a, and 45% in Spring 8b, and in the ultimate deformation capacity, which is 2.5% drift in Spring 8a and 4% drift in Spring 8b. They also differ in the presence of a residual strength plateau, which does not exist in Spring 8a, but does in Spring 8b.

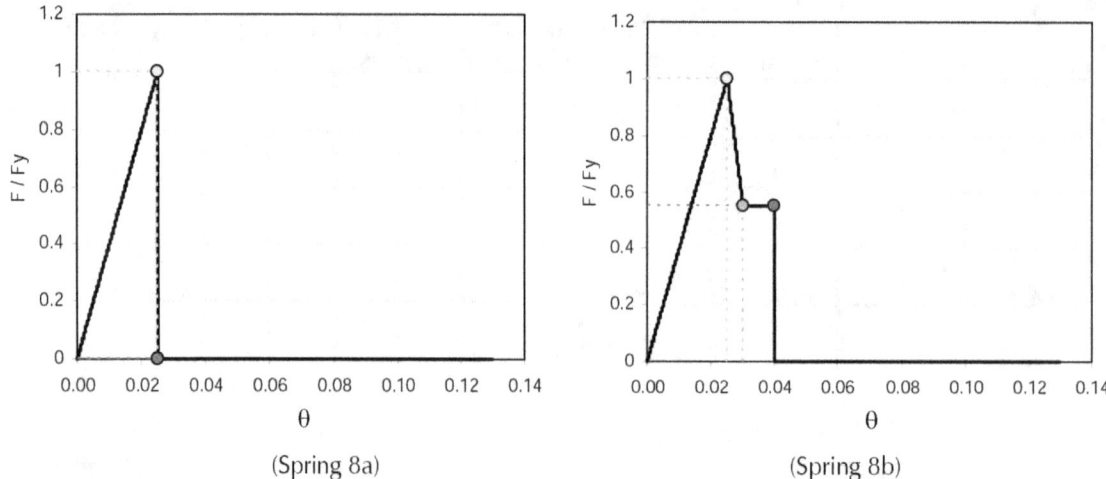

(Spring 8a)   (Spring 8b)

Figure 3-33   Force-displacement capacity boundaries for Spring 8a and Spring 8b.

The hysteretic behaviors of Spring 8a and Spring 8b, both with and without cyclic degradation, are shown in Figure 3-34 and Figure 3-35. In each figure, the initial force-displacement capacity boundary (before cyclic degradation) is overlaid onto the hysteretic plots.

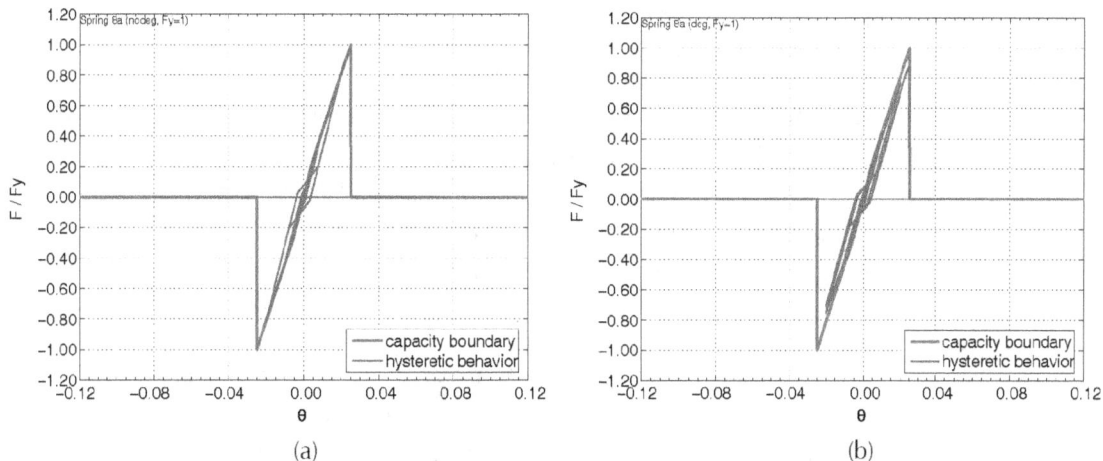

Figure 3-34   Initial force-displacement capacity boundary overlaid onto hysteretic behaviors for Spring 8a: (a) without cyclic degradation; and (b) with cyclic degradation.

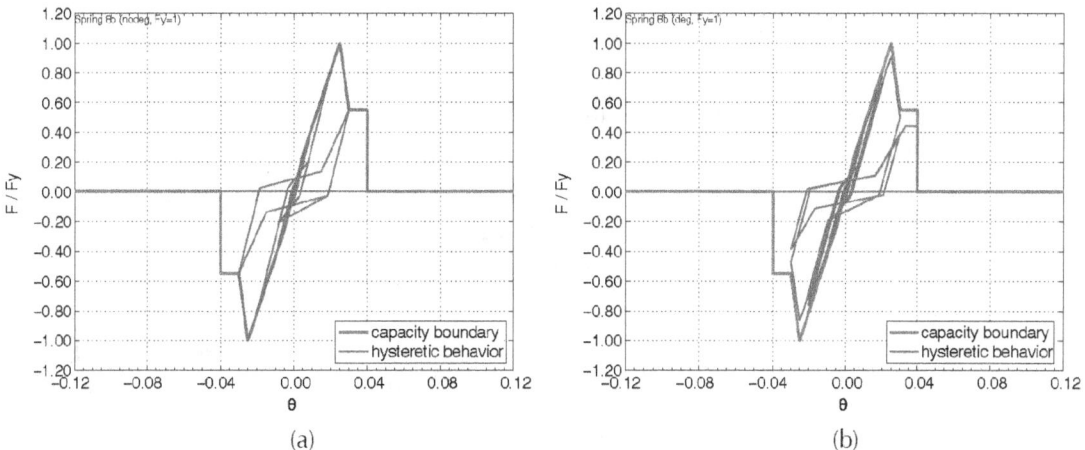

Figure 3-35   Initial force-displacement capacity boundary overlaid onto hysteretic behaviors for Spring 8b: (a) without cyclic degradation; and (b) with cyclic degradation.

3.3   Multiple Spring Models

Multiple spring models were used to represent the behavior of more complex structural systems containing subsystems with different hysteretic and force-displacement capacity boundary characteristics linked by rigid diaphragms. Multi-spring SDOF systems were developed by placing individual springs in parallel. Combinations were performed in a manner consistent with combinations that would be encountered in real structural systems. For each such combination, variations in the relative contribution of individual springs to the initial stiffness and maximum lateral strength over a range of periods were considered.

3.3.1 Multi-Spring Combinations of Single-Spring Systems

Of the numerous combinations possible, only assemblages consisting of two springs in parallel were considered in this investigation. Furthermore, only springs including cyclic degradation were considered in multi-spring combinations. This was done to limit the number of possible permutations under consideration, but also because, in general, realistic systems experiencing strong in-cycle degradation will also experience cyclic degradation.

Two-spring assemblages consisting of a lateral-force-resisting system (Springs 2, 3, 4, 5, 6, or 7), working in combination with a gravity frame system (Springs 1a, 1b, 8a, or 8b), were used. For example, a combination of Spring 2a with Spring 1a would be representative of a non-ductile moment frame system with a typical gravity frame back-up system in parallel.

In general, it is not realistic to assume that the contribution of each subsystem to the peak lateral strength of the combined system would be equal. In most cases, the lateral-force-resisting system in a building would be expected to be stronger and stiffer than the gravity system. For this reason, systems were combined using an additional parameter, N, as a multiplier on the contribution of lateral-force-resisting springs in the combined system. Multi-spring systems then carry a designation of "NxJa+1a" or "NxJa+1b" where "N" is the peak strength multiplier (N = 1, 2, 3, 5, or 9), "J" is the lateral-force-resisting spring number (J = 2, 3, 4, 5, 6, or 7), and 1a or 1b is the gravity system identifier. Using this designation, "3x2a+1a" would identify a system consisting of a multiple of three non-ductile moment frame springs (Spring 2a) in combination with a single gravity system spring (Spring 1a).

To investigate potential period-dependency, multi-spring systems were tuned to center the resulting periods of vibration for each set of "NxJa" lateral-force-resisting systems approximately around T=1.0s (representing relatively stiff systems) and T=2.0s (representing relatively flexible systems). This was accomplished by assuming two different story masses of M=8.87 tons or M=35.46 tons, respectively.

In summary the following series of multi-spring systems were investigated:

- Series 1:   NxJa + 1a (M=8.87 ton; relatively stiff)
- Series 2:   NxJb + 1a (M=35.46 ton; relatively flexible)
- Series 3:   NxJa + 1b (M=8.87 ton; relatively stiff)
- Series 4:   NxJb + 1b (M=35.46 ton; relatively flexible)

Multi-spring combinations using Spring 8a and Spring 8b were created and analyzed, however, the resulting behavior was not substantially different from other systems analyzed. As a result, this data was not investigated in detail, and information on these combinations has not been provided. As part of the series of investigations, each "NxJa" lateral-force-resisting system was analyzed without the 1a or 1b gravity system in order to compare results both with and without the contribution of the back-up system. A representative force-displacement capacity boundary from each multi-spring system is shown in Figure 3-36 through Figure 3-41.

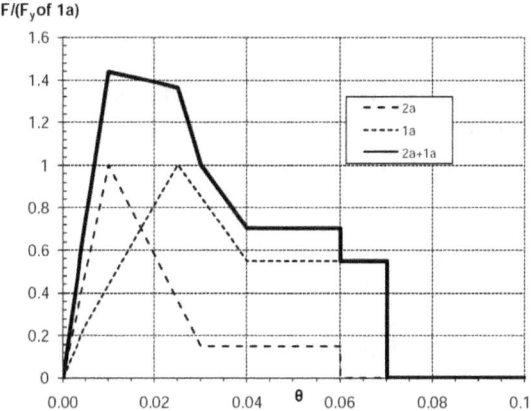

Figure 3-36   Combined force-displacement capacity boundary for spring 2a +1a (normalized by the strength of Spring 1a).

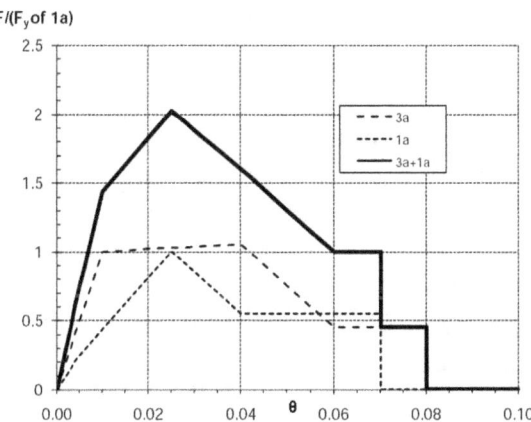

Figure 3-37   Combined force-displacement capacity boundary for spring 3a +1a (normalized by the strength of Spring 1a).

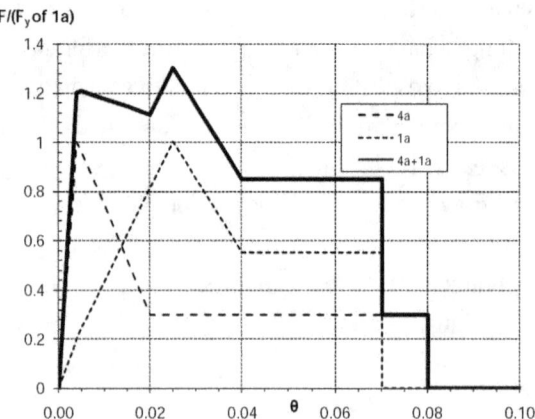

Figure 3-38   Combined force-displacement capacity boundary for spring
                4a +1a (normalized by the strength of Spring 1a).

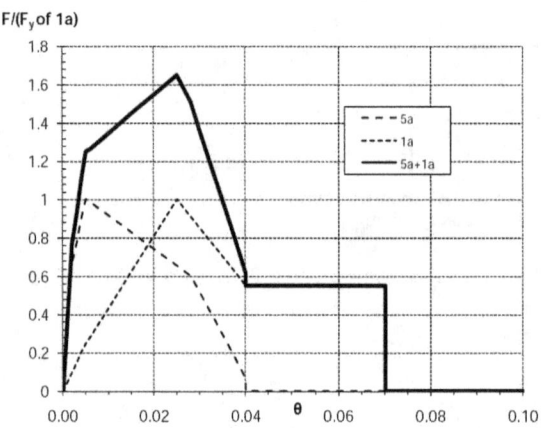

Figure 3-39   Combined force-displacement capacity boundary for spring
                5a +1a (normalized by the strength of Spring 1a).

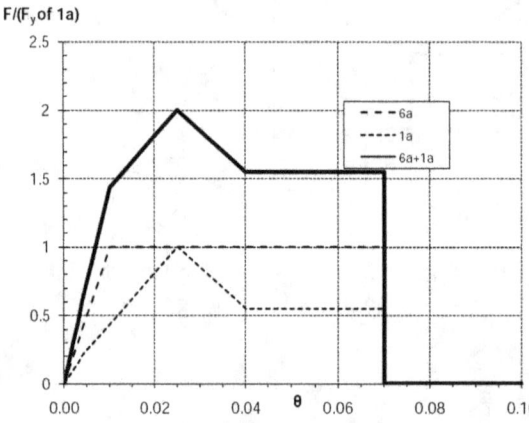

Figure 3-40   Combined force-displacement capacity boundary for spring
                6a +1a (normalized by the strength of Spring 1a).

**3: Development of Single-Degree-of-Freedom Models
        for Focused Analytical Studies**   **FEMA P440A**

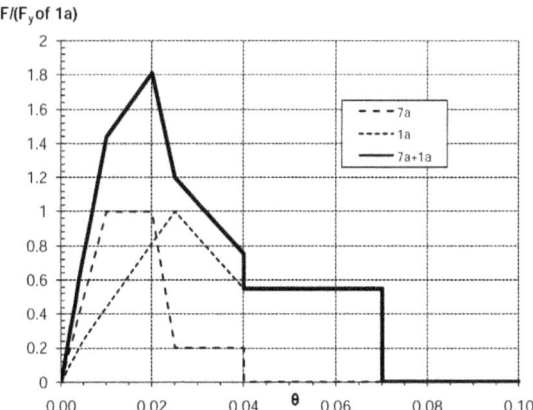

Figure 3-41   Combined force-displacement capacity boundary for spring
7a +1a (normalized by the strength of Spring 1a).

Each multi-spring combination was subjected to an ATC-24 type loading
protocol with a degrading force-displacement capacity boundary (cyclic
degradation). The resulting hysteretic behaviors for the combination of
Nx2a+1a for (N = 1, 2, 3, 5, and 9) are shown in Figure 3-42 through Figure
3-44. In addition, individual Spring 2a is shown in Figure 3-44 for
comparison. In each figure, the initial combined force-displacement capacity
boundary (before cyclic degradation) is overlaid onto the hysteretic plots.

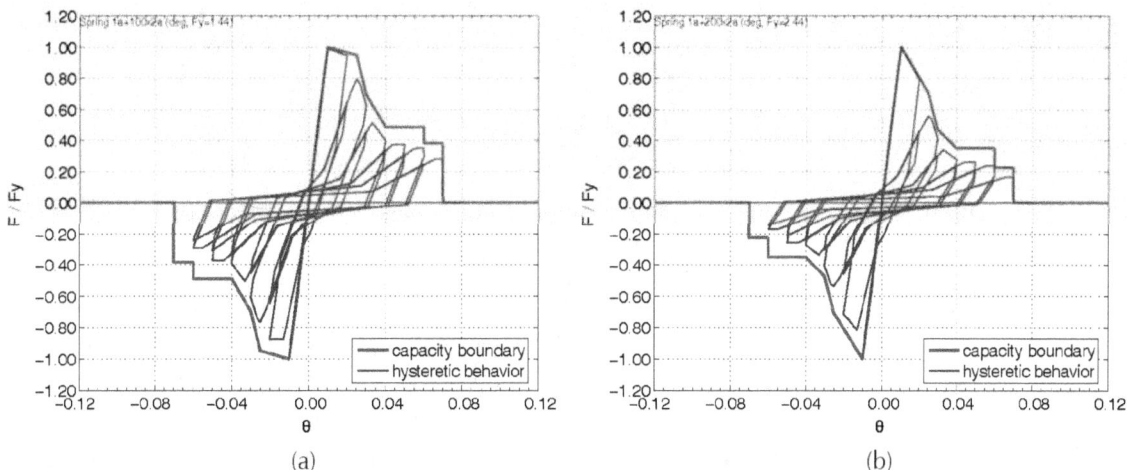

Figure 3-42   Initial force-displacement capacity boundary overlaid onto hysteretic behavior for:
(a) Spring 1x2a+1a; and (b) Spring 2x2a+1a; both with cyclic degradation.

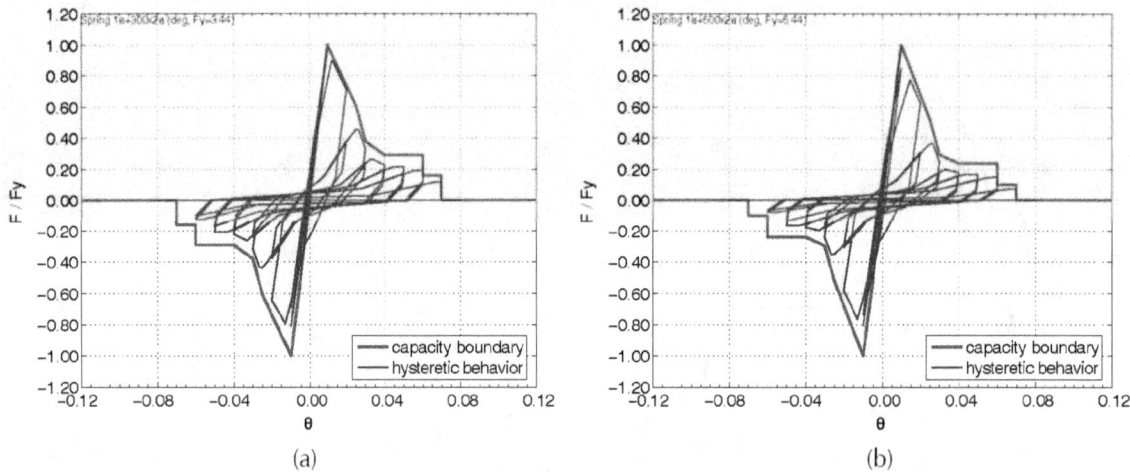

Figure 3-43   Initial force-displacement capacity boundary overlaid onto hysteretic behavior for: (a) Spring 1x3a+1a; and (b) Spring 5x2a+1a; both with cyclic degradation.

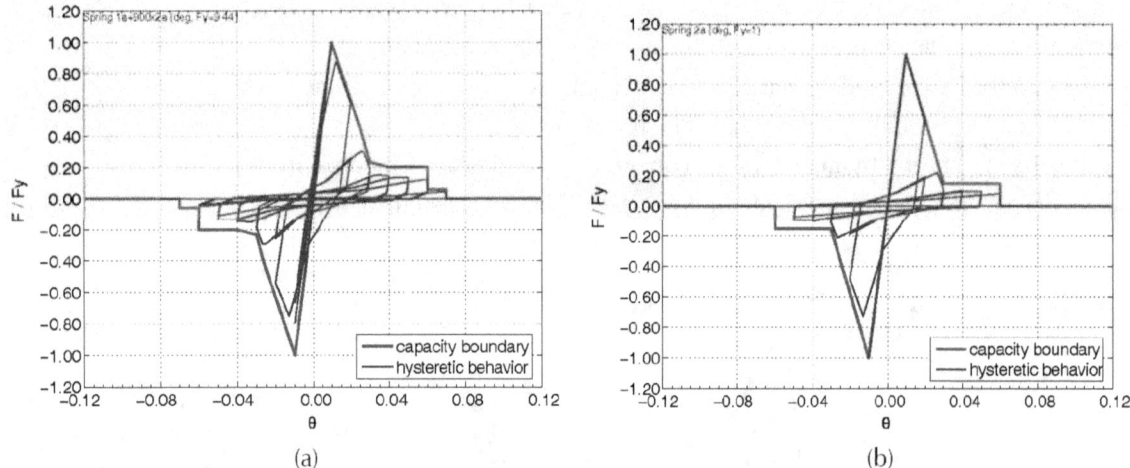

Figure 3-44   Initial force-displacement capacity boundary overlaid onto hysteretic behavior for: (a) Spring 9x2a+1a; and (b) individual Spring 2a; both with cyclic degradation.

As might be expected, the more the multiplier "N" for Spring 2a increases, the more the combined system resembles Spring 2a itself (Figure 3-44), and the more the behavior of the combined system would be expected to be dominated by the characteristics of the lateral-force-resisting spring component. Conversely, for lower multiples of "N", the characteristics of the gravity system are more visible in the combined system properties (Figure 3-42), and would be expected to play a more significant role in the behavior of the combined system.

**3: Development of Single-Degree-of-Freedom Models
for Focused Analytical Studies**

Chapter 4
Results from Single-Degree-of-Freedom Focused Analytical Studies

This chapter summarizes the results of focused analytical studies on single-spring and multi-spring systems, compares results to recommendations contained in FEMA 440, and explains the development of a new equation measuring the potential for lateral dynamic instability.

4.1 Summary of Analytical Results

There were 160 single-spring systems (eight basic spring types, "a" and "b" versions of each, with and without cyclic degradation, tuned to five different periods of vibration). Each system was subjected to incremental dynamic analysis using 56 ground motion records scaled to multiple levels of increasing intensity. This resulted in over 600,000 nonlinear response history analyses on single-spring systems.

There were 600 multi-spring systems (six lateral-force-resisting springs, "a" and "b" versions of each, five relative strength multipliers, five different gravity spring combinations, tuned with two different story masses). Each system was subjected to incremental dynamic analysis using 56 ground motion records scaled to multiple levels of increasing intensity. This resulted in over 2,000,000 nonlinear response history analyses on multi-spring systems.

In total, results from over 2.6 million nonlinear response history analyses were available for review. Given the large volume of analytical data, customized algorithms were developed for post-processing, statistical analysis, and visualization of results. Results are summarized in the sections that follow. More complete sets of data are presented in the appendices. A customized visualization tool that was developed to view results of multi-spring studies, along with all available data, is included on the CD accompanying this report. Use of the visualization tool is described in Appendix C and Appendix D.

4.2 Observations from Single-Spring Studies

This section summarizes the results from nonlinear dynamic analyses of single-spring systems. Results from these studies were used to:

- identify predominant characteristics of median incremental dynamic analysis (IDA) curves for these systems,

- demonstrate a relationship between IDA curves and features of the force-displacement capacity boundaries, and

- qualitatively determine the effects of different degrading behaviors on the dynamic stability of structural systems.

Only selected results are presented here. Quantile (16th, 50th and 84th percentile) IDA curves for each of the single-spring systems are provided in Appendix B. The horizontal axis for all single-spring IDA results is the maximum story drift ratio, θ_{max}, in radians.

4.3 Characteristics of Median IDA Curves

Individual incremental dynamic analysis (IDA) curves for single ground motion records are very sensitive to dynamic interaction between the properties of the system and the characteristics of the ground motion. Quantile (16th, 50th and 84th percentile) IDA curves, however, are much more stable and provide better information on the central tendency (median) and variability (dispersion) in system response. In general, median IDA curves exhibit the following characteristics (Figure 4-1):

- An initial linear segment corresponding to linear-elastic behavior in which in lateral deformation demand is proportional to ground motion intensity, regardless of the characteristics of the system or the ground motion. This segment extends from the origin to the onset of yielding.

- A second curvilinear segment corresponding to inelastic behavior in which lateral deformation demand is no longer proportional to ground motion intensity. As intensity increases, lateral deformation demands increase at a faster rate. This segment corresponds to softening of the system, or reduction in stiffness (reduction in the slope of the IDA curve). In this segment, the system "transitions" from linear behavior to eventual dynamic instability. Although a curvilinear segment is always present, in some cases the transition can be relatively long and gradual, while in other cases it can be very short and abrupt.

- A final linear segment that is horizontal, or nearly horizontal, in which infinitely large lateral deformation demands occur at small increments in ground motion intensity. This segment corresponds to the point at which a system becomes unstable (lateral dynamic instability). For SDOF systems, this point corresponds to the ultimate deformation capacity at which the system loses all lateral-force-resisting capacity.

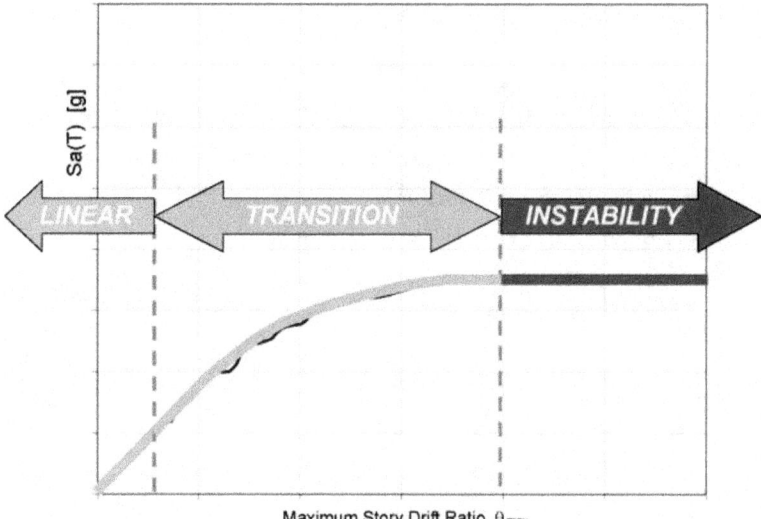

Figure 4-1   Characteristic segments of a median IDA curve.

In some systems, the initial linear segment can be extended beyond yield into
the inelastic range (Figure 4-2). In this segment lateral deformation demand
is approximately proportional to ground motion intensity, which is consistent
with the familiar equal-displacement approximation for estimating inelastic
displacements. The range of lateral deformation demands over which the
equal-displacement approximation is applicable depends on the
characteristics of the force-displacement capacity boundary of the system and
the period of vibration.

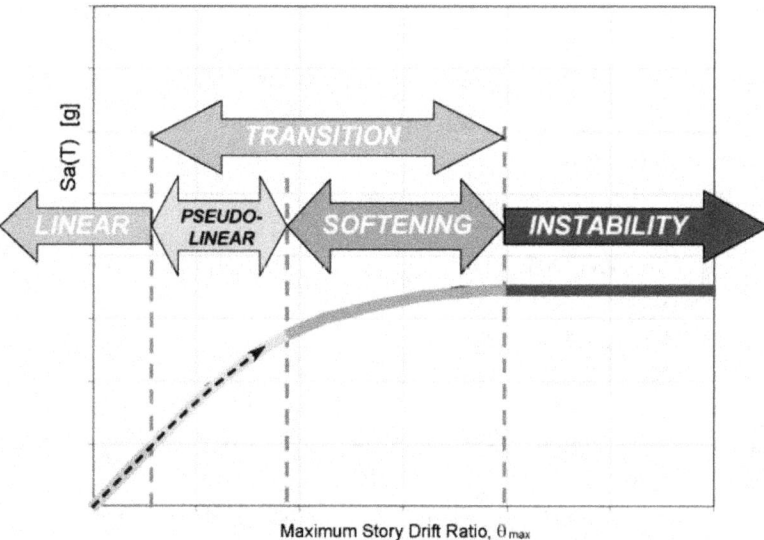

Figure 4-2   Characteristic segments of a median IDA curve with a pseudo-
                  linear segment.

4.3.1 Dependence on Period of Vibration

Figure 4-3 shows the force-displacement capacity boundary and resulting IDA curves for Spring 3a with different periods of vibration. Each system is tuned to a different lateral strength and stiffness so results are compared using the normalized intensity measure $R = S_a(T,5\%)/S_{ay}(T,5\%)$. Intensities larger than $R = 1.0$ mean the system is behaving inelastically.

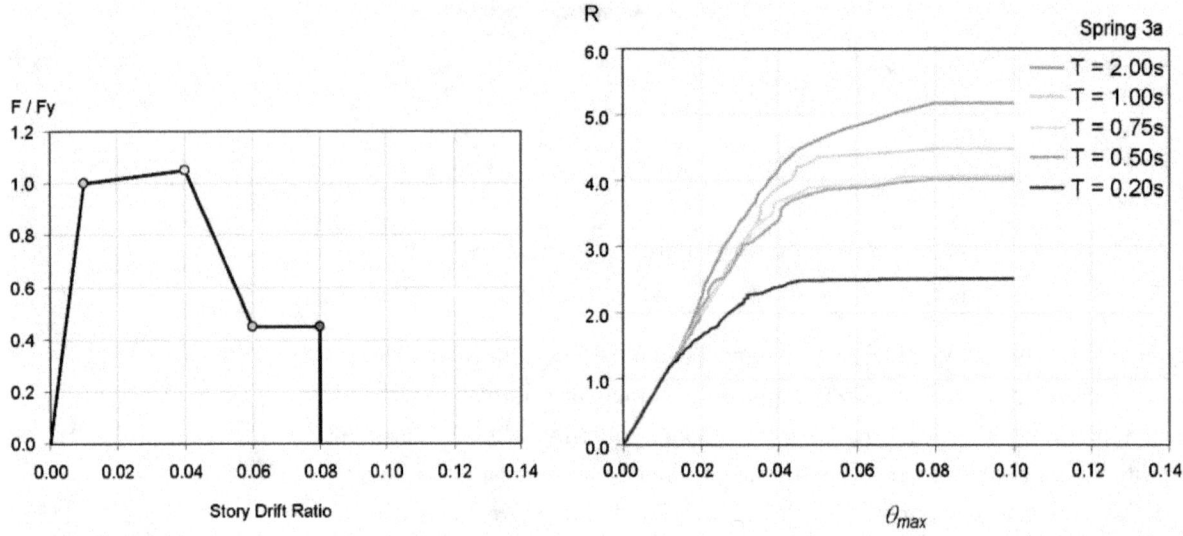

Figure 4-3   Force-displacement capacity boundary and median IDA curves for Spring 3a with various periods of vibration.

In general, moderate and long period systems with zero or positive post-yield stiffness in the force-displacement capacity boundary follow the equal displacement trend well into the nonlinear range. For systems with periods longer than 0.5s, Spring 3a exhibits an extension of the initial linear segment well beyond the yield drift of 0.01. In contrast, the short period system (T=0.2s) diverges from the initial linear segment just after yielding, even at deformations within the strength-hardening segment of the force-displacement capacity boundary (drifts between 0.01 and 0.04).

4.3.2 Dispersion in Response

Nonlinear response is sensitive to the characteristics of the ground motion record, and will vary from one ground motion to the next, even when scaled to the same intensity. For a given level of ground motion intensity, the lateral deformation demand can be significantly smaller or significantly larger than the value shown on median IDA curves. As the level of ground motion intensity increases, the dispersion in response tends to increase.

Figure 4-4 shows three quantile IDA curves for Spring 3b with period of vibration of 2.0s. The 50% (median) IDA curve indicates that, for a given level of ground motion intensity (S_a), half of all deformation demands are smaller and half are larger than values along this curve. Because the distribution of demands is lognormally distributed, the dispersion about the median is not symmetric. The upper (16%) curve in the figure indicates that, for a given level of ground motion intensity, 16% of all lateral deformation demands are to the left of this curve while 84% are to the right. This means that lateral deformation demands along this curve have an 84% chance of being exceeded. Similarly the lower (84%) curve corresponds to lateral deformation demands with a 16% chance of being exceeded.

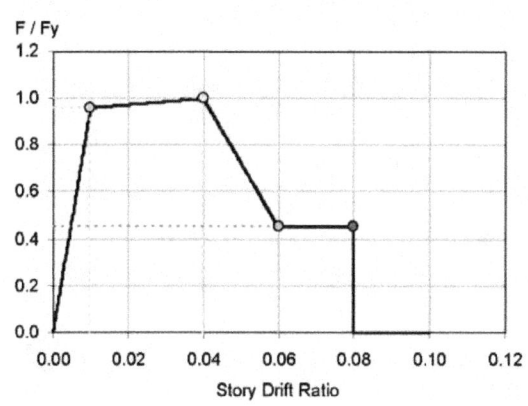 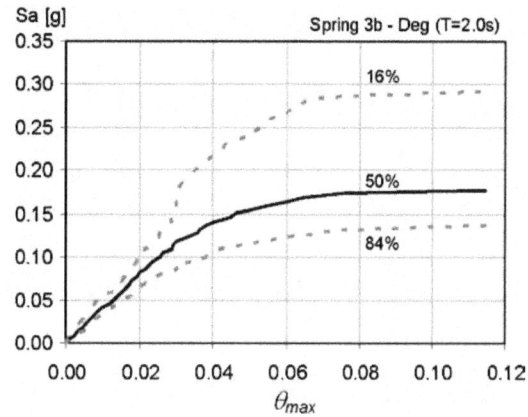

Figure 4-4   Force-displacement capacity boundary and 16th, 50th and 84th percentile IDA curves for Spring 3b with a period of vibration T=2.0s.

4.4   Influence of the Force-Displacement Capacity Boundary

Comparisons between force-displacement capacity boundaries and median IDA curves show a strong correlation between the shape of the resulting curves and key features of the force-displacement capacity boundary, such as post-yield behavior and onset of degradation, slope of degradation, ultimate deformation capacity, and presence of cyclic degradation.

Figure 4-5 shows the force-displacement capacity boundary and resulting median IDA curve for Spring 3b with a period of 2.0s. With a positive post-yield slope, delayed onset of degradation, and robust residual strength plateau with an extended maximum deformation capacity, the resulting IDA curve includes both linear and pseudo-linear segments and a gradual transition to lateral dynamic instability.

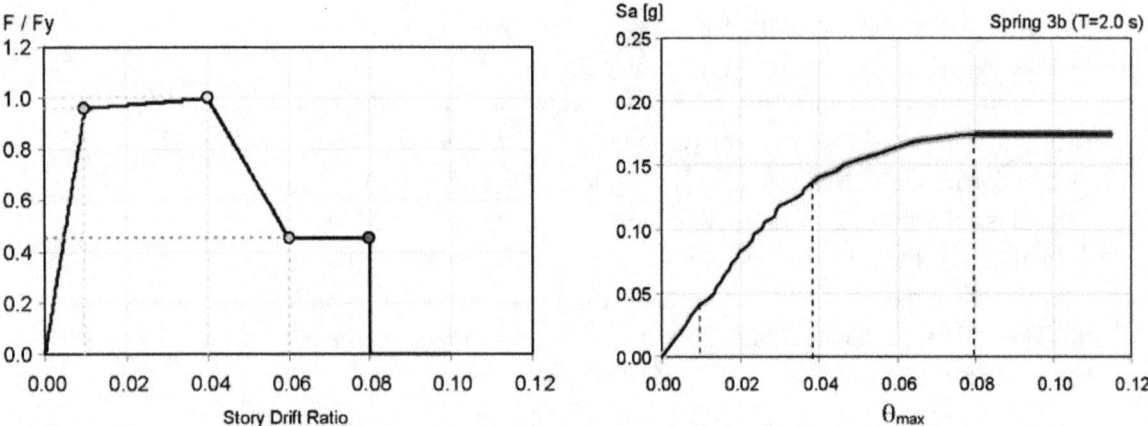

Figure 4-5   Force-displacement capacity boundary and median IDA curve for Spring 3b with a period of vibration T=2.0s.

Figure 4-6 shows the force-displacement capacity boundary and resulting median IDA curve for Spring 2a with a period of 2.0s. With the onset of degradation occurring immediately after yielding, the shape of the resulting IDA curve changes. The pseudo-linear segment disappears, but with the presence of a residual strength plateau, the transition segment remains somewhat gradual until lateral dynamic instability.

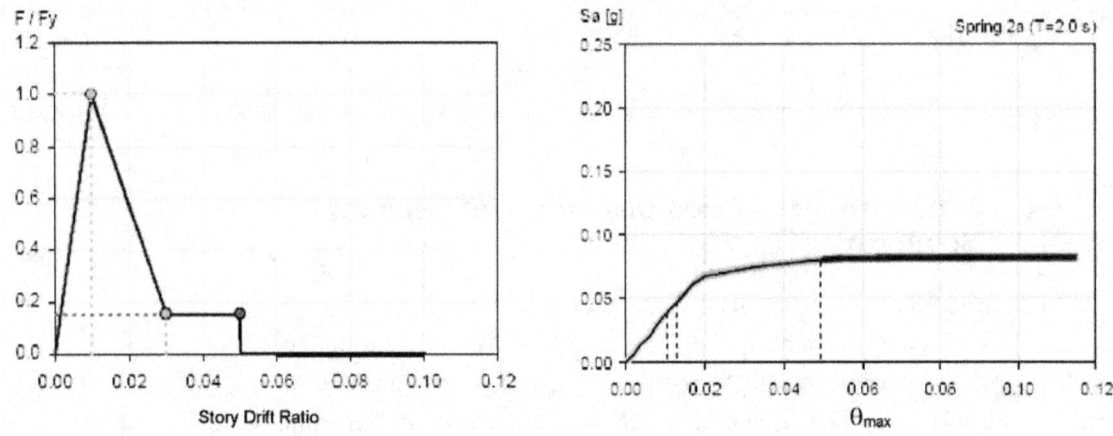

Figure 4-6   Force-displacement capacity boundary and median IDA curve for Spring 2a with a period of vibration T=2.0s.

Figure 4-7 shows the force-displacement capacity boundary and resulting median IDA curve for Spring 6a with a period of 2.0s. With a broad yielding plateau, the pseudo-linear segment extends well into the inelastic range. Without a residual strength plateau, however, the system abruptly transitions into lateral dynamic instability.

**4: Results from Single-Degree-of-Freedom
Focused Analytical Studies**

FEMA P440A

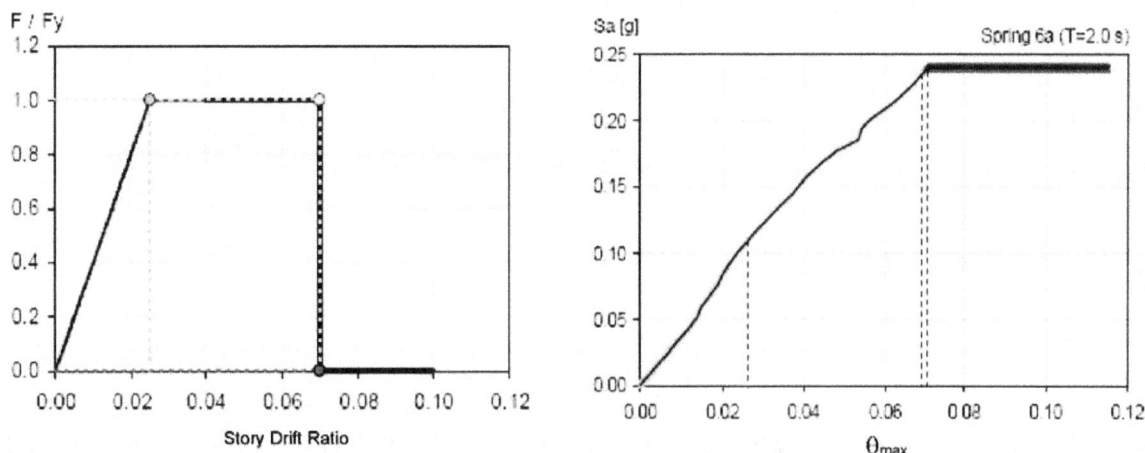

Figure 4-7   Force-displacement capacity boundary and median IDA curve for Spring 6a with a period of vibration T=2.0s.

Figure 4-8 shows the force-displacement capacity boundary and resulting median IDA curve for Spring 8a with a period of 2.0s. With severe strength degradation occurring immediately after yielding, and the absence of a residual strength plateau, the system abruptly transitions from linear elastic behavior directly into lateral dynamic instability with little or no transition.

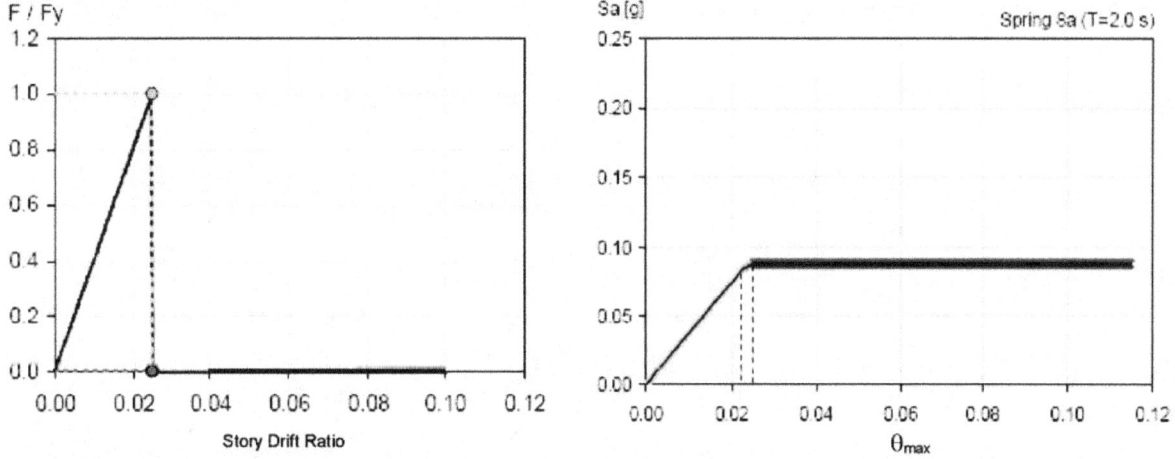

Figure 4-8   Force-displacement capacity boundary and median IDA curve for Spring 8a with a period of vibration T=2.0s.

These observed relationships suggest that dynamic response is directly influenced by the features of a force-displacement capacity boundary. Figure 4-9 shows how the characteristic segments of a median IDA curve relate to these features. Note that the relationship depicted in this idealized graphical representation is dependent upon the period of the system, as described in Section 4.3.1.

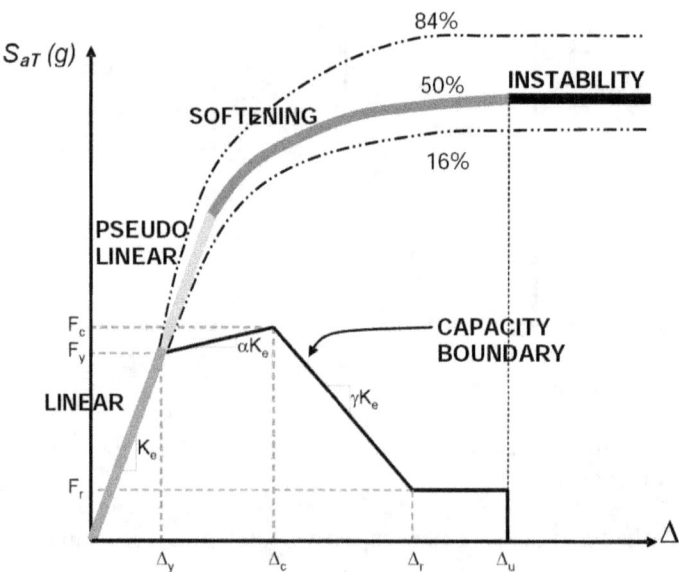

Figure 4-9   Relationship between IDA curves and the features of a typical force-displacement capacity boundary

For low levels of ground motion intensity, the initial linear segment of the IDA curve is controlled by the effective stiffness of the system, K_e. Since the response is linear there is no dispersion evident in this segment. As the intensity increases the system reaches its yield point, F_y, Δ_y. Systems with a non-negative post-elastic stiffness, αK_e, will likely exhibit a pseudo-linear segment. Beyond yield, dispersion appears in the nonlinear response due to ground motion variability, and the 16th and 84th percentile IDA curves begin to diverge from the median curve.

The extent of the pseudo-linear segment depends on the initial post elastic stiffness, αK_e, and ends prior to reaching the strength hardening limit, F_c, Δ_c (also known as the capping point). For systems that exhibit negative stiffness, γK_e, immediately after yielding, the pseudo-linear segment may be very short or non-existent. Also, for short-period systems, the pseudo-linear segment can be very short, even if the system has positive post-yield stiffness.

As the ground motion intensity increases further, deformation demands increase at a faster rate, the IDA curve begins to flatten, and the curvilinear softening segment emerges. Dispersion between the quantile curves also increases. Beyond the strength hardening limit, F_c, Δ_c, degradation occurs, and the softening increases at a faster rate. The presence of a residual

4: Results from Single-Degree-of-Freedom Focused Analytical Studies

strength plateau, F_r, Δ_r, can extend the softening segment and delay the eventual transition into lateral dynamic instability. The point at which instability occurs corresponds to the ultimate deformation capacity, Δ_u, at which the system loses all lateral force resistance.

This relationship suggests that it is possible to estimate the nonlinear dynamic behavior of a system based on knowledge of the characteristics of the force-displacement capacity boundary of the system. The influence that important features of the force-displacement capacity boundary have on nonlinear response is explained in more detail in the sections that follow.

4.4.1 Post-Yield Behavior and Onset of Degradation

The three systems shown in Figure 4-10 have the same elastic stiffness, same yield strength, but different post-yield characteristics. The force-displacement capacity boundary of Spring 2a experiences strength degradation immediately after yielding. In contrast, Spring 3a has a moderate yielding plateau before the onset of similar strength degradation, while Spring 6a has elastic-perfectly-plastic behavior up to the ultimate deformation capacity.

These three systems have the same elastic behavior, but at drift ratios larger than 0.02, their relative potential for in-cycle strength degradation, and their resulting collapse behaviors, are all very different. Key parameters related to the observed change in response are the post-yield slope and the strength hardening limit (capping point). The presence of a non-negative post-yield slope and any delay before the onset of degradation reduces potential in-cycle strength degradation and improves the collapse capacity of a system.

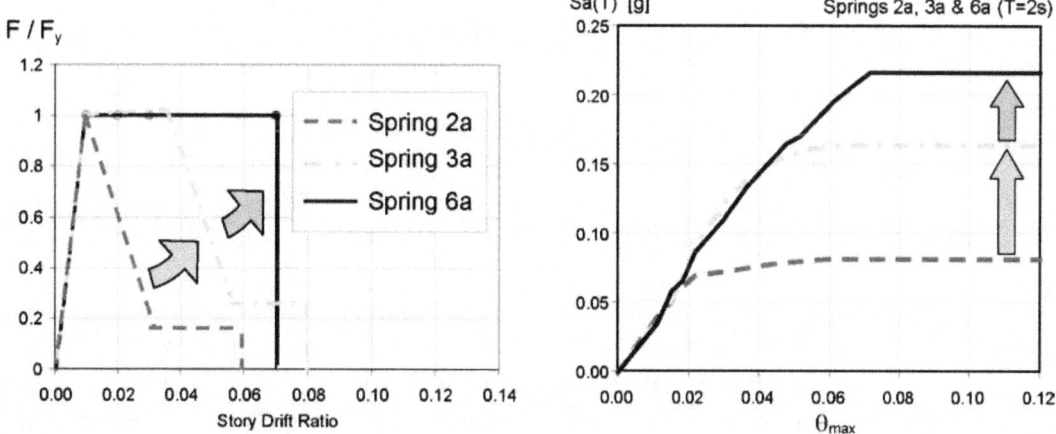

Figure 4-10   Effect of post-yield behavior on the collapse capacity of a system (Springs 2a, 3a and 6a with T=2.0s).

4.4.2 Slope of Degradation

Figure 4-11 shows the force-displacement capacity boundaries of Spring 2a and Spring 2b along with the corresponding IDA curves. These two systems have the same elastic stiffness, same yield strength, but differ in the negative slope of the strength-degrading segment and, therefore, in their potential for in-cycle strength degradation. They also have the same ultimate deformation capacity, but Spring 2b has a shorter residual strength plateau than Spring 2a because of the different slope.

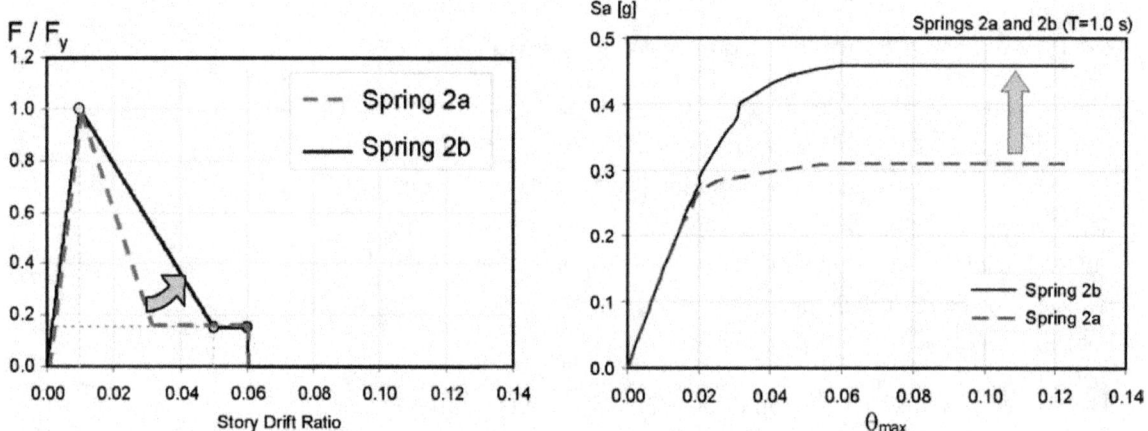

Figure 4-11   Effect of slope of degradation on the collapse capacity of a system (Springs 2a and 2b with T=1.0s).

The two systems have the same elastic behavior, but their response at drift demands larger than 0.01 is very different. Spring 2a, with a steeper degrading slope, likely experiences in-cycle strength degradation and reaches its collapse capacity relatively early, while Spring 2b, with a more shallow degrading slope, reaches a collapse capacity that is approximately 50% larger.

Figure 4-12 shows the force-displacement capacity boundaries of Spring 5a and Spring 5b along with the corresponding IDA curves. As in the case of Springs 2a and 2b, these two systems differ in the negative slope of the strength-degrading segments. They also differ in the presence of a residual strength plateau, which exists in Spring 5a, but not in Spring 5b.

As shown in the figure, the median IDA curves are similar up to 0.005 drift, at which both systems reach their peak strength. Beyond this point, the curves diverge as a result of the change in negative slope. Spring 5a, with steeper degrading slopes, reaches its collapse capacity sooner, while Spring 5b, with more shallow degrading slopes, reaches a higher collapse capacity.

**4: Results from Single-Degree-of-Freedom
Focused Analytical Studies**

The key parameter related to the observed change in response is the negative slope of the strength-degrading segment. In both examples, the change in negative slope changed the magnitude of potential in-cycle strength degradation, and overshadowed any changes in the residual strength plateau, as long as the ultimate deformation capacity remained the same.

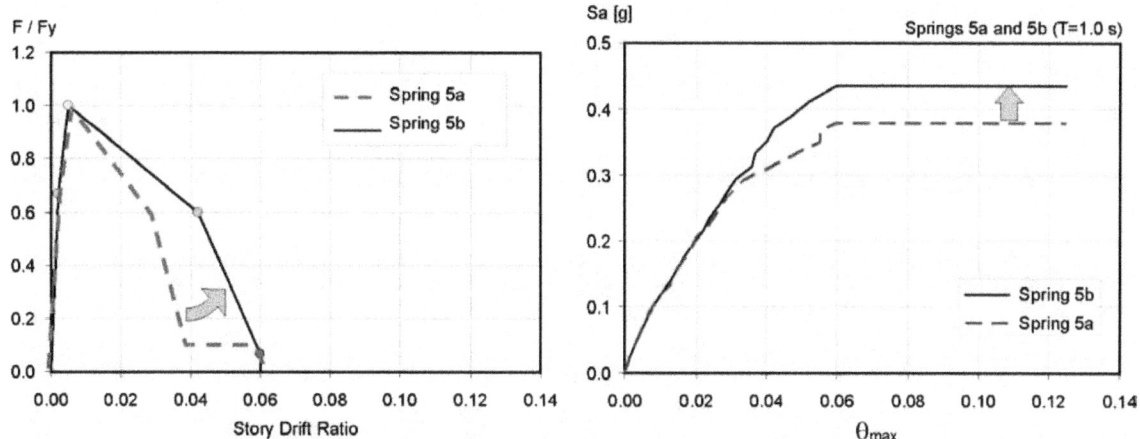

Figure 4-12   Effect of slope of degradation on the collapse capacity of a system (Springs 5a and 5b with T=1.0s).

4.4.3   Ultimate Deformation Capacity

Figure 4-13 shows the force-displacement capacity boundaries and corresponding IDA curves for Springs 1a and 1b. Figure 4-14 shows the force-displacement capacity boundaries and corresponding IDA curves for Springs 6a and 6b. These spring systems have very different post-yield behaviors, one with strength degradation (Springs 1a and 1b) and the other with elasto-plastic behavior (Springs 6a and 6b). In both cases, the "b" versions of each spring have higher ultimate deformation capacities.

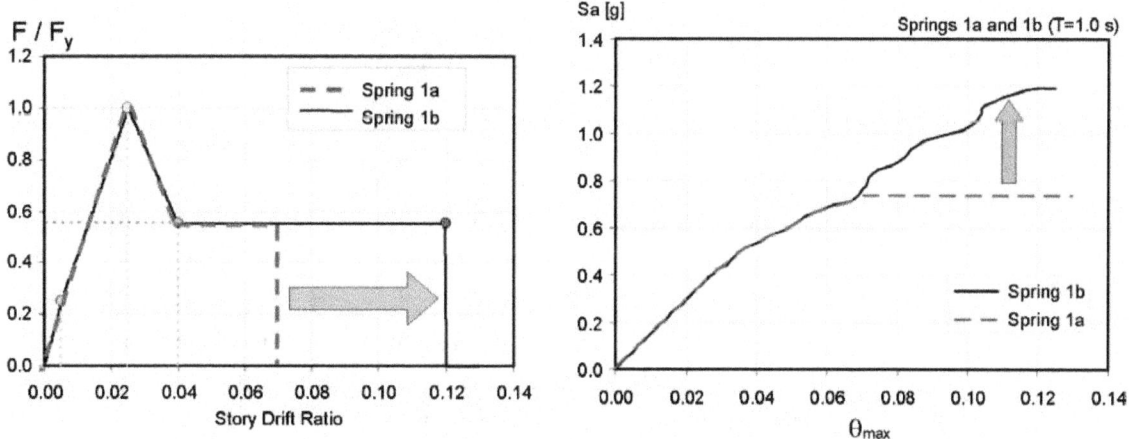

Figure 4-13   Effect of ultimate deformation capacity on the collapse capacity of a system (Springs 1a and 1b with T=1.0s).

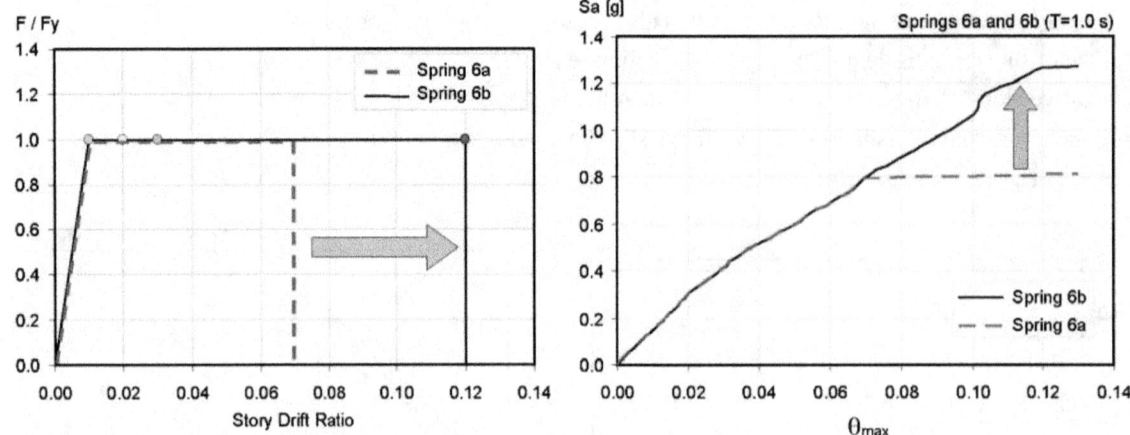

Figure 4-14   Effect of ultimate deformation capacity on the collapse capacity of a system (Springs 6a and 6b with T=1.0s).

In both examples, increasing the ultimate deformation capacity resulted in more than a 50% increase in collapse capacity. The key parameter related to the observed change in response is the increment in the ultimate deformation capacity. Observed changes in collapse capacity resulting from increases in the ultimate deformation capacity were insensitive to the other characteristics of the post-yield behavior of the springs.

4.4.4 Degradation of the Force-Displacement Capacity Boundary (Cyclic Degradation)

In general, most components will exhibit some level of cyclic degradation. To investigate the effects of cyclic degradation, the "a" and "b" versions of each spring (except Spring 6) were analyzed with both a constant force-displacement capacity boundary and a degrading force-displacement capacity boundary.

Consistent with observations from past studies, comparison of results between springs both with and without cyclic degradation show that the effects of cyclic degradation (as measured by gradual movement of the capacity boundary) are relatively unimportant in comparison with in-cycle degradation (as measured by the extent and steepness of negative slopes in the capacity boundary). This trend is illustrated for Spring 3b in Figure 4-15, but can be observed in the results for many spring systems. Although the system without cyclic degradation has a higher median collapse capacity, the difference is not very large. For the single-spring systems studied, the difference between median collapse capacity with and without cyclic degradation is shown in Appendix B. In general, this difference was typically less than 10%.

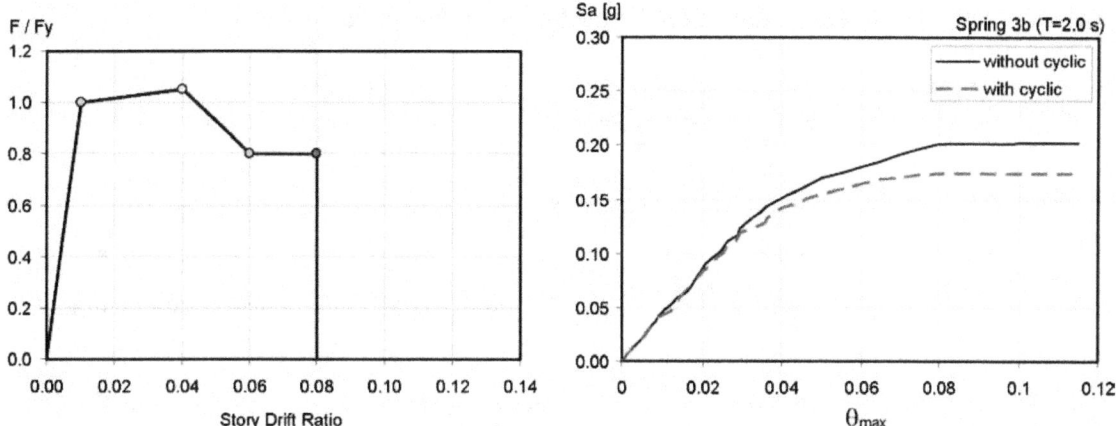

Figure 4-15   Effect of degradation of the force-displacement capacity boundary on the collapse capacity of a system (Spring 3b, T=2.0s, with and without cyclic degradation).

This observation has two important exceptions. First, the effect of cyclic degradation increases as the level of in-cycle degradation increases. Systems such as Spring 2b with a steep negative slope in the capacity boundary, indicating a strong potential for severe in-cycle strength degradation, showed as much as 30% difference in median collapse capacity between systems with and without cyclic degradation (Figure 4-16). Second, the effect of cyclic degradation increases as the period of vibration decreases. The short period (T=0.5s) versions of each spring showed more influence from cyclic degradation than the corresponding longer period (T=1.0s or T=2.0s) versions. This can be seen in the plots in Appendix B.

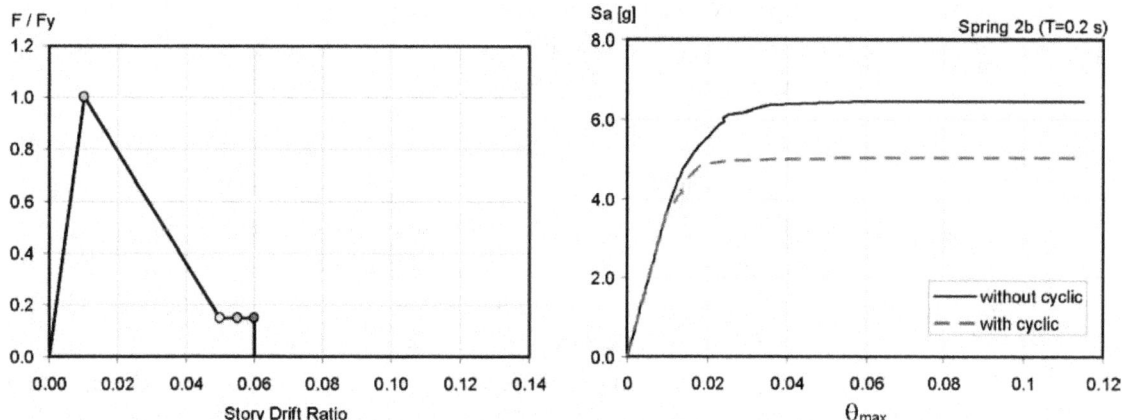

Figure 4-16   Effect of degradation of the force-displacement capacity boundary on the collapse capacity of a system (Spring 2b, T=0.2s, with and without cyclic degradation).

4.5   Observations from Multi-Spring Studies

This section summarizes the results from nonlinear dynamic analyses of multi-spring systems. Results from these studies were used to qualitatively:

- understand the influence of key features of the combined force-displacement capacity boundary on the nonlinear response of multi-spring systems,

- determine the effects of lateral strength on the dynamic stability of multi-spring systems, and

- determine the effects of secondary systems on the dynamic stability of multi-spring systems.

Only selected results are presented here. Combinations Nx2a+1a and Nx3a+1a, for N = 1, 2, 3, 5, or 9, are used to highlight trends observed to be generally applicable for the set of multi-spring combinations studied in this investigation. Results for each combination, plotted versus normalized and non-normalized intensity measures, are provided in Appendix C and Appendix D.

4.5.1   Normalized versus Non-Normalized Results

Two intensity measures were used in conducting incremental dynamic analyses. One was the 5% damped spectral acceleration at the fundamental period of vibration of the oscillator, $S_a(T,5\%)$. While generally appropriate for single-degree-of-freedom systems, this measure does not allow comparison between systems having different periods of vibration. For this reason, a normalized intensity measure, $R = S_a(T,5\%)/S_{ay}(T,5\%)$ was also used, where $S_{ay}(T,5\%)$ is the intensity that causes first yield to occur in the system.

In order to compare the response of different spring systems, it is necessary to plot the IDA curves from several springs in a single figure using a common intensity measure. This can be done in two ways. One way is to plot them using the normalized intensity measure, $R = S_a(T,5\%)/S_{ay}(T,5\%)$. First yield occurs at a normalized intensity of one, and increasing values of $S_a(T,5\%)/S_{ay}(T,5\%)$ represent increasing values of ground motion intensity with respect to the intensity required to initiate yielding in the system. Normalized plots provide a measure of system capacity relative to the yield intensity, and are useful for comparing results across different spring types when evaluating the influence of the key features of the force-displacement capacity boundary on the response of the system.

A second way to compare results is to plot them using an absolute (non-normalized) intensity measure that is somewhere in the middle of the range that would be suitable for the systems being plotted (e.g., T=1.0s). When evaluating the effects of increasing or decreasing the relative contribution of one subsystem with respect to another, use of a single absolute intensity measure allows comparison of results based on the relative strengths of different systems.

Since each method has advantages for viewing results and drawing comparisons, results for multi-spring systems were plotted using both the normalized intensity measure, $S_a(T,5\%)/S_{ay}(T,5\%)$, and non-normalized intensity measures, $S_a(1s,5\%)$ for stiff systems and $S_a(2s,5\%)$ for flexible systems. Results for normalized intensity measures, IM = $S_a(T,5\%)/S_{ay}(T,5\%)$, are provided in Appendix C, and results for non-normalized intensity measures, IM = $S_a(1s,5\%)$ or $S_a(2s,5\%)$, are provided in Appendix D. The horizontal axis in all cases is the maximum story drift ratio, θ_{max}, in radians.

4.5.2 Comparison of Multi-Spring Force-Displacement Capacity Boundaries

Figure 4-17 shows the force-displacement capacity boundaries for multi-spring systems Nx2a+1a and Nx3a+1a, normalized by the yield strength, F_y, of the combined system. Figure 4-18 shows the force-displacement capacity boundaries for the same two systems, normalized by the strength of the weakest system. Depending on the normalizing parameter used along the vertical axis, the resulting curves look very different.

In Figure 4-17, the use of a normalized base shear, F/F_y or S_a/S_{ay}, along the vertical axis allows for a better qualitative comparison of the relative shapes of the force-displacement capacity boundaries, without the added complexity caused by the different yield strengths of the systems. In this figure, it is easier to see how increasing the multiplier "N" on the lateral-force-resisting spring causes the combined system to more closely resemble the lateral spring itself (i.e., as "N" increases from 1 to 9, the combination Nx2a+1a begins to look more like Spring 2a).

Figure 4-17, however, is misleading with regard to the relative strengths of the combined systems. In normalizing to the yield strength of the combined system, higher values of yield strength will reduce the plotted values by a larger ratio, so curves for higher strength systems will plot below curves for lower strength systems in F/F_y coordinates.

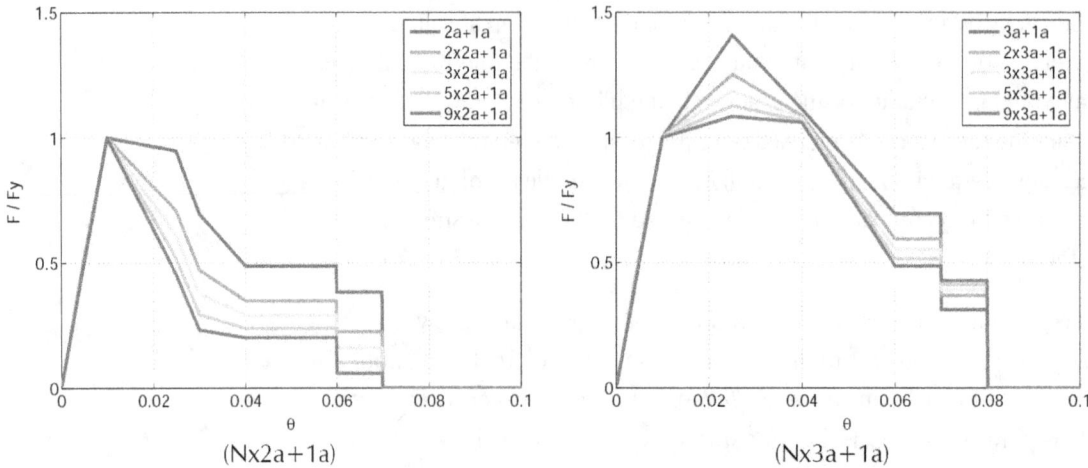

Figure 4-17   Force-displacement capacity boundaries for multi-spring systems Nx2a+1a and Nx3a+1a, normalized by the yield strength, Fy, of the combined system.

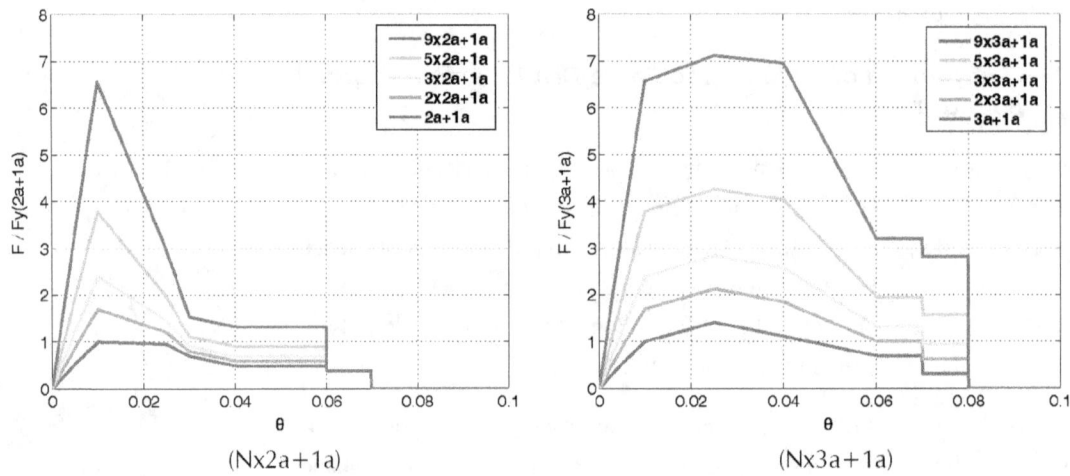

Figure 4-18   Force-displacement capacity boundaries for multi-spring systems Nx2a+1a and Nx3a+1a, normalized by the yield strength of the weakest system

In Figure 4-18, normalizing to the strength of the weakest system allows for a better comparison of the relative strength between the systems. In this figure it is easier to see how increasing the multiplier "N" on the lateral-force-resisting spring increases the strength of the combined system.

4.5.3   Influence of the Combined Force-Displacement Capacity Boundary in Multi-Spring Systems

Regardless of the normalizing parameter, Figure 4-17 and Figure 4-18 show how the combined force-displacement capacity boundaries change as the relative contributions of the springs vary. Results from single-spring studies demonstrated the influence of key features of the force-displacement capacity boundary on the nonlinear dynamic response of a single-spring system. Results from multi-spring studies followed the same relationships. Multi-spring systems in which the combined force-displacement capacity boundary had more favorable features (e.g., delayed onset of degradation, more gradual slope of degradation, higher residual strength, and higher ultimate deformation capacity) performed better.

Figure 4-19 shows median IDA curves plotted versus the normalized intensity measure $R = S_a(T,5\%)/S_{ay}(T,5\%)$ for multi-spring systems Nx2a+1a and Nx3a+1a with a mass of 8.87 tons, representing a series of relatively stiff systems. As "N" increases, the yield strength of the combined system increases, and each system has a correspondingly shorter period of vibration.

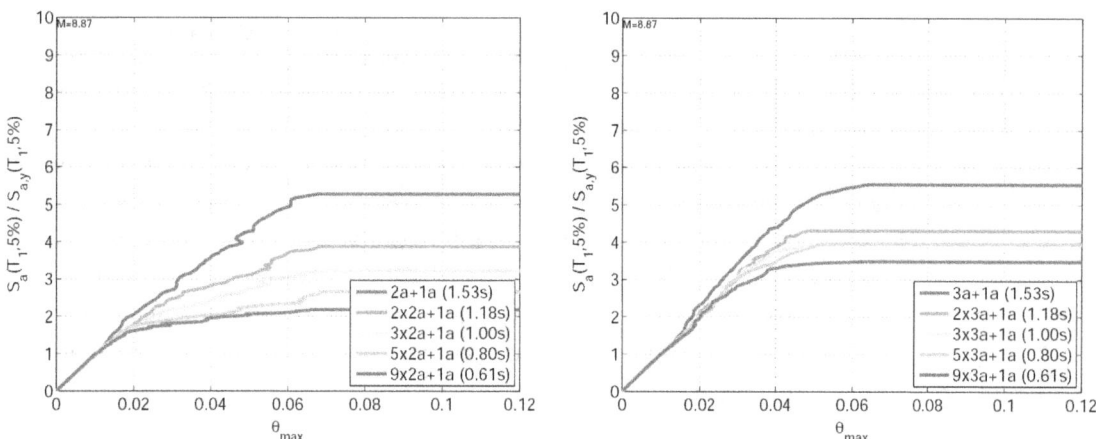

Figure 4-19   Median IDA curves plotted versus the normalized intensity measure Sa(T,5%)/Say(T,5%) for systems Nx2a+1a and Nx3a+1a with a mass of 8.87 tons.

Figure 4-20 shows median IDA curves for the same two systems with a mass of 35.46 tons, representing a series of relatively flexible systems. Because each system has a different period of vibration, normalized plots are used to qualitatively compare IDA curves between systems. Normalized curves, however, can be somewhat misleading with regard to the effect of changing "N" in the different spring combinations. The plotting positions in Figure 4-19, for example, are not an indication of the absolute collapse capacity of each system. Rather, they are a measure of collapse capacity relative to the intensity required to initiate yielding. Systems with high yield strengths may actually collapse at higher absolute intensities than systems with lower yield strengths, but because of the normalization to yield intensity, they plot out at lower ratios.

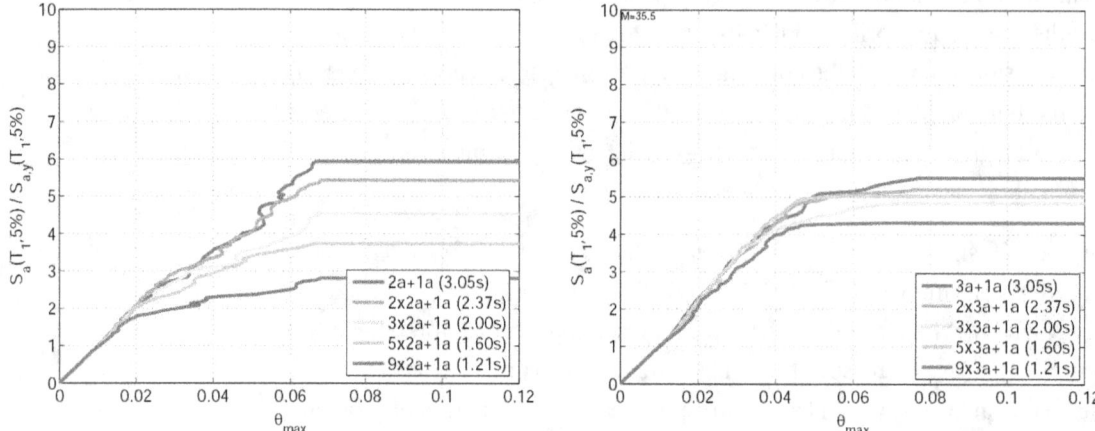

Figure 4-20   Median IDA curves plotted versus the normalized intensity measure
$S_a(T,5\%)/S_{ay}(T,5\%)$ for systems Nx2a+1a and Nx3a+1a with a mass of 35.46 tons.

For system Nx2a+1a, Figure 4-19 and Figure 4-20 show that as "N" increases, collapse capacity decreases. The reason for this can be seen in the combined force-displacement capacity boundaries for system Nx2a+1a shown in Figure 4-17. Because of the characteristics of Spring 2a, combinations with higher multiples of "N" have steeper negative slopes. As was the case with single-spring systems, steeper slopes in the strength-degrading segment of the force-displacement capacity boundary result in lower collapse capacities.

For system Nx3a+1a, the results are the same, but less pronounced. Similar to system Nx2a+1a, the force-displacement capacity boundaries shown in Figure 4-17 for system Nx3a+1a with higher multiples of "N" have steeper negative slopes, but the differences are less significant.

Figure 4-19 and Figure 4-20 also show that, in general, combinations with systems that have more favorable characteristics result in higher median collapse capacities relative to yield intensity. For example, in Figure 4-19, system 9x2a+1a exhibits a median collapse capacity that is approximately 2.3 times the yield intensity while system 9x3a+1a exhibits a median collapse capacity that is approximately 3.5 times the yield intensity. The reason for this can be seen by comparing the combined force-displacement capacity boundaries for systems Nx2a+1a and Nx3a+1a shown in Figure 4-17. The post-yield characteristics of system Nx3a+1a are more favorable in terms of the post-yield slope, onset of degradation, and slope of degradation, resulting in better performance.

A more direct illustration of this behavior can be seen by comparing combinations using the "a" and "b" versions of primary spring components. Figure 4-21 shows the median IDA curves for systems Nx3a+1a and

**4: Results from Single-Degree-of-Freedom
Focused Analytical Studies**

Nx3b+1a. By definition, the "b" version of each spring was created to have more favorable characteristics than the "a" version of the same spring, with all other parameters being equal. As shown in the figure, the curves for system Nx3b+1a outperform all corresponding combinations of Nx3a+1a in terms of collapse capacity relative to yield intensity, for all values of "N" from 1 to 9.

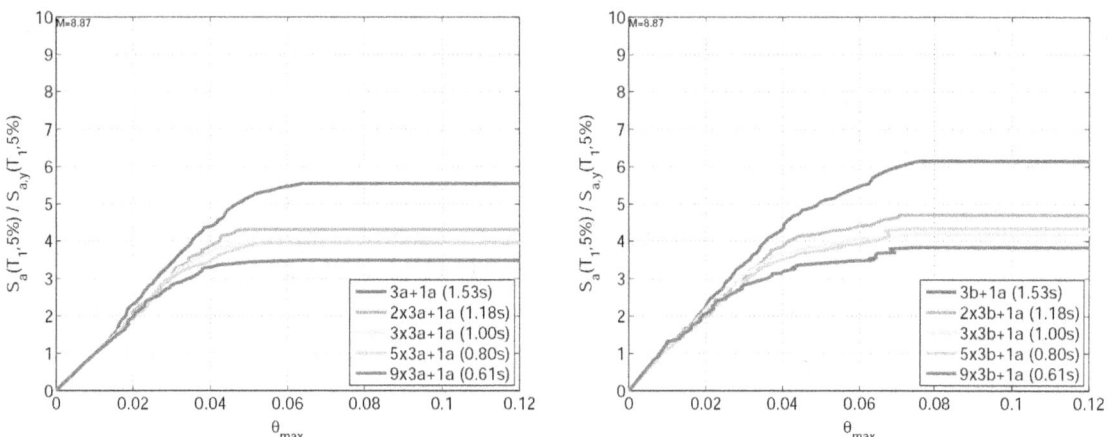

Figure 4-21   Median IDA curves plotted versus the normalized intensity measure $S_a(T,5\%)/S_{ay}(T,5\%)$ for systems Nx3a+1a and Nx3b+1a with a mass of 8.87 tons.

4.5.4   Effects of the Lateral Strength of Multi-Spring Systems

Plotting of results using absolute (non-normalized) intensity measures allows for comparison of results based on the relative strengths of different systems. Non-normalized intensity measures of $S_a(1s,5\%)$ for stiff systems and $S_a(2s,5\%)$ for flexible systems were used to identify the effects of the lateral strength of the multi-spring system on the lateral dynamic stability of the system.

Figure 4-22 shows median IDA curves for multi-spring systems Nx2a+1a and Nx3a+1a tuned with a mass of 8.87 tons. They are plotted versus $S_a(1s,5\%)$, which is an intensity measure keyed to a period of T=1.0s, located in the middle of the range of periods for the relatively stiff set of multi-spring systems. Figure 4-23 shows median IDA curves for same set of systems Nx2a+1a and Nx3a+1a, tuned with a mass of 35.46 tons. In this figure, the curves are plotted versus $S_a(2s,5\%)$, which is keyed to a period of T=2.0s, located in the middle of the range of periods for the relatively flexible set of multi-spring systems.

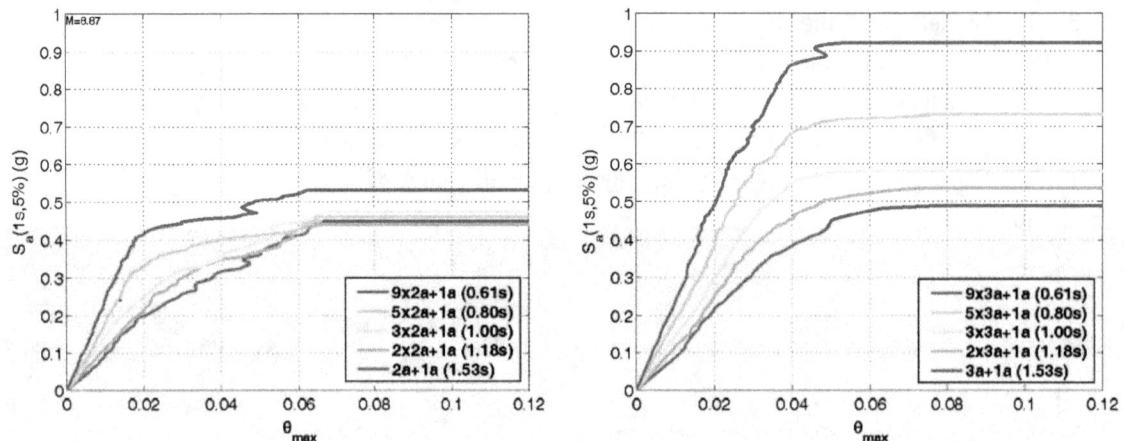

Figure 4-22   Median IDA curves plotted versus the common intensity measure $S_a(1s,5\%)$ for systems Nx2a+1a and Nx3a+1a with a mass of 8.87 tons.

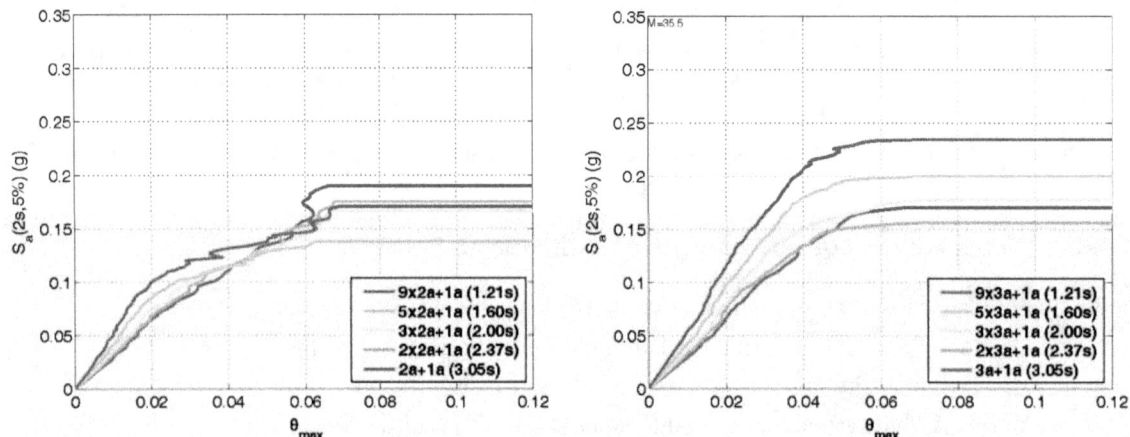

Figure 4-23   Median IDA curves plotted versus the common intensity measure $S_a(2s,5\%)$ for systems Nx2a+1a and Nx3a+1a with a mass of 35.46 tons.

In comparing non-normalized plots of IDA curves for various multi-spring combinations, the following observations were made:

- Increases in the lateral strength of a system change the intensity that initiates yielding in the system as well as the intensity at collapse (lateral dynamic instability). The incremental change in collapse capacity, however, is less than proportional to the increase in yield strength.

- The effectiveness of increasing the lateral strength of a system is a function of the shape of the force-displacement capacity boundary. Incremental changes in yield strength are more effective for ductile systems than they are for systems with less ductile behavior.

4: Results from Single-Degree-of-Freedom Focused Analytical Studies

- The effectiveness of increasing the lateral strength of a system is also a function of the period of system. Incremental changes in yield strength are more effective for stiff systems than they are for flexible systems.

These effects can be observed by comparing the combined force-displacement capacity boundaries in Figure 4-18 with the resulting IDA curves in Figure 4-22 and Figure 4-23. Figure 4-22 shows that as "N" increases, the yield intensity increases significantly, however, increases in intensity at lateral dynamic instability are not as significant. For example, Figure 4-18 shows that the yield strength of system 9x3a+1a is approximately 6.5 times higher than the yield strength of system 3a+1a, but Figure 4-22 shows that the collapse capacity is only about two times higher.

Comparing results between systems Nx2a+1a and Nx3a+1a in Figure 4-22 shows that increases in collapse capacity that do occur as a result of changes in lateral strength are more pronounced for the more ductile Spring 3a than they are for the less ductile Spring 2a. For example, the increase in collapse capacity for system Nx3a+1a, as "N" increases from 1 to 9, is a factor of approximately 2.0. For system Nx2a+1a the corresponding increase in collapse capacity is a factor of approximately 1.25.

Comparing results between Figure 4-22 and Figure 4-23 shows that as the period increases, the increment in collapse capacity caused by a change in lateral strength decreases. For example, the increase in collapse capacity shown in Figure 4-22 for the relatively stiff combinations of system Nx3a+1a is a factor of approximately 2.0. The increase in collapse capacity shown in Figure 4-23 for the relatively flexible combinations of system Nx3a+1a is a factor of approximately 1.3.

4.5.5 Effects of Secondary System Characteristics

The contribution of a secondary ("gravity") system acting in parallel with a primary lateral-force-resisting system always results in an improvement in post-yield performance, especially close to collapse. This result was observed both qualitatively and quantitatively (i.e., both in normalized and non-normalized coordinates).

The improvement is larger when considering secondary systems with larger ultimate deformation capacities. Figure 4-24 shows median IDA curves plotted versus the normalized intensity measure $R = S_a(T,5\%)/S_{ay}(T,5\%)$ for multi-spring systems Nx2a+1a and Nx2a+1b with a mass of 8.87 tons. In the figure it can be seen that combinations with Spring 1b (with a larger ultimate deformation capacity) perform significantly better than combinations with

Spring 1a. This result was observed in combinations with all lateral-force-resisting springs.

Near collapse, secondary systems with larger deformation capacities have an even greater influence, even if the lateral strength is small compared to that of the primary system. This can be observed by comparing differences between systems 9x2a+1a and 9x2a+1b in Figure 4-24. Even though the relative contribution of Spring 1 in these combinations is small, the resulting collapse capacity is increased significantly.

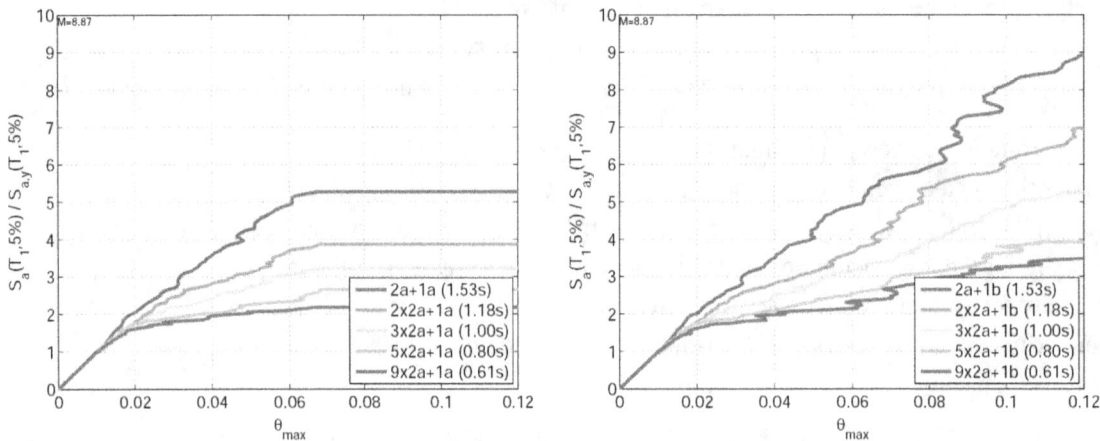

Figure 4-24   Median IDA curves plotted versus the normalized intensity measure $S_a(T,5\%)/S_{ay}(T,5\%)$ for systems Nx2a+1a and Nx2a+1b with a mass of 8.87 tons.

The contribution of the secondary system is more noticeable and significant in systems where the primary lateral resisting system is less ductile. Figure 4-25 shows median IDA curves plotted versus the normalized intensity measure $R = S_a(T,5\%)/S_{ay}(T,5\%)$ for multi-spring systems Nx2a+1a and Nx3a+1a with a mass of 8.87 tons.

Comparing the systems in Figure 4-25 shows a much wider spread between the median IDA curves for system Nx2a+1a than the curves for system Nx3a+1a. This means that the behavior of Spring 2a is more heavily influenced by the combination with Spring 1a than Spring 3a. The reason for this can be explained by the relative contributions of each spring to the combined force-displacement capacity boundaries in Figure 4-17.

Spring 2a, which represents a non-ductile moment frame system, has less favorable post-yield behavior in its force-displacement capacity boundary than does Spring 3a, which represents a ductile moment frame system. As such, Spring 2a is more favorably impacted by the characteristics of Spring 1a, and combinations with Spring 1a result in greater changes in performance. However, as "N" increases from 1 to 9, system Nx2a+1a

4: Results from Single-Degree-of-Freedom
Focused Analytical Studies

becomes more like Spring 2a, and the positive influences of Spring 1a diminish.

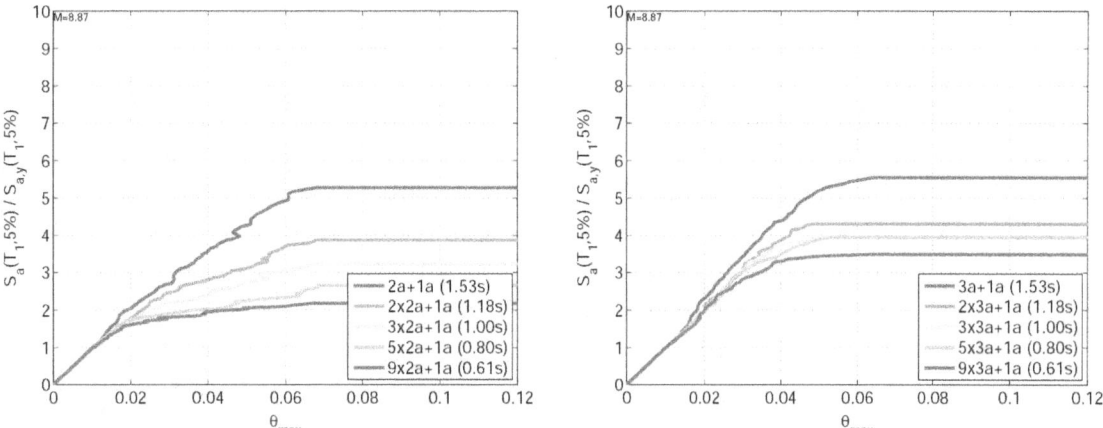

Figure 4-25   Median IDA curves plotted versus the normalized intensity measure Sa(T,5%)/Say(T,5%) for systems Nx2a+1a and Nx3a+1a with a mass of 8.87 tons.

4.6   Comparison with FEMA 440 Limitations on Strength for Lateral Dynamic Instability

In FEMA 440, a minimum strength requirement (maximum value of R) was developed as an approximate measure of the need to further investigate the potential for lateral dynamic instability caused by in-cycle strength degradation and P-delta effects. The recommended limitation is shown in Equation 4-1, with terms defined in Equation 4-2 and Equation 4-3, and illustrated in Figure 4-26:

$$R_{max} = \frac{\Delta_d}{\Delta_y} + \frac{\alpha_e^{-t}}{4} \qquad (4\text{-}1)$$

where

$$t = 1 + 0.15 \ln T \qquad (4\text{-}2)$$

and

$$\alpha_e = \alpha_{P-\Delta} + \lambda \left(\alpha_2 - \alpha_{P-\Delta} \right) \qquad (4\text{-}3)$$

for $0 < \lambda < 1.0$.

In-cycle strength degradation caused by P-delta is represented by $\alpha_{P-\Delta}$. The effects from all other sources of cyclic and in-cycle strength and stiffness degradation are represented by the term $\left(\alpha_2 - \alpha_{P-\Delta} \right)$. At the time, it was apparent that modeling rules specified the use of hysteretic envelopes idealized from cyclic test results and would, consequently, overestimate actual in-cycle losses. For this reason, these effects were reduced by factor λ, which was less than 1.0.

Base shear

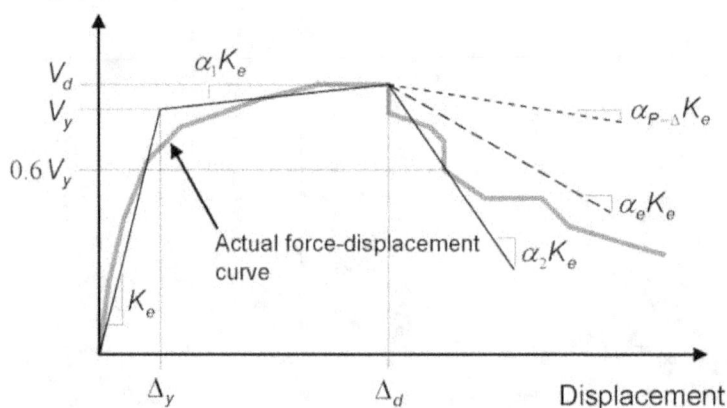

Figure 4-26　　Idealized force-displacement curve for nonlinear static analysis (from FEMA 440).

According to FEMA 440, the idealized force-displacement relationship (Figure 4-26) and the factor λ were based on judgment, and significant variability should be expected in the value predicted using the equation for R_{max}. As such, R_{max} was intended only for identification of cases where further investigation using nonlinear response history analysis should be performed, and not as an accurate measure of the strength required to avoid lateral dynamic instability.

To further investigate correlation between the FEMA 440 equation for R_{max} and lateral dynamic instability, the results of this equation were compared to quantile IDA curves for selected multi-spring systems included in this investigation. In making this comparison, parameters in the FEMA 440 equation for R_{max} were estimated from multi-spring force-displacement capacity boundaries idealized as shown in Figure 4-27.

Results from this comparison indicate that values predicted by the FEMA 440 equation for R_{max} are variable, but generally plot between the median and 84th percentile results for lateral dynamic instability of the systems investigated. The trends observed in this comparison indicate that an improved equation, in a form similar to R_{max}, could be developed as a more accurate and reliable (less variable) predictor of lateral dynamic instability for use in current nonlinear static analysis procedures.

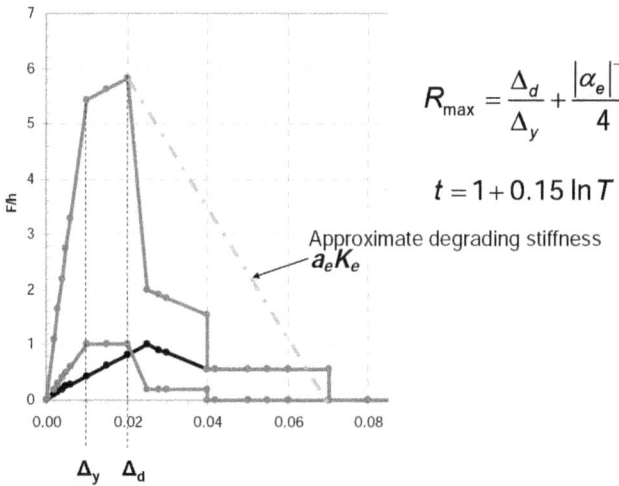

$$R_{max} = \frac{\Delta_d}{\Delta_y} + \frac{|\alpha_e|^{-t}}{4}$$

$$t = 1 + 0.15 \ln T$$

Approximate degrading stiffness
$a_e K_e$

Figure 4-27   Idealization of multi-spring force-displacement capacity boundaries to estimate effective negative stiffness for use in the FEMA 440 equation for R_{max}.

4.6.1 Improved Equation for Evaluating Lateral Dynamic Instability

An improved estimate for the strength ratio at which lateral dynamic instability might occur was developed based on nonlinear regression of the extensive volume of data generated during this investigation. In performing this regression, results were calibrated to the median response of the SDOF spring systems studied in this investigation.

A median-targeted strength ratio for lateral dynamic instability, R_{di}, is defined as:

$$R_{di} = \left(\frac{\Delta_c}{\Delta_y} \right)^a + b \frac{T_e}{3|\gamma|} + \frac{F_r}{F_c} \left(\frac{\Delta_u - \Delta_r}{\Delta_y} \right) \sqrt[3]{T_e} \qquad (4\text{-}4)$$

where T_e is the effective fundamental period of vibration of the structure, Δ_y, Δ_c, Δ_r, and Δ_u are displacements corresponding to the yield strength, F_y, capping strength, F_c, residual strength, F_r, and ultimate deformation capacity at the end of the residual strength plateau, as shown in Figure 4-28. Parameters a and b are functions given by:

$$a = 1 - \exp(-dT_e) \qquad (4\text{-}5)$$

$$b = 1 - \left(\frac{F_r}{F_c} \right)^2 \qquad (4\text{-}6)$$

where the parameter d is a constant equal to 4 for systems with stiffness degradation, and 5 for systems without stiffness degradation. The parameter γ is the ratio of the post-capping slope (degrading stiffness) to the initial effective slope (elastic stiffness).

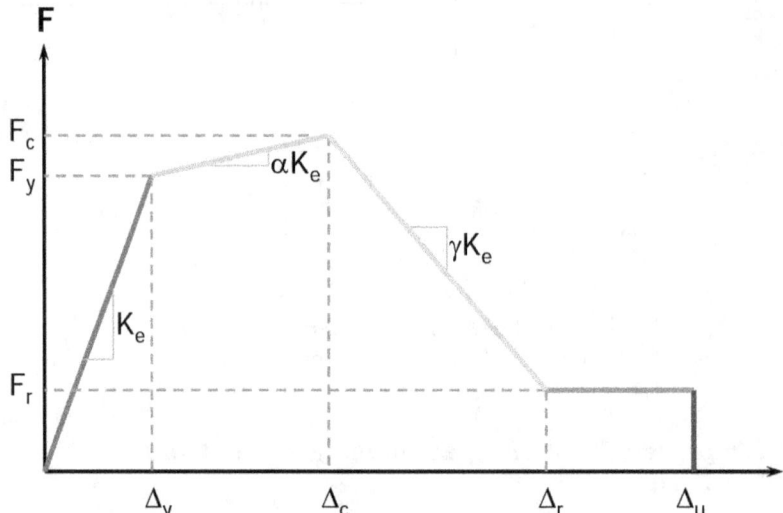

Figure 4-28   Simplified force-displacement boundary for estimating the median collapse capacity associated with lateral dynamic instability.

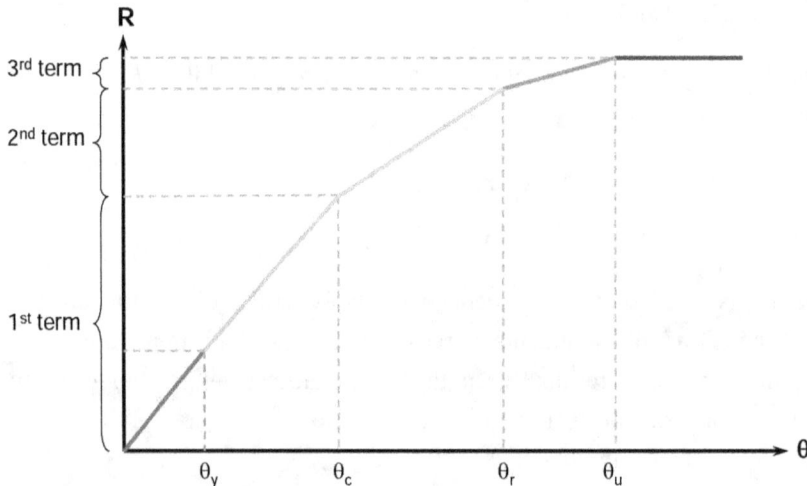

Figure 4-29   Relationship between Equation 4-4 and the segments of a typical IDA curve.

The three terms in Equation 4-4 relate to the segments of a typical force-displacement capacity boundary (Figure 4-28) and typical IDA curve (Figure 4-29). The first term provides an estimate of the median ground motion intensity corresponding to the end of the pseudo-linear segment of an IDA curve (i.e., intensity at the onset of degradation). The second term provides an estimate of the increment in ground motion intensity required to push the structure onto the residual strength plateau. The third term provides an estimate of the increment in ground motion intensity required produce lateral dynamic instability (collapse).

As developed, the equation for R_{di} is intended to be a more reliable (less variable) predictor of median response at lateral dynamic instability. The resulting equation was compared to the FEMA 440 equation for R_{max} and overlaid onto results for selected multi-spring systems. With few exceptions, Figure 4-30 through Figure 4-35 show that the equation for R_{di} consistently predicts median response over a range of system types and periods of vibration.

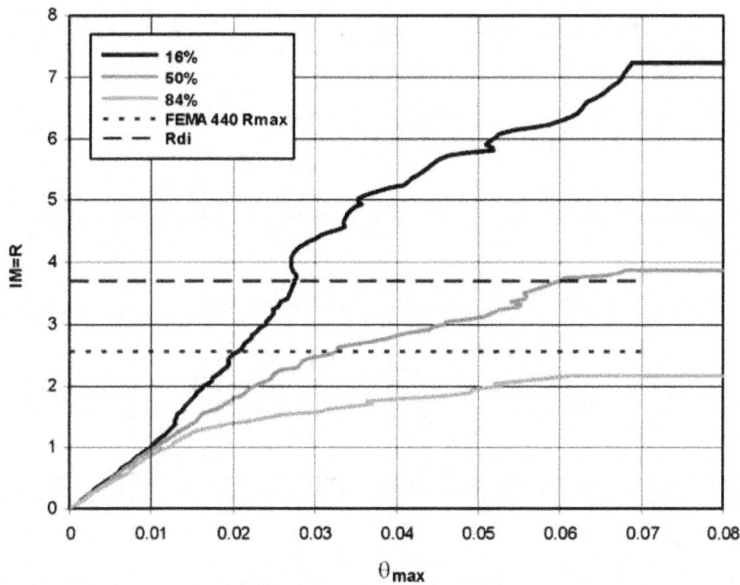

Figure 4-30   Comparison of R_{di} with FEMA 440 R_{max} and IDA results for system 2x2a+1a with T=1.18s.

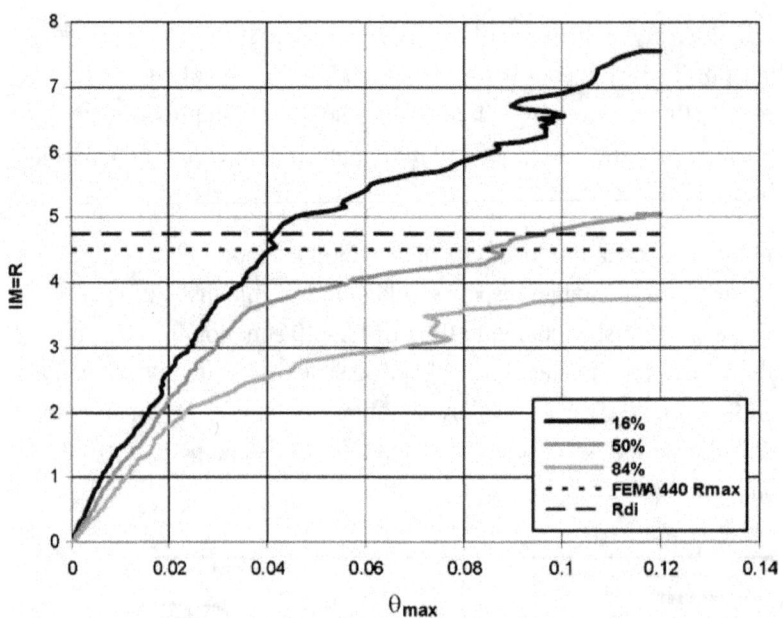

Figure 4-31   Comparison of R_{di} with FEMA 440 R_{max} and IDA results for system 3x3b+1b with T=1.0s.

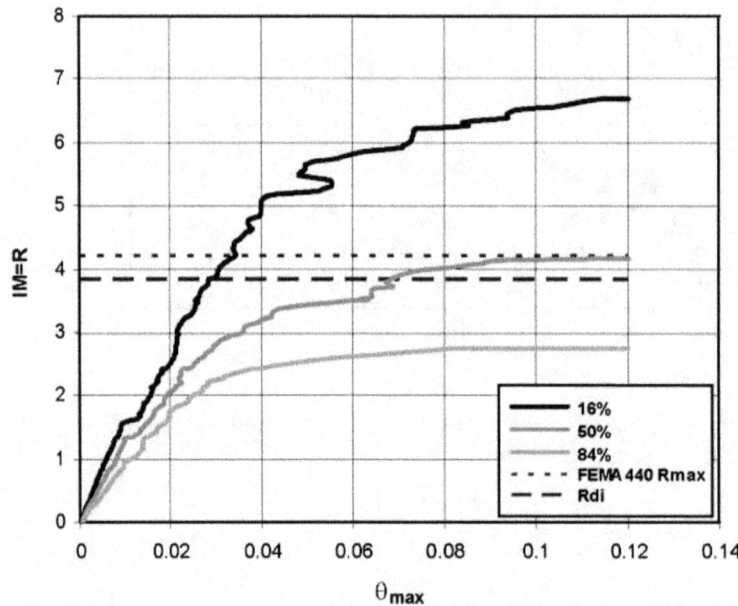

Figure 4-32   Comparison of R_{di} with FEMA 440 R_{max} and IDA results for system 9x3b+1b with T=0.61s.

4: Results from Single-Degree-of-Freedom Focused Analytical Studies

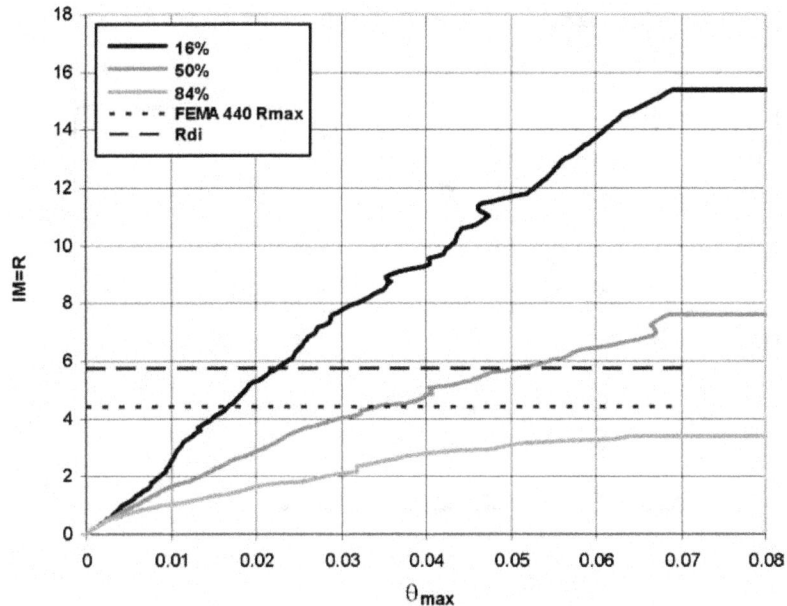

Figure 4-33   Comparison of R_{di} with FEMA 440 R_{max} and IDA results for system 5x5a+1a with T=1.15s.

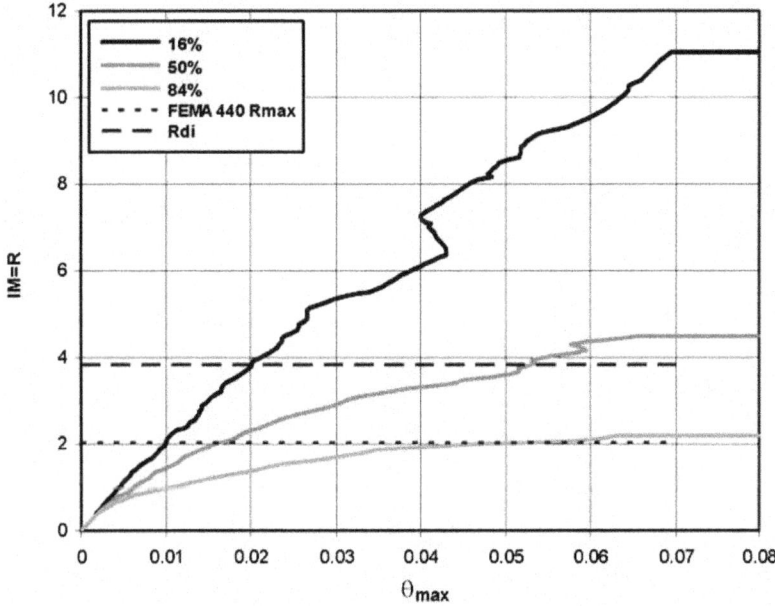

Figure 4-34   Comparison of R_{di} with FEMA 440 R_{max} and IDA results for system 5x5a+1a with T=0.58s.

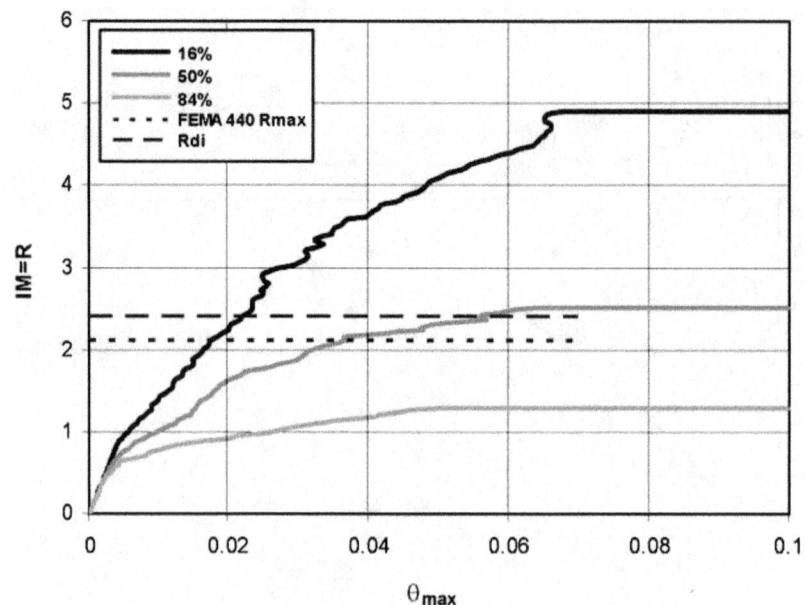

Figure 4-35   Comparison of R_{di} with FEMA 440 R_{max} and IDA results for system 9x5a+1a with T=0.34s.

4: Results from Single-Degree-of-Freedom
Focused Analytical Studies

Chapter 5
Findings, Conclusions, and Recommendations

This chapter summarizes the findings, conclusions, and recommendations resulting from the literature review and focused analytical studies of this investigation. Information from other chapters is collected and repeated here for ease of reference. In this chapter, findings have been grouped into the following categories:

- Findings related to improved understanding of nonlinear degrading response and judgment in implementation of nonlinear analysis results in engineering practice.

- Recommended improvements to current nonlinear analysis procedures

- Suggestions for further study

From the literature review, it is apparent that in-cycle strength and stiffness degradation are real phenomena that have been observed and documented to cause instability in individual components. Focused analytical studies have shown that larger assemblies of components of mixed hysteretic behavior experience similar negative stiffness that can lead to lateral dynamic instability. These studies have been able to link nonlinear dynamic response to major characteristics of component and system degrading behavior.

These studies have also confirmed many of the conclusions regarding degradation and lateral dynamic instability presented in FEMA 440: (1) in-cycle strength degradation is a significant contributor to dynamic instability; (2) cyclic degradation can increase the potential for dynamic instability, but its effects are far less significant in comparison with in-cycle degradation; and (3) an equation, such as R_{max}, could be used as an indicator of potential lateral dynamic instability for use in current nonlinear static analysis procedures.

5.1 Findings Related to Improved Understanding and Judgment

This section summarizes observations and conclusions related to improved understanding of nonlinear degrading response and judgment in implementation of nonlinear analysis results in engineering practice. Findings, and practical ramifications for engineering practice, are summarized in the sections that follow.

5.1.1 Sidesway Collapse versus Vertical Collapse

Lateral dynamic instability is manifested in structural systems as sidesway collapse caused by loss of lateral-force-resisting capacity. Most sidesway collapse mechanisms can be explicitly simulated in nonlinear response history analyses. It should be noted, however, that inelastic deformation of structural components can result in shear and flexural-shear failures in members, and failures in joints and connections, which can lead to an inability to support vertical loads (vertical collapse) long before sidesway collapse can be reached.

5.1.1.1 Practical Ramifications

Behavior of real structures can include loss of vertical-load-carrying capacity at lateral displacements that are significantly smaller than those associated with sidesway collapse. Use of the findings of this investigation with regard to lateral dynamic instability (sidesway collapse) in engineering practice should include consideration of possible vertical collapse modes that could be present in the structure under consideration.

5.1.2 Relationship between Loading Protocol, Cyclic Envelope, and Force-Displacement Capacity Boundary

Historically, the term "backbone curve" has referred to many different things. For this reason, two new terms have been introduced to distinguish between different aspects of hysteretic behavior. These are the *force-displacement capacity boundary*, and *cyclic envelope*.

5.1.2.1 Force-Displacement Capacity Boundary

A force-displacement capacity boundary defines the maximum strength that a structural member can develop at a given level of deformation, resulting in an effective "boundary" for the strength of a member in force-deformation space (Figure 5-1). In many cases, the force-displacement capacity boundary corresponds to the monotonic force-deformation curve.

A cyclic load path cannot cross a force-displacement capacity boundary. If a member is subjected to increasing deformation and the boundary is reached, then the strength that can be developed in the member is limited and the response must continue along the boundary (in-cycle strength degradation). Only displacement excursions intersecting portions of the capacity boundary with a negative slope will result in in-cycle strength degradation.

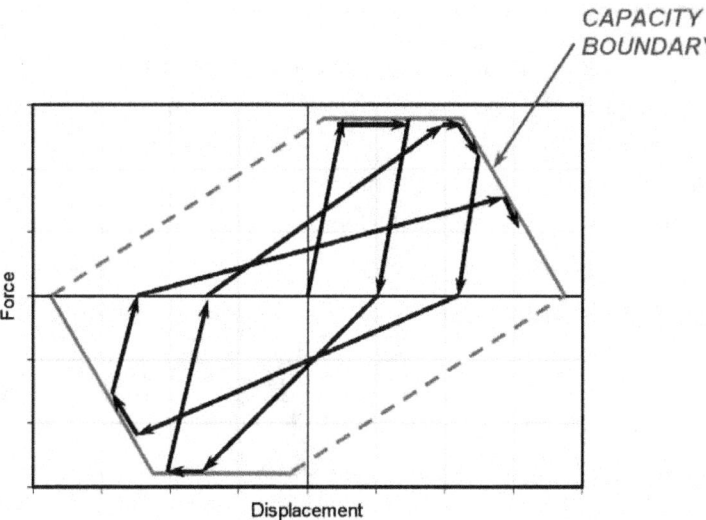

Figure 5-1   Example of a force-displacement capacity boundary.

5.1.2.2   Cyclic Envelope

A cyclic envelope is a force-deformation curve that envelopes the hysteretic behavior of a component or assembly that is subjected to cyclic loading (Figure 5-2).

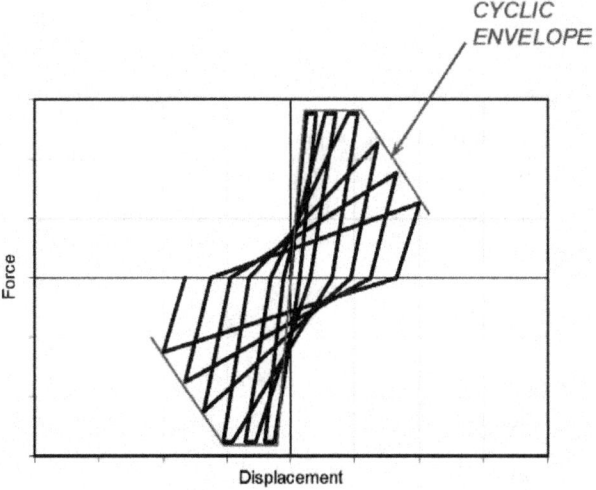

Figure 5-2   Example of a cyclic envelope.

The characteristics of the cyclic envelope are strongly influenced by the points at which unloading occurs in a test, and are therefore strongly influenced by the loading protocol that was used in the experimental program. Nominally identical specimens loaded with different loading protocols will have different cyclic envelopes depending on the number of cycles used in the loading protocol, the amplitude of each cycle, and the sequence of the loading cycles, as illustrated in Section 2.2.3.

Under lateral deformations that are less than or equal to those used to generate the cyclic envelope, differences between the cyclic envelope and the force-displacement capacity boundary are of no consequence. However, under larger lateral displacements these differences will affect the potential for in-cycle degradation to occur, and will significantly affect system behavior and response (Figure 5-3).

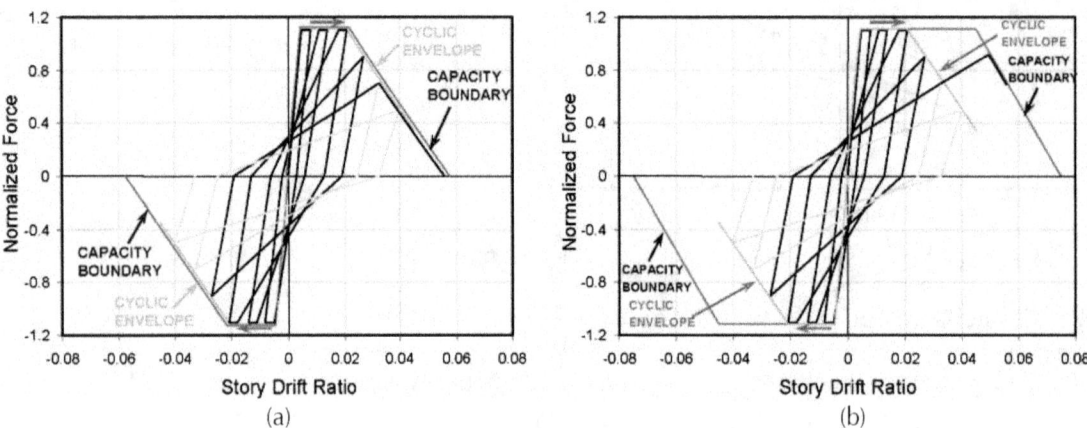

Figure 5-3   Comparison of hysteretic behavior when the force-displacement capacity boundary is: (a) equal to the cyclic envelope, and (b) extends beyond the cyclic envelope.

Constraining nonlinear hysteretic behavior within the limits of a cyclic envelope that does not capture the full range of permissible force-deformation response, as defined by the force-displacement capacity boundary, will result in overly pessimistic predictions of the nonlinear dynamic response of a system.

5.1.2.3   Practical Ramifications

Nonlinear component parameters should be based on the force-displacement capacity boundary, which is different from a cyclic envelope. Determining the force-displacement capacity boundary from test results using a single cyclic loading protocol can result in significant underestimation of the actual capacity for force-deformation response and subsequent overestimation of nonlinear displacement demands.

5.1.3 Characteristics of Median IDA Curves

Observed relationships between IDA curves and degrading component characteristics suggest that dynamic response is directly influenced by the features of a force-displacement capacity boundary. This relationship, which is dependent upon the period of vibration of the system, is depicted in the idealized graphical representation of Figure 5-4.

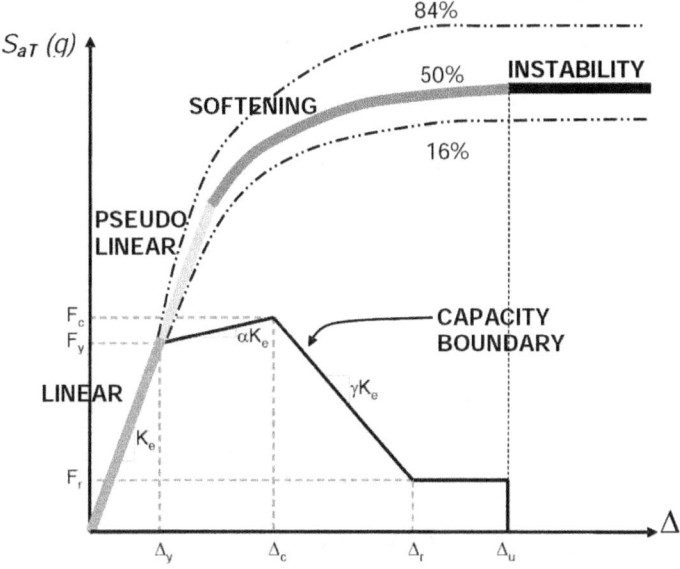

Figure 5-4   Relationship between IDA curves and the features of a typical force-displacement capacity boundary.

In general, median IDA curves were observed to exhibit the following characteristics:

- An initial linear segment corresponding to linear-elastic behavior in which in lateral deformation demand is proportional to ground motion intensity, regardless of the characteristics of the system or the ground motion. This segment extends from the origin to the onset of yielding.

- A second curvilinear segment corresponding to inelastic behavior in which lateral deformation demand is no longer proportional to ground motion intensity. As intensity increases, lateral deformation demands increase at a faster rate. This segment corresponds to softening of the system, or reduction in stiffness (reduction in the slope of the IDA curve). In this segment, the system "transitions" from linear behavior to eventual dynamic instability. Although a curvilinear segment is always present, in some cases the transition can be relatively long and gradual, while in other cases it can be very short and abrupt.

- A final linear segment that is horizontal, or nearly horizontal, in which infinitely large lateral deformation demands occur at small increments in ground motion intensity. This segment corresponds to the point at which a system becomes unstable (lateral dynamic instability). For SDOF systems, this point corresponds to the ultimate deformation capacity at which the system loses all lateral-force-resisting capacity.

In some systems, the initial linear segment can be extended beyond yield into the inelastic range. In this pseudo-linear segment, lateral deformation demand is approximately proportional to ground motion intensity, which is consistent with the familiar equal-displacement approximation for estimating inelastic displacements. The range of lateral deformation demands over which the equal-displacement approximation is applicable depends on the characteristics of the force-displacement capacity boundary of the system and the period of vibration.

5.1.3.1 Practical Ramifications

The observed relationships support the conclusion that it is possible to estimate nonlinear dynamic response based on knowledge of the characteristics of the force-displacement capacity boundary.

5.1.4 Dependence on Period of Vibration

In general, moderate and long period systems with zero or positive post-yield stiffness in the force-displacement capacity boundary follow the equal displacement trend well into the nonlinear range, as shown for Spring 3a in Figure 5-5. For systems with periods longer than 0.5s, Spring 3a exhibits an extension of the initial linear segment well beyond the yield drift of 0.01. In contrast, the short period system (T=0.2s) diverges from the initial linear segment just after yielding, even at deformations within the strength-hardening segment of the force-displacement capacity boundary (drifts between 0.01 and 0.04 in the figure).

5.1.4.1 Practical Ramifications

It is important to consider the dependence on period of vibration in conjunction with the effects of other parameters identified in this investigation. The generalized effect of any one single parameter can be misleading.

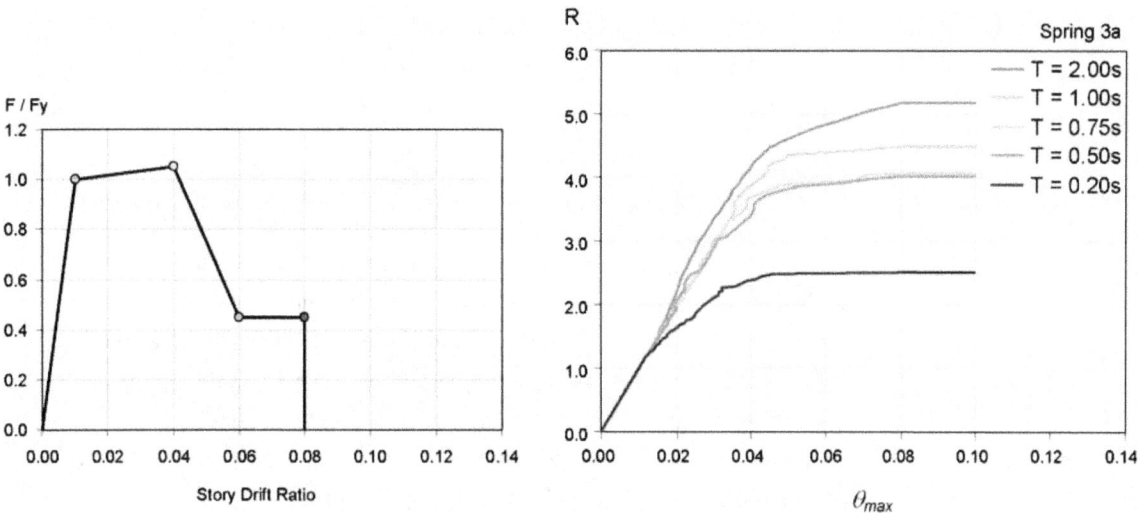

Figure 5-5   Force-displacement capacity boundary and median IDA curves for Spring 3a with various periods of vibration.

5.1.5   Dispersion in Response

Nonlinear response is sensitive to the characteristics of the ground motion record, and will vary from one ground motion to the next, even when scaled to the same intensity (Figure 5-6). For a given level of ground motion intensity, the lateral deformation demand can be significantly smaller or significantly larger than the value shown on median IDA curves, as indicated by the 16th and 84th percentile curves in the figure. As the level of ground motion intensity increases, the dispersion in response tends to increase.

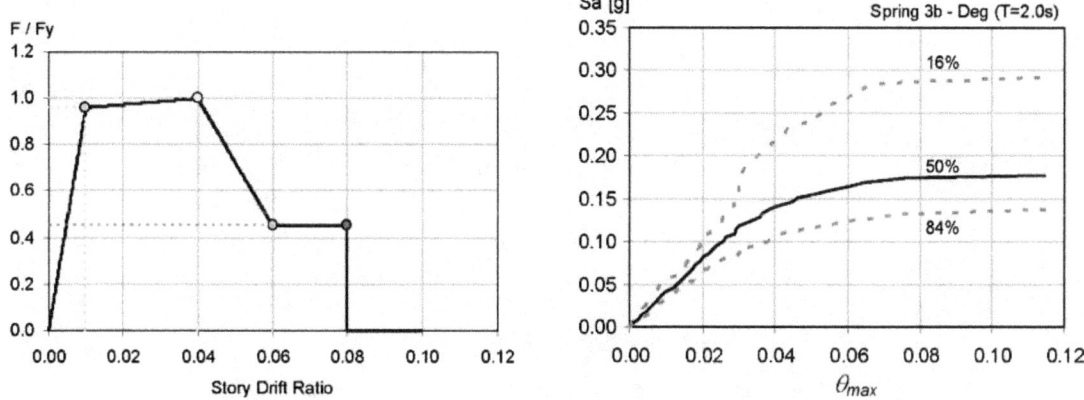

Figure 5-6   Force-displacement capacity boundary and 16th, 50th and 84th percentile IDA curves for Spring 3b with a period of vibration T=2.0s.

5.1.5.1   Practical Ramifications

It is important to recognize the level of uncertainty that is inherent in nonlinear analysis, particularly regarding variability in response due to ground motion uncertainty. It may not be sufficient to rely on median (50%) estimates of response for certain design or evaluation quantities of interest, unless the intensity of the ground motion is associated with an appropriately rare probability of exceedance.

5.1.6 Influence of the Force-Displacement Capacity Boundary

Key features of a force-displacement capacity boundary that were observed to influence the shape of median IDA curves included post-yield behavior and onset of degradation, slope of degradation, ultimate deformation capacity, and presence of cyclic degradation. Systems in which the force-displacement capacity boundary had more favorable post-yield characteristics (e.g., delayed onset of degradation, more gradual slope of degradation, higher residual strength, and higher ultimate deformation capacity) were observed to perform better.

5.1.6.1 Post-Yield Behavior and Onset of Degradation

The presence of a non-negative post-yield slope and delay before the onset of degradation reduced potential in-cycle strength degradation and significantly improved the collapse capacity of a system (Figure 5-7).

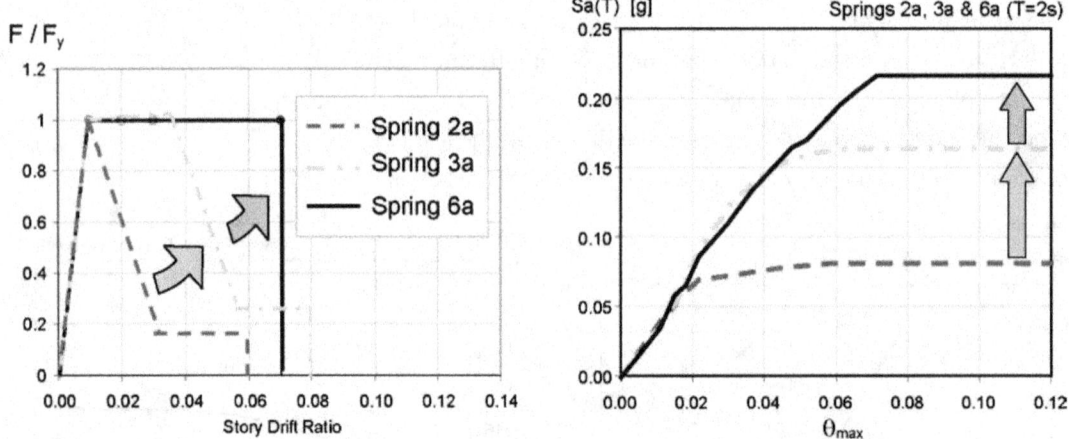

Figure 5-7   Effect of post-yield behavior on the collapse capacity of a system (Springs 2a, 3a and 6a with T=2.0s).

5.1.6.2 Slope of Degradation

Differences in the negative slope of the strength-degrading segment significantly affected the collapse capacity of a system. Systems with more shallow degrading slopes reached higher collapse capacities than systems with steeper degrading slopes (Figure 5-8). Changes in negative slope changed the magnitude of potential in-cycle strength degradation, and overshadowed any changes in other parameters (e.g., the residual strength plateau), as long as the ultimate deformation capacity remained the same.

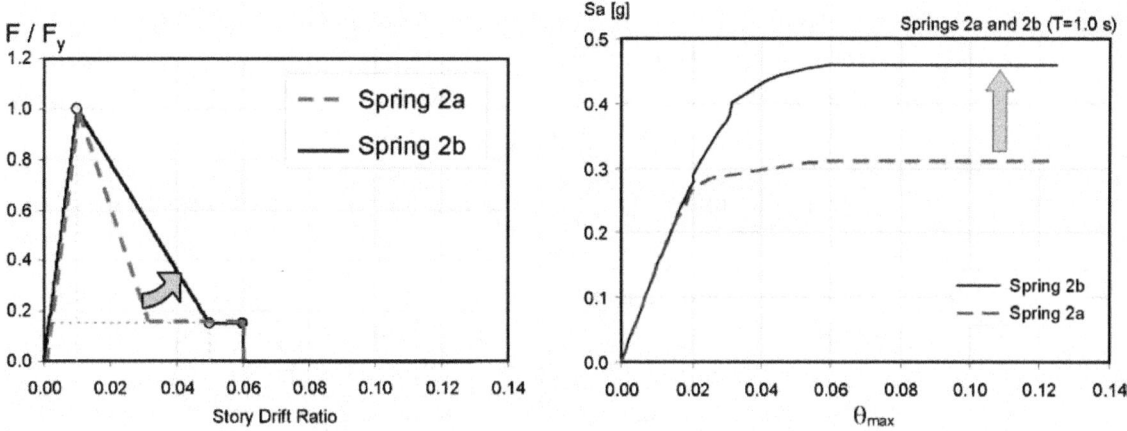

Figure 5-8   Effect of slope of degradation on the collapse capacity of a system (Springs 2a and 2b with T=1.0s).

5.1.6.3   Ultimate Deformation Capacity

Increasing the ultimate deformation capacity resulted in significant increases in collapse capacity (Figure 5-9). The key parameter related to the observed change in response is the increment in the ultimate deformation capacity. Observed changes in collapse capacity resulting from increases in the ultimate deformation capacity were insensitive to other characteristics of the post-yield behavior of the springs.

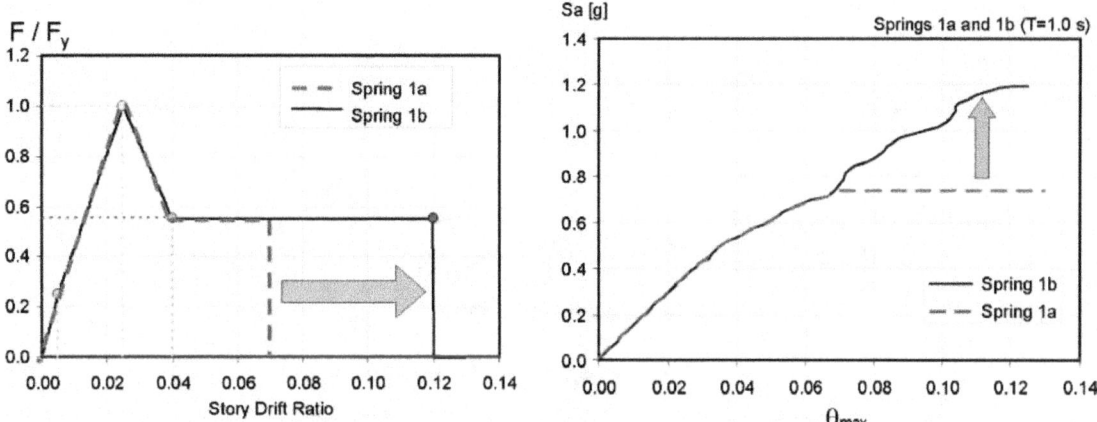

Figure 5-9   Effect of ultimate deformation capacity on the collapse capacity of a system (Springs 1a and 1b with T=1.0s).

5.1.6.4   Practical Ramifications

Observed relationships between selected features of the force-displacement capacity boundary and the resulting characteristics of median IDA curves support the conclusion that the nonlinear dynamic response of a system can be correlated to the parameters of the force-displacement capacity boundary of that system. Of particular interest is the relationship between global deformation demand and the intensity of the ground motion at lateral dynamic instability (collapse). Results indicate that it is possible to use nonlinear static procedures to estimate the potential for lateral dynamic instability of systems exhibiting in-cycle degradation.

5.1.7 Cyclic Degradation of the Force-Displacement Capacity Boundary

In general, most components will exhibit some level of cyclic degradation. Consistent with observations from past studies, comparison of results between springs both with and without cyclic degradation show that the effects of cyclic degradation (as measured by gradual movement of the capacity boundary) are relatively unimportant in comparison with in-cycle degradation (as measured by the extent and steepness of negative slopes in the capacity boundary). This trend is illustrated for Spring 3b in Figure 5-10, but can be observed in the results for many spring systems in Appendix B.

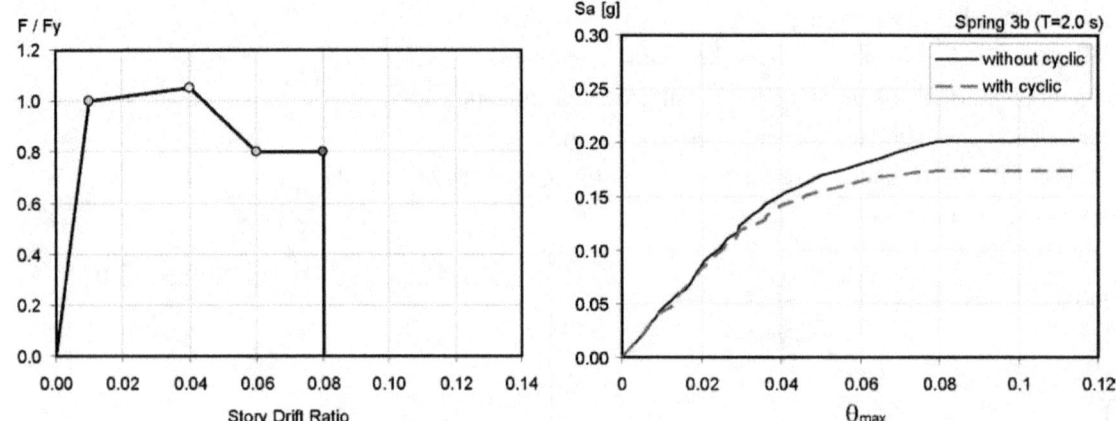

Figure 5-10   Effect of degradation of the force-displacement capacity boundary on the collapse capacity of a system (Spring 3b, T=2.0s, with and without cyclic degradation).

5.1.7.1 Practical Ramifications

In most cases the effects of in-cycle strength degradation dominate the nonlinear dynamic behavior of a system. This suggests that in many cases the effects of cyclic degradation can be neglected. Instead, the focus should be on more accurately characterizing the force-displacement capacity boundary, which controls the onset of in-cycle degradation (where it occurs).

Two situations in which the effects of cyclic degradation were observed to be important include: (1) short period systems; and (2) systems with very strong in-cycle strength degradation effects (very steep and very large drops in lateral strength). In these cases, the effects of cyclic degradation can be important and should be considered.

5: Findings, Conclusions, and Recommendations   FEMA P440A

5.1.8 Effects of Secondary System Characteristics

The contribution of a secondary ("gravity") system acting in parallel with a primary lateral-force-resisting system always resulted in an improvement in nonlinear response, especially close to collapse. This result was observed both qualitatively and quantitatively (i.e., both in normalized and non-normalized coordinates).

The improvement was larger when considering secondary systems with larger ultimate deformation capacities, even if the lateral strength of the secondary system was small in comparison to that of the primary system. This result is illustrated in Figure 5-11, and is supported by results described in Section 5.1.6.3. In the figure, as the system combination ratio increases, the relative combination of the secondary system diminishes, yet the resulting collapse capacities for combinations with Spring 1b (larger ultimate deformation capacity) are significantly higher than combinations with Spring 1a (smaller ultimate deformation capacity).

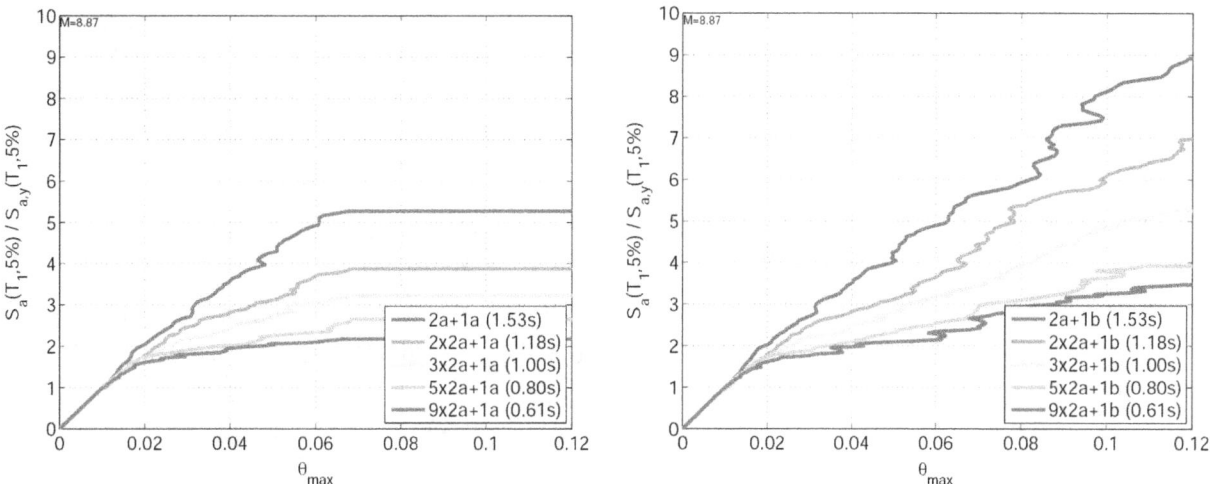

Figure 5-11   Median IDA curves plotted versus the normalized intensity measure $S_a(T,5\%)/S_{ay}(T,5\%)$ for systems Nx2a+1a and Nx2a+1b with a mass of 8.87 tons.

5.1.8.1 Practical Ramifications

Consideration of the contribution of secondary ("gravity") systems acting in parallel with primary lateral resisting systems is important and should be included in nonlinear modeling for collapse simulation. For seismic retrofit of existing structures, this suggests that adding a relatively weak (but ductile) system in parallel with the primary system could substantially increase collapse capacity and delay the onset of lateral dynamic instability. The introduction of such a secondary system could be significantly less complicated and less expensive than direct improvements to the strength, stiffness and deformation capacity of the primary system.

5.1.9 Effects of Lateral Strength

Increasing the lateral strength of a system was observed to increase collapse capacity, with the following limitations:

- Increases in the lateral strength of a system changed the intensity that initiated yielding in the system and the intensity at collapse (lateral dynamic instability). The incremental change in collapse capacity, however, was less than proportional to the increase in yield strength (Figure 5-12).

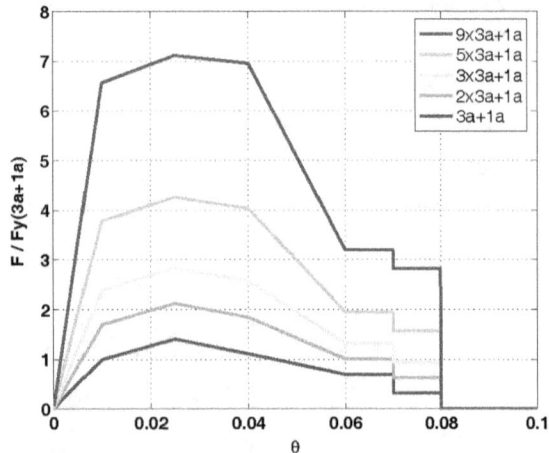 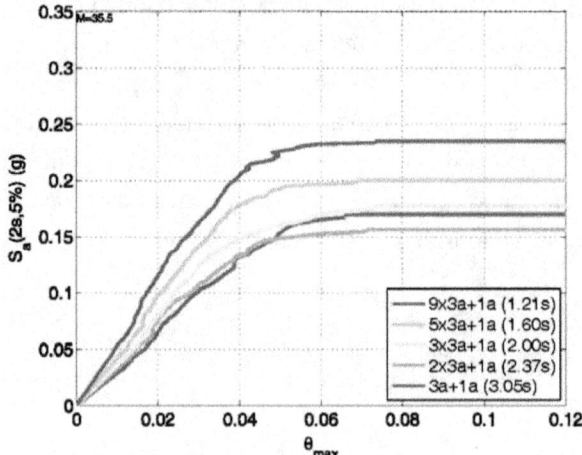

Figure 5-12   Force-displacement capacity boundaries and median IDA curves plotted versus the common intensity measure $S_a(2s,5\%)$ for system Nx3a+1a with a mass of 35.46 tons.

- The effectiveness of increasing the lateral strength of a system was a function of the shape of the force-displacement capacity boundary. Incremental changes in yield strength were more effective for ductile systems than they were for systems with less ductile behavior.

- The effectiveness of increasing the lateral strength of a system was also a function of the period of system. Incremental changes in yield strength were more effective for stiff systems than they were for flexible systems.

5.1.9.1 Practical Ramifications

Increasing the lateral strength of a system can improve collapse behavior, but will not result in equal increases in collapse capacity. The effectiveness of seismic retrofit strategies that involve increasing the lateral strength will depend on the characteristics of the force-displacement capacity boundary of the existing system as well as the period of vibration.

5.2 Recommended Improvements to Current Nonlinear Analysis Procedures

Prevailing practice for performance-based seismic design is based on the FEMA 273 *NEHRP Guidelines for the Seismic Rehabilitation of Buildings* (FEMA, 1997) and its successor documents, FEMA 356 *Prestandard and Commentary for the Seismic Rehabilitation of Buildings* (FEMA, 2000), and ASCE/SEI Standard 41-06 *Seismic Rehabilitation of Existing Buildings* (ASCE, 2006b). Recommendations contained in FEMA 440 *Improvement of Nonlinear Static Seismic Analysis Procedures* (FEMA, 2005) were incorporated into the developing ASCE/SEI Standard 41-06 in 2005. ASCE/SEI Standard 41-06 Supplement No. 1 was published in 2007. Together these resource documents form the basis of nonlinear analysis in current engineering practice. This section summarizes recommended clarifications and improvements to current nonlinear analysis procedures as characterized in these documents.

5.2.1 Current Nonlinear Static Procedures

The Coefficient Method is one method of estimating maximum inelastic displacements of a system. The process begins with the generation of an idealized force-deformation curve (i.e., static pushover curve) relating base shear to roof displacement. From this curve, an effective period, T_e, is obtained, and the maximum global displacement (target displacement) for a specified level of ground motion intensity is estimated using Equation 5-1:

$$\delta_t = C_0 C_1 C_2 S_a \frac{T_e^2}{4\pi^2} g \qquad (5\text{-}1)$$

In this expression the first three terms are coefficients that modify the elastic displacement of the system. C_0 is the first mode participation factor. This coefficient essentially converts from spectral ordinates to roof displacement. This C_1 coefficient (Equation 5-2) increases elastic displacements in short period systems, essentially accounting for exceptions to the equal displacement approximation.

$$C_1 = 1 + \frac{R-1}{aT_e^2} \qquad (5\text{-}2)$$

where:

$$R = \frac{S_a}{V_y / W} \cdot C_m \qquad (5\text{-}3)$$

and C_m is the effective mass factor to account for higher mode mass participation effects. The C_2 coefficient (Equation 5-4) increases elastic displacements in short period and weak systems to account for stiffness degradation, hysteretic pinching, and cyclic strength degradation.

$$C_2 = 1 + \frac{1}{800}\left(\frac{R-1}{T_e}\right)^2 \qquad (5\text{-}4)$$

Importantly, C_2 does not account for displacement amplification due to in-cycle strength degradation, which can result in lateral dynamic instability. In-cycle strength degradation is addressed by a minimum strength requirement (maximum value of R) used as a trigger for the need to further investigate the potential for lateral dynamic instability using nonlinear response history analysis. The minimum strength requirement in current nonlinear analysis procedures was described in Section 4.6 (and is repeated in the equations that follow):

$$R_{max} = \frac{\Delta_d}{\Delta_y} + \frac{\alpha_e^{-t}}{4} \qquad (5\text{-}5)$$

where

$$t = 1 + 0.15 \ln T \qquad (5\text{-}6)$$

and

$$\alpha_e = \alpha_{P-\Delta} + \lambda\left(\alpha_2 - \alpha_{P-\Delta}\right) \qquad (5\text{-}7)$$

for $0 < \lambda < 1.0$.

Values of R (Equation 5-3) are compared to R_{max}. Systems in which $R < R_{max}$ are deemed to meet the minimum strength requirement to avoid lateral dynamic instability, and nonlinear response history analysis is not required.

5.2.2 Clarification of Terminology and Use of the Force-Displacement Capacity Boundary for Component Modeling

For nonlinear analysis, ASCE/SEI 41-06 specifies component modeling and acceptability criteria based on the conceptual force-displacement relationship ("backbone") depicted in Figure 5-13. Since the term "backbone curve" has been used to refer to many different things, its definition related to nonlinear component modeling is not clear. In Section 2.8 of the standard, it is permitted to derive modeling parameters and acceptance criteria using experimentally obtained cyclic response characteristics from subassembly testing. So defined, the standard can be interpreted to condone the use of cyclic envelopes from component tests to generate the necessary force-displacement relationships.

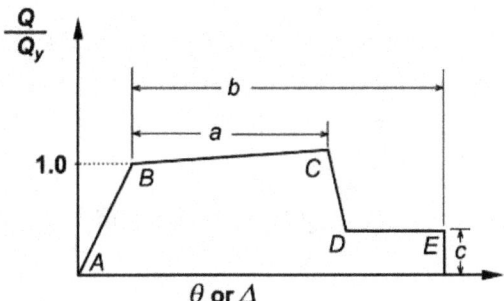

Figure 5-13   Conceptual force-displacement relationship ("backbone") used in ASCE/SEI 41-06 (adapted from FEMA 356).

The use of a cyclic envelope, as opposed to a force-displacement capacity boundary, has been shown to result in underestimation of the actual capacity for force-deformation response and subsequent overestimation of nonlinear deformation demands. In some cases the resulting conservatism can be very large.

For this reason, introduction and use of two new terms are recommended to distinguish between different aspects of hysteretic behavior. These are the *force-displacement capacity boundary*, and *cyclic envelope*, defined in Section 5.1.2. Important conceptual differences between the force-displacement capacity boundary and a loading protocol-specific cyclic envelope should be clarified in future revisions to ASCE/SEI 41, and the use of an appropriate force-displacement capacity boundary should be specified for characterizing component hysteretic behavior.

Proper definition of the hysteretic behavior in a component model requires an understanding of: (1) the initial force-displacement capacity boundary; and (2) how the force-displacement capacity boundary degrades under cyclic loading. The ideal method for establishing an initial force-displacement capacity boundary is through monotonic testing. Once the initial force-displacement capacity boundary is defined, degradation parameters should be established based on results from cyclic tests.

There is no recognized testing protocol that incorporates realistic consideration of the force-displacement capacity boundary. The use of several cyclic loading protocols is desirable to ensure that the degradation parameters are properly identified and the calibrated component model is general enough to represent response under any type of loading.

In Commentary Section C6.3.1.2.2 of ASCE/SEI 41-06 Supplement No. 1, it is suggested that the sudden drop from Point C to Point D (in Figure 5-10) can be overly pessimistic, and that a more gradual slope from Point C to Point E might be more realistic for concrete components. Some experimental results suggest that such an adjustment could be applicable for other types of components. If the actual monotonic curve is not available, or cannot be estimated, use of a force-displacement capacity boundary with this alternate slope can be considered.

5.2.3   Improved Equation for Evaluating Lateral Dynamic Instability

In comparison with results for selected multi-spring systems in this investigation, the FEMA 440 equation for R_{max} was shown to predict values that are variable, but generally fall between the median and 84th percentile results for lateral dynamic instability. This result suggests that the current equation for R_{max} would be conservative if used in conjunction with a capacity boundary generated from a pushover analysis. It could be very conservative if the pushover curve was based on component modeling parameters determined using a cyclic envelope rather than a force-displacement capacity boundary.

The trends observed in this comparison indicate that an improved equation, in a form similar to R_{max}, could be developed as a more accurate and reliable (less variable) predictor of lateral dynamic instability for use in current nonlinear static analysis procedures. An improved estimate for the strength ratio at which lateral dynamic instability might occur was developed based on nonlinear regression of the extensive volume of data generated during this investigation. In performing this regression, results were calibrated to the median response of the SDOF spring systems studied in this investigation.

A median-targeted minimum strength requirement (maximum value of R) for lateral dynamic instability, R_{di}, is proposed in Equation 5-8:

$$R_{di} = \left(\frac{\Delta_c}{\Delta_y}\right)^a + b\frac{T_e}{3|\gamma|} + \frac{F_r}{F_c}\left(\frac{\Delta_u - \Delta_r}{\Delta_y}\right)\sqrt[3]{T_e} \qquad (5\text{-}8)$$

where T_e is the effective fundamental period of vibration of the structure, Δ_y, Δ_c, Δ_r, and Δ_u are displacements corresponding to the yield strength, F_y, capping strength, F_c, residual strength, F_r, and ultimate deformation capacity at the end of the residual strength plateau, as shown in Figure 5-14. Parameters a and b are functions given by:

$$a = 1 - \exp(-dT_e) \qquad (5\text{-}9)$$

$$b = 1 - \left(\frac{F_r}{F_c} \right)^2 \qquad\qquad (5\text{-}10)$$

The parameter d is a constant equal to 4 for systems with stiffness degradation, and 5 for systems without stiffness degradation. The parameter γ is the ratio of the post-capping slope (degrading stiffness) to the initial effective slope (elastic stiffness).

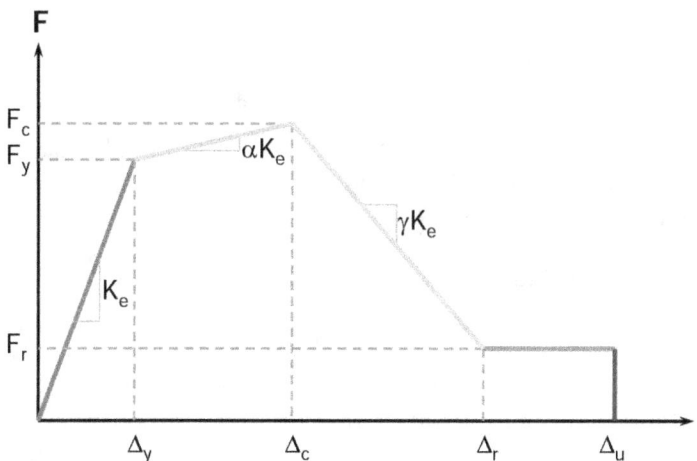

Figure 5-14   Simplified force-displacement boundary for estimating the median collapse capacity associated with dynamic instability.

5.2.3.1 Practical Ramifications

Since R_{di} has been calibrated to median response, use of this equation could eliminate some of the conservatism built into the current R_{max} limitation on use of nonlinear static procedures. Calibrated using the extensive volume of data generated during this investigation, use of this equation could improve the reliability of current nonlinear static procedures with regard to cyclic and in-cycle degradation.

In conjunction with a pushover curve used as a system force-displacement capacity boundary, the equation for R_{di} could be used to determine if a system is susceptible to lateral dynamic instability for a specified level of spectral acceleration, S_{aT}. Similar to R_{max}, use of R_{di} would involve comparison with R (Equation 5-3). If $R < R_{di}$ the system could be deemed satisfactory without additional nonlinear dynamic analysis. This capability is, of course, limited to systems for which the assumption of SDOF behavior is appropriate (i.e., MDOF effects are not significant).

Calculated values of R_{di} should be viewed carefully with respect to the intensity measure (S_{aT}) considered. Collapse limit states (i.e., lateral dynamic instability) should be evaluated for intensities associated with rare ground motions (long return periods). Evaluation of collapse limit states at lower ground motion intensities leaves open the possibility that collapse could occur during events in which those intensities are exceeded.

In addition, the development of the proposed equation for R_{di} targeted median response, which was intentionally less conservative than the level at which the FEMA 440 equation for R_{max} appeared to be predicting. Median response implies a fifty percent chance of being above or below the specified value. Use of R_{di} in engineering practice should consider whether or not a median predictor represents an appropriate level of safety against the potential for lateral dynamic instability. If needed, a reduction factor could be applied to Equation 5-8 to reduce the resulting values of R_{di} and ultimately achieve a correspondingly higher level of safety.

5.2.4 Simplified Nonlinear Dynamic Analysis Procedure

From empirical relationships for characteristic segments of IDA curves for many systems, Vamvatsikos and Cornell (2006) suggested that static pushover curves could be used to estimate nonlinear dynamic response. The open source software tool, *Static Pushover 2 Incremental Dynamic Analysis* (SPO2IDA), was created as a product of that research, and can be obtained at http://www.ucy.ac.cy/~divamva/software.html. A Microsoft Excel version of the SPO2IDA application has also been provided on the CD accompanying this report.

As the name suggests, SPO2IDA transforms static pushover (SPO) curves to incremental dynamic analysis (IDA) plots. It utilizes a large database of IDA results to fit representative 16th, 50th, and 84th percentile IDA curves to a given idealized single-degree-of-freedom (SDOF) oscillator subjected to a static pushover analysis. The relationships between force-displacement capacity boundaries and IDA curves observed in this investigation are consistent with this notion.

Focused analytical studies comparing force-displacement capacity boundaries to incremental dynamic analysis results led to the concept of a simplified nonlinear dynamic analysis procedure. In this procedure, a nonlinear static analysis is used to generate an idealized force-deformation curve (i.e., static pushover curve). The resulting curve is then used as a force-displacement capacity boundary to constrain the hysteretic behavior of an equivalent SDOF oscillator. This SDOF oscillator is then subjected to incremental dynamic analysis (or approximate IDA results are obtained using SPO2IDA).

The concept of a simplified nonlinear dynamic analysis procedure is described in the steps outlined below.

- **Develop an analytical model of the system.**

  Models can be developed in accordance with prevailing practice for seismic evaluation, design, and rehabilitation of buildings described in ASCE/SEI 41-06. Component properties should be based on force-displacement capacity boundaries, rather than cyclic envelopes.

- **Perform a nonlinear static pushover analysis.**

  Subject the model to a conventional pushover analysis in accordance with prevailing practice. Lateral load increments and resulting displacements are recorded to generate an idealized force-deformation curve.

- **Conduct an incremental dynamic analysis of the system based on an equivalent SDOF model.**

  The idealized force-deformation curve is, in effect, a system force-displacement capacity boundary that can be used to constrain a hysteretic model of an equivalent SDOF oscillator. This SDOF oscillator is then subjected to incremental dynamic analysis to check for lateral dynamic instability and other limit states of interest. Alternatively, an approximate incremental dynamic analysis can be accomplished using the idealized force-deformation curve and SPO2IDA.

- **Determine probabilities associated with limit states of interest.**

  Results from incremental dynamic analysis can be used to obtain response statistics associated with limit states of interest in addition to lateral dynamic instability. SPO2IDA can also be used to obtain median, 16th, and 84th percentile IDA curves relating displacements to intensity. Using the fragility relationships described in Appendix E in conjunction with a site hazard curve, this information can be converted into annual probabilities of exceedance for each limit state. Probabilistic information in this form can be used to make enhanced decisions based on risk and uncertainty, rather than on discrete threshold values of acceptance.

The procedure is simplified because only a SDOF oscillator is subjected to nonlinear dynamic analysis. Further simplification is achieved through the use of SPO2IDA, which avoids the computational effort associated with incremental dynamic analysis. This simplified procedure has several advantages over nonlinear static analysis procedures: (1) lateral dynamic instability is investigated explicitly; (2) results include the effects of record-to-record variability in ground motion; (3) response can be characterized probabilistically; and (4) uncertainty can be considered explicitly. Results can be investigated for any limit state that can be linked to the demand parameter of interest (e.g. roof displacement).

Use of the procedure is explained in more detail in the example application contained in Appendix F.

5.3 Suggestions for Further Study

This section summarizes suggestions for further study that will expand the application of results to more complex systems, fill in gaps in existing knowledge, and enhance future practice.

5.3.1 Application of Results to Multiple-Degree-of-Freedom Systems

Multi-story buildings are more complex dynamic systems whose seismic response is more difficult to estimate than that of SDOF systems. Recent studies have suggested that it may be possible to estimate the collapse capacity of multiple-degree-of-freedom (MDOF) systems through dynamic analysis of equivalent SDOF systems.

As part of the focused analytical work, preliminary studies of MDOF systems were performed as summarized in Appendix G. These studies investigated the use of nonlinear static analyses combined with incremental dynamic analyses of equivalent SDOF systems to evaluate dynamic instability of multi-story buildings ranging in height from 4 to 20 stories. Preliminary results indicate that many of the findings for SDOF systems in this investigation (e.g., the relationship between force-displacement capacity boundary and IDA curves; the equation for R_{di}) may be applicable to MDOF systems. More detailed study of the application of these results to MDOF systems is recommended as a result of this investigation, and additional MDOF investigations are planned under a project funded by the National Institute of Standards and Technology (NIST).

5.3.2 Development of Physical Testing Protocols for Determination of Force-Displacement Capacity Boundaries

Important conceptual differences exist between force displacement capacity boundaries and loading protocol-specific cyclic envelopes. Proper definition of hysteretic behavior in a component model requires an understanding of the initial force-displacement capacity boundary and how that boundary degrades under cyclic loading. The use of several loading protocols is desirable, but there is no recognized testing procedure that accomplishes this.

The loading protocol for experimental investigations described in Section 2.8 of ASCE/SEI 41-06 is not specific enough to produce a true force-displacement capacity boundary. For a set of identical specimens, necessary testing could conceivably include: (1) monotonic loading to get an initial capacity boundary; (2) multiple symmetric cyclic loading cases to calibrate cyclic degradation; (3) high frequency or long duration cyclic loading cases to check for fracture or fatigue; (4) cyclic loading followed by a monotonic push to more clearly observe changes due to cyclic degradation; and (5) unsymmetrical cyclic loading. Development of a specification for physical testing protocols necessary to generate appropriate force-displacement capacity boundaries is recommended.

5.3.3 Development and Refinement of Tools for Approximate Nonlinear Dynamic Analysis

Nonlinear dynamic analysis has obvious advantages over nonlinear static analysis procedures. Disadvantages are related to increased computational effort. Studies have shown that the characteristics of nonlinear dynamic response can be estimated through simplified approximate relationships based on the results of static pushover analyses.

Software tools such as SPO2IDA have the capability to estimate dynamic response without the computational effort associated with incremental dynamic analysis. This approximation facilitates the use of dynamic analysis results to supplement and inform more simplified analysis procedures (e.g., nonlinear static procedures). Development and refinement of similar approximate tools for performing nonlinear dynamic analyses is recommended.

5.4 Concluding Remarks

Using FEMA 440 as a starting point, this investigation has advanced the understanding of degradation and lateral dynamic instability by:

- Investigating and documenting currently available empirical and theoretical knowledge on nonlinear cyclic and in-cycle strength and stiffness degradation, and their affects on the stability of structural systems

- Supplementing and refining the existing knowledge base with focused analytical studies

This investigation has resulted in an extensive collection of available research on component modeling of degradation, and a database of analytical results from over 2.6 million nonlinear response history analyses documenting the effects of a variety of parameters on the overall response of SDOF systems with degrading components.

Results have confirmed conclusions regarding degradation and dynamic instability presented in FEMA 440, provided updated information on modeling to differentiate between cyclic and in-cycle strength and stiffness degradation, and linked nonlinear dynamic response to major characteristics of component and system degrading behavior. This information has resulted in:

- an improved understanding of nonlinear degrading response and the practical ramifications of this information for engineering practice

- recommendations to better account for nonlinear degrading response in the context of current nonlinear analysis procedures

- suggestions for further study

Results from this investigation will ultimately improve the modeling of structural components considering cyclic and in-cycle degrading behavior, improve the characterization of lateral dynamic instability, and reduce the conservatism in current analysis procedures making it more cost-effective to strengthen existing buildings for improved seismic resistance in the future.

Appendix A

Detailed Summary of Previous Research

This appendix contains a detailed summary of the development of hysteretic models for nonlinear analysis. It also contains summaries of publications that were extensively reviewed for guidance on appropriately targeting and scoping focused analytical studies.

A.1 Summary of the Development of Hysteretic Models

A.1.1 Non-Deteriorating Models

Early studies that incorporated nonlinear behavior in seismic response of structures assumed the structure to have an elastoplastic hysteretic behavior or a bilinear hysteretic behavior (e.g., Berg and Da Deppo, 1960; Penzien 1960a, 1960b; Iwan 1961). These might be perfectly plastic with no post-elastic stiffness or with some strain hardening. More accurate models were also developed with smooth rounded transitions from elastic to plastic regions. (Ramberg and Osgood, 1943; Pinto and Guiffre, 1970; Menegotto and Pinto, 1973). Ramberg-Osgood and Giuffre-Menegotto-Pinto models continue to be used today for modeling non-degrading structures such as steel moment-frame structures when fracture and buckling do not occur, and have recently been used successfully to model the hysteretic behavior of buckling-restrained braces (e.g., Lin *et al.*, 2004).

Other examples of non-degrading smooth hysteretic models commonly used are the Bouc-Wen model (Bouc, 1967a, 1967b; Wen, 1976, 1989) and the Ozdemir model (1976). Unlike the Ramberg-Osgood and Giuffre-Menegotto-Pinto models in which the force-deformation relationship is described by an algebraic equation, in the Bouc-Wen and Ozdemir models the force-displacement relationship and the force-deformation characteristics are described by a differential equation. These models are relatively easy to implement and are capable of describing, relatively well, non-degrading hysteretic behavior. An extension of smooth models to a three-dimensional tensorial idealization of Prager's model was developed by Casciati and Faravelli (1985, Casciati, 1989). Although models based on differential equations with smooth loading curves, such as the Bouc-Wen or Ozdemir models, are relatively easy to implement, they generally exhibit a local violation of Drucker's stability postulate. In particular, Thyagarajan and Iwan (1990) concluded that the Wen-Bouc model tends to exhibit a pronounced drift, particularly when post-yield stiffness is small.

A.1.2 Piecewise Linear Deteriorating Models

Many structural materials and structural elements will exhibit some level of degradation of stiffness or strength or both, or may also exhibit other phenomena such as pinching, when subjected to cyclic reversed loading, this is especially true for reinforced concrete elements subjected to several large cyclic reversals. Deterioration can be the result of, for example, cracking, crushing, rebar buckling, crack or gap opening and closing, loss of bond, and interaction with high shear or axial stresses. The level of degradation depends, on the one hand, on the characteristics of the structural element such as properties of the materials, geometry, level of detailing, and type and characteristics of the connections, and on the other hand, on the loading history (e.g., loading intensity on each cycle, number of cycles, and sequence of loading cycles).

One of the earliest attempts to model deterioration of structural elements subjected to cyclic reversals was conducted by Jacobsen (1958) who proposed a behaviorist model to study the response of connections to cyclic loading. His model consisted of a combination of sliding blocks arranged in series which experienced frictional forces of different amplitudes and which were joined by Hookean springs with different elastic stiffnesses. Although this model allowed the mathematical description of observed static behavior, earthquake response of deteriorating structures was not studied until 1962, when Hisada proposed a degrading model (Hisada, Nakagawa and Izumi, 1962) for studying earthquake response of degrading structures.

Concerned with the stiffness degradation observed in reinforced concrete elements, Clough and Johnston (Clough, 1966; Clough and Johnston, 1966) developed a degrading model which incorporated stiffness degradation after reloading. In this model unloading occurred with a stiffness equal to the initial stiffness but reloading was aimed towards the largest excursion in previous cycles. They used the model to study the response of SDOF systems subjected to four recorded acceleration time histories. In particular, this study computed ratios of maximum deformation of elastoplastic systems to maximum deformation of stiffness-degrading systems. When evaluating these ratios they wrote "*these ratios demonstrate conclusively that there is no significant difference between the yield amplitudes generated in the two materials. The ratios vary between 0.8 and 1.2 except in a few cases.*" They concluded that "*earthquake ductility requirements in the degrading stiffness systems are not materially different from those observed in ordinary elastoplastic structures, except for structures having a period of vibration less than ½ second.*" Based on their study they also concluded that "*the ductility required in the members of reinforced concrete frame buildings will be about the same as is required in equivalent steel frame buildings.*"

An unrealistic feature of the Clough model when experiencing large load reversals followed by small load reversals was pointed out by Mahin and Bertero (1976) and by Riddell and Newmark (1979) who showed that after a small unloading the model would unrealistically reload toward the point of maximum deformation. They modified the model to reload along the same unloading branch until the reloading branch was reached and then aim toward the point of peak deformation. Mahin and Bertero (1976) also made the model more versatile by incorporating a positive post-yield stiffness and variable unloading stiffness as a function of the peak deformation. The model proposed by Mahin and Bertero (1972), which is often referred to as the modified-Clough model, has been incorporated in several general nonlinear analysis programs and has been used extensively to model the behavior of flexurally controlled reinforced concrete elements.

An early model proposed for nonlinear analysis of reinforced concrete is the Takeda model (Takeda, Sozen and Nielsen, 1970). This model incorporated some of the features of the Clough model but also added other features such as a trilinear loading curve to incorporate pre-cracking and post-cracking stiffnesses, a variable unloading stiffness which was a function of the peak deformation, and improved hysteretic rules for inner cyclic loops. This model has also been incorporated in several general analysis programs and has been extensively used in earthquake engineering to study the seismic response of reinforced concrete structures.

A slight modification to the Takeda model was proposed by Otani and Sozen (1972) who replaced the trilinear initial loading segments of the Takeda model by a bilinear relationship. The resulting model is known as bilinear Takeda model. Otani (1981) compared the response of six different hysteretic models (Ramberg-Osgood, degrading bilinear, modified-Clough, bilinear Takeda, Takeda and degrading trilinear) when subjected to horizontal components of the 1940 El Centro and the 1954 Taft records. He concluded that "*maximum response amplitudes are not as sensitive to details in the differences in hysteretic rules of these models.*"

Other early models developed specifically for reinforced concrete structures include the model developed by Nielsen and Imbeault (1970) who proposed a degrading bilinear system whose stiffness would change only when a prior maximum displacement was exceeded, and the degrading model proposed by Anagnostopoulos (1972) which combines Nielsen's degrading bilinear model and the Clough model. Models where reloading is aimed at the point of maximum deformation in prior cycles are sometimes also referred to as "peak-oriented" models (Rahnama and Krawinkler, 1993; Medina and Krawinkler, 2004). The Clough model, the modified-Clough, Takeda, bilinear Takeda and Anagnostopoulos models are all peak-oriented models.

Iwan developed a general class of stiffness-degrading and pinching models (Iwan, 1973, 1977, 1978). Similar to prior models developed by him (Iwan, 1966, 1967) and by Jacobsen (1958), this model consisted of a collection of linear elastic and Coulomb slip elements. He then studied the response of a wide range of stiffness-degrading and pinching models when subjected to an ensemble of 12 accelerograms recorded in various earthquakes (Iwan and Gates, 1979a, 1979b). After comparing the response of the various degrading and non-degrading systems they noted "*despite the quite different load deformation characteristic the overall effect for a given ductility is nearly the same. This is a rather surprising result which may be useful in design, for it implies that it may not be necessary to know the precise details of the load-deflection behavior of a structure in order to make a reasonably accurate estimate of its response.*" In another study aimed at estimating inelastic spectra from elastic spectra using equivalent linear methods, Iwan (1980) concluded that "*the differences in hysteretic behavior considered herein appear to have only a secondary effect on the accuracy of the results.*"

Chopra and Kan (1973) studied the effects of stiffness degradation on ductility requirements of two idealized multistory buildings, one having a period of vibration of 0.5 s and the other 2.0 s. They concluded that "*stiffness degradation has little influence on ductility requirements for flexible buildings, but it leads to increased ductility requirements for stiff buildings.*"

Riddell and Newmark (Riddell and Newmark, 1979; Newmark and Riddell, 1980) studied the influence of hysteretic behavior on inelastic spectral ordinates. They considered an elastoplastic system, a bilinear system and a stiffness degrading system. They compared average inelastic spectral ordinates for the three systems for ground motions scaled to peak ground acceleration, peak ground velocity and peak ground displacement in the acceleration, velocity and displacement-controlled spectral regions, respectively. They arrived at similar conclusions to those of Clough or those of Iwan; in particular they concluded that *"the ordinates of the average spectra do not vary significantly when various nonlinear models are used."* They also noted *"It is particularly significant that, on the average, the stiffness degradation phenomenon is not as critical as one might expect"* and concluded that *"the use of the elastoplastic idealization provides, in almost every case, a conservative estimate of the average response to a number of earthquake ground motions."*

Mahin and Bertero (1981) used the modified Clough model to also study the difference in response of elastoplastic systems and stiffness-degrading systems. They noted that *"ductility demands for a stiffness degrading system subjected to a particular ground motion can differ significantly from those obtained for elastoplastic systems in some period ranges. However, it appears that, on average, the differences are generally small."* Similar conclusions were also reached by Powel and Row (1976) who studied the influence of analysis assumptions on computed inelastic response of three different types of reinforced concrete ten-story buildings. They concluded that *"degrading stiffness appears to have no substantial influence on interstory drift demands."* Nassar and Krawinkler (1991), used the modified Clough model to study the difference of strength reduction factors associated to increasing levels of ductility demands in bilinear and stiffness degrading models. In their report they wrote *"... except for very short period systems, the stiffness-degrading models allow higher reduction factors than the bilinear model, for systems without strain hardening. This difference diminishes with strain hardening. This is a very interesting result in that it suggests that the stiffness degrading model behaves "better" than the bilinear model, i.e., it has a smaller inelastic strength demand for the same ductility ratio."*

Other piece-wise models that incorporate degradation include the Park and Ang mechanistic model (Park and Ang, 1985) and the three-parameter model (Park, Reinhorn and Kunnath, 1987; Valles et al., 1996). The three-parameter model includes strain hardening, variable unloading stiffness, pinching and cyclic load degradation (that is, decreasing yielding strength as a function of maximum deformation, hysteretic energy demand, or a combination of the two). The model was further improved in Kunnath and Reinhorn (1990). Rahnama and Krawinkler (1993) developed a general piece-wise linear

---

hysteretic model which was incorporated into a SDOF analysis program referred to as SNAP (SDOF Nonlinear Analysis Program). The model has a bilinear skeleton relationship and includes variable unloading stiffness, peak-oriented stiffness degradation at reloading, pinching, cyclic strength deterioration as a function of hysteretic energy demands, and also the capability of accelerating the degradation of loading stiffness beyond the peak-oriented degradation.

They used this model to study the influence of hysteretic behavior of SDOF systems subjected to 15 recorded ground motions recorded at firm sites during California earthquakes to study constant-ductility strength-reduction factors. Their results confirmed observations of Nassar and Krawinkler (1991) and of previous investigators who noted that the effect of stiffness degradation was small, on average, leading to smaller displacement demands except for short-period structures where displacement demands in systems with stiffness degradation were larger than those in bilinear systems. In their study they also noted that cyclic strength deterioration increased displacement demands but that the increase was not large unless the strength deteriorates to a small value, and noted that further research was needed before quantitative conclusions could be drawn.

Rahnama and Krawinkler also studied the effect of in-cyclic degradation by considering a negative post-elastic slope (that is, negative strain hardening) in bilinear and degrading models. They concluded that "*ratios of reduction factors for degrading and bilinear systems become significantly larger than 1.0 when negative hardening is present, particularly if the periods of vibration are short and the ductility demands are high.*"

Gupta and Krawinkler (1998, 1999) investigated the effects of pinching and stiffness degradation in SDOF and MDOF structures using the hysteretic model previous developed by Rahnama (Rahnama and Krawinkler, 1993) which was incorporated in the DRAIN-2DX analysis program (Allahabadi and Powell, 1988). They concluded that "*for SDOF systems, pinching leads to a relatively small amplification of the displacement response for systems with medium and long periods, regardless of the yielding strength. For short period structures, which are subject to a larger number of cycles, the displacement amplification increases significantly.*" They also noted that "*the effect of the pinched force-deformation relationship on the displacement ratio is not very sensitive to the severity of the ground motion.*" For MDOF structures they concluded that "*pinching of the force-deformation characteristics of inelastic systems has a global (roof) drift similar to that observed in SDOF systems.*" They also investigated the effect of negative post-yield stiffness in SDOF and MDOF systems. They concluded that "*for SDOF systems a negative post-yield stiffness (which could represent P-Δ effects) has a large effect on the displacement demand for systems with bilinear characteristics. The effect increases rapidly with an increase in the negative stiffness ratio α, with decrease in the yield strength of the system, and a decrease in the period. Dynamic instability, caused by attainment of zero lateral resistance, is a distinct possibility and was observed under several of the ground motion records.*" For systems with negative strain hardening they noted that the pinching model exhibits better behavior than the bilinear model.

Recently, Ruiz-Garcia and Miranda (2005) examined the effect of hysteretic behavior on maximum deformations of SDOF systems subjected to an ensemble of 240 ground motions recorded in California. They considered seven different types of hysteretic behavior: elastoplastic, bilinear, modified Clough, Takeda, origin-oriented, moderate degrading, and severely degrading. The modified Clough, Takeda and origin-oriented models only exhibit stiffness degradation while the moderate degrading and severely degrading systems exhibit both stiffness and cyclic strength degradation. They found that the effect of positive post-yield stiffness was relatively small except for systems with very short periods of vibration (T<0.2s). When subjected to firm soil records they found that the effects of hysteretic behavior were relatively small for structures with periods of vibration larger than about 0.7s.

The same authors used the modified Clough model to examine the effect of stiffness degradation on single-degree-of-freedom systems subjected to ground motions recorded on soft soil sites (Miranda and Ruiz-Garcia, 2002, Ruiz-Garcia and Miranda, 2004, 2006b). They concluded that the effects of stiffness degradation were larger for structures on soft soil sites than those observed for structures on firm sites. In particular, they concluded that for structures with periods of vibration shorter than the predominant period of the ground motion, the lateral displacement demands in stiffness degrading systems on average are 25% larger than those of non-degrading systems and that, in order to control lateral deformations to comparable levels of those in non-degrading structures, stiffness-degrading structures in this spectral region need to be designed for higher lateral forces.

A model similar to the one developed by Krawinkler and his coworkers but with additional capabilities to model connection fracture was developed and incorporated into DRAIN-2DX by Shi and Foutch (Shi and Foutch, 1997; Foutch and Shi, 1998). They studied the influence of hysteretic behavior on the seismic response of buildings by considering seven different hysteretic behaviors, which included non-degrading and degrading models, and nine steel moment-resisting frame models of buildings with three, six, and nine stories. They concluded that "*Hysteresis type has only a minimum effect on ductility demands of structures.*" When evaluating ratios of deformations of degrading to bilinear behavior they noted that "*For the non-pinching hysteresis models, the maximum ratios of ductility demand to the bilinear hysteresis model range from 1.10 to 1.15 when the period of the structure is less than 1.0 second. For pinching hysteresis types the maximum ratios are on the order of 1.25 to 1.30.*"

Gupta and Kunnath (1998) arrived at similar conclusions. More recently Medina and Krawinkler (2004) studied the effects of hysteretic behavior (i.e., bilinear, peak oriented and pinching) in the evaluation of peak deformation demands and their distribution over the height for regular frame structures over a wide range of stories (from 3 to 18) and fundamental periods (from 0.3 s. to 3.6 s.). The study did not consider monotonic in-cycle deterioration. The ground motions used were those with frequency content characteristic of what they referred to as "ordinary ground motions" (that is, no near-fault or soft soil effects). They concluded that "*the degree of stiffness degradation is important for the seismic performance evaluation of regular frames because systems with a large degree of stiffness degradation tend to exhibit larger peak drift demands and a less uniform distribution of peak drifts over the height.*"

Based on the general class of non-degrading and degrading models developed by Iwan (1966, 1967, 1973), Mostaghel (1998, 1999) developed a general hysteretic model by providing an analytical description (that is, with differential equations) of physical models consisting of a series of linear springs, dashpots, and sliders. The model includes the effects of pinching, stiffness degradation, and load deterioration. He showed that complex multi-linear hysteretic behavior can be obtained by solving $(2n-1)$ differential equations where n is the number of linear segments in the model.

A.1.3 Smooth Deteriorating Hysteretic Models

Degradation and pinching have also been incorporated in smooth hysteretic models. Some examples of degrading smooth models are the Baber model (Baber and Wen, 1981; Baber and Noori 1985, 1986) which extends the Bouc-Wen model to include stiffness degradation and pinching.

More recently, Sivaselvan and Reinhorn (1999, 2000) developed a versatile hysteretic model that is conceptually based on a general class of non-degrading and degrading models developed by Iwan (1966, 1967, 1973) but extended the model developed by Mostaghel (1998, 1999) to include smooth curvilinear segments. Stiffness degradation is incorporated using a pivot rule analogous to the one incorporated in the three-parameter model (Park, Reinhorn and Kunnath, 1987). Cyclic strength degradation is modeled by reducing the capacity in the backbone curve while pinching is achieved by adding an additional slip-lock spring in series with the main smooth hysteretic spring, which is similar to the Bouc-Wen model. The hysteretic behavior is then described by the solution of four time-independent differential equations which are solved using Runge-Kutta's method. Although they provided specific rules for controlling stiffness degradation, cyclic degradation, and pinching, they showed that other rules could be implemented as well. This hysteretic model has been incorporated in recent versions of IDARC (Valles et al., 1996).

A.1.4   Hysteretic Models for Steel Braces

Experimental research on the behavior of steel braces has shown that their behavior under severe cyclic loading is complicated and not fully understood. Cyclic nonlinear behavior of steel brace members is complex as a result of various phenomena occurring in the braces and their connections, such as yielding in tension, buckling in compression, post-buckling deterioration of compressive load capacity, deterioration of axial stiffness with cycling, low-cycle fatigue fractures at plastic hinge regions, Bauschinger effect, and buckling and fracture in the gusset plates. As in the models previously described, element models for steel braces can be classified as either phenomenological models in which the load-deformation behavior of steel braces is described through a series of hysteretic rules that try to reproduce behavior observed experimentally, or material-based models such as finite element models and fiber element models where the steel brace is discretized into small elements and the overall behavior of the brace is obtained from uniaxial, biaxial or triaxial material behavior of the material.

A significant amount of both experimental and analytical work on the behavior of steel bracing has been conducted at the University of Michigan under the direction of Professors Goel and Hanson. One of the first analytical models for predicting the force-deformation behavior of axially-loaded members with intermediate slenderness ratios was developed by Higginbotham and Hanson (1976). Prathuangsit, Goel, and Hanson, (1978) proposed a model with rotational end springs to simulate the end restraint resulting from the flexural rigidity of the connections of axially loaded (that is, bracing) members. They showed that members with balanced strength connections, (that is, which form plastic hinges simultaneously at midspan and at the ends), have more efficient compressive load and energy dissipation capacities than members of the same length and same cross-sectional properties with unbalanced strength connections. They concluded that the hysteresis behavior of a balanced strength member can be represented adequately by that of a pin-connected member of the same cross section and same effective slenderness ratio

Jain, Goel, and Hanson tested 17 tube specimens and eight angle specimens under repeated axial loading (Jain, Goel and Hanson, 1976, 1978a, b). The objective of this experimental investigation was to quantify the reduction in maximum compressive loads and increase in member length, and to study the influence of the buckling mode and the shape of the cross-section on the hysteretic behavior and dissipation of energy through the hysteretic cycles. They concluded that local buckling and shape of the cross section can have a significant influence on the hysteretic behavior of axially loaded steel members. Based on their experimental results they developed a hysteresis model for steel tubular members that included a reduction in compressive strength and an increase in member length with the number of cycles (Jain and Goel, 1978, 1980) which was then incorporated in the DRAIN-2D analysis program. This model has been extensively used by investigators at the University of Michigan and elsewhere to study the seismic response of concentrically braced steel frames.

Astaneh-Asl and Goel investigated the behavior of double-angle bracing members subjected to out-of-plane buckling due to severe cyclic load reversals (Astaneh-Asl et al. 1982; Astaneh-Asl and Goel 1984). Nine full-size test specimens were subjected to severe inelastic axial deformations. Test specimens were made of back-to-back A36 steel angle sections connected to the end gusset plates by fillet welds or high-strength bolts. Five of the test specimens were designed according to current design procedures and code requirements. These specimens experienced fracture in gusset plates and stitches during early cycles of loading. Based on observations and analysis of the behavior of the specimens, new design procedures were proposed for improved ductility and energy dissipation capacity of double-angle bracing members. Goel and El-Tayem (1986) investigated the behavior of angle cross-bracing.

Gugerili and Goel (1982) tested nine commercially available wide-flange shapes and structural tubes with different slenderness and width-to-thickness ratios in order to investigate the effects of cross section and slenderness ratio. A general rule was developed for transitioning between compression and tension mechanism lines of constant shape and elastic segments. The new rule also included the effects of residual elongation. The theoretical model was used as a basis for a developing a semi-empirical model for predicting experimental hysteresis loops more accurately than previous models. This model included the decrease in compressive strength with cumulative plastic hinge deformation.

Tang and Goel (1987) developed a procedure to predict the fracture life of bracing members. Their empirically based procedure was refined with an energy approach using Jain's hysteresis model and was then also incorporated in the the DRAIN-2D program. Based on the experimental results of previous researchers Hassan and Goel (1991) formulated a refined and practical hysteresis model for bracing the members of concentrically (chevron) braced steel structures subjected to severe earthquakes. A more recent model was implemented in a Structural Nonlinear Analysis Program (SNAP) (Rai et al, 1996), which eliminates the brace once it is estimated that it has fractured (while this program has the same name, it is different from the program developed by Krawinkler and his coworkers).

Professors Popov and Mahin at the University of California at Berkeley have also conducted a significant amount of experimental and analytical research on steel braces. Zayas, Popov, and Mahin, (1979, 1980) conducted a series of experimental tests on scaled tubular brace members subjected to severe inelastic cyclic loading. The tubular brace specimens considered were one-sixth scale models of braces of the type used for offshore platforms. A method of predicting the reduction in buckling load was presented. Maison and Popov (1980) performed an experimental and analytical investigation of the behavior of structural steel frames with K-braces subjected to severe cyclic loadings simulating seismic effects. They developed an empirical brace model to analyze steel frames with K-braces.

Black and Popov, E.P. (1980, 1981) conducted an experimental study on 24 commercially available steel struts, commonly used as bracing members. They investigated the effects of loading patterns, end conditions, cross-sectional shapes, and slenderness ratios on the hysteresis response of members. A large variety of shapes were tested, including wide flanges, structural tees, double-angles, a double-channel, and thick-walled and thin-walled square and round tubes. Two types of boundary conditions were considered, pinned-pinned and fixed-pinned, with effective slenderness ratios of 40, 80, and 120. An explanation was provided regarding the fundamental mechanisms responsible for the observed degradation in the buckling load capacity during inelastic cycling.

Ikeda and Mahin (1984) developed an analytical model for simulation of the inelastic buckling behavior of steel braces. In their model, buckling is simulated by the use of predefined straight-line segments and simple rules regarding factors such as buckling load deterioration and plastic growth. The model, based on the approach of Maison and Popov (1980) overcomes some of the limitations of earlier models. They proposed a systematic method for selecting input parameters along with several rules governing the values of certain parameters. The model was subsequently refined in 1986 by combining analytical formulations describing plastic hinge behavior with empirical formulas that are based on a study of experimental data. Analytical expressions for the axial force versus axial deformation behavior of braces were derived as solutions of the basic beam-column equation based on specified assumptions. While analytical expressions form the basis of this model, several empirical behavioral characteristics were implemented in the modeling to achieve better representation of observed cyclic inelastic behavior.

Khatib and Mahin (1987, 1988) conducted an analytical investigation on concentrically braced frames. Their study stressed that concentrically braced steel frames designed by conventional methods may exhibit several undesirable modes of behavior. In particular, they showed that chevron-braced frames have an inelastic cyclic behavior that is often characterized by a rapid redistribution of internal forces, a deterioration of strength, a tendency to form soft stories, and fracture due to excessive deformation demand. They identified parameters having a significant influence on these phenomena and provided recommendations for preferable ranges of brace slenderness, approaches for designing beams, and a simplified capacity design approach for proportioning columns and connections. They also proposed several alternative brace configurations with improved behavior including zipper bracing systems which incorporate vertical linkage elements in a conventional chevron-braced frame to decrease concentrations of interstory drift demands.

More recently Uriz and Mahin (2004, Uriz, 2005) conducted experimental testing of a nearly full-size, two-story Special Concentric Braced Frame

(SCBF) specimen. They also conducted an analytical investigation using the same reliability framework used to assess Special Moment Resisting Frame (SMRF) structures during the FEMA/SAC Steel Project in order to assess the confidence with which SCBFs might achieve the seismic performance expected of new SMRF construction.

Other institutions have also been actively involved in research on the behavior of steel braces. For example, Nonaka (1973, 1977) at Kyoto University, conducted elastic, perfectly plastic analyses of a bar under repeated axial loading. The bar was taken as a one-dimensional continuum with both ends simply supported. His analysis considered the plastic interaction for the combined action of bending and axial deformation, based on a piecewise-linear yield condition. With a number of simplifying assumptions, a closed form solution was derived that can describe the hysteretic behavior of a bar, such as a structural brace or a truss member, under any given history of tension and/or compression or of corresponding displacements. Nonaka's closed form solution was later extended by Shibata (1982) for a bar of ideal I-section with bilinear stress-strain relationship. For a bar of arbitrary solid cross-section with a piecewise linear stress-strain relationship, an incremental load-displacement relationship was also obtained in analytical form.

At the University of Canterbury in New Zealand, Remennikov and Walpole (1997a, 1997b) developed an analytical model for the inelastic response analysis of braced steel structures. Their model combines the analytical formulation of plastic hinge behavior with empirical formulas developed on the basis of experimental data. The brace is modeled as a pin-ended member, with a plastic hinge located at the midspan and braces with other end conditions are handled using the effective length concept. Step-wise regression analysis is employed to approximate the plastic conditions for the steel UC section. Verification of the brace model is performed on the basis of quasi-static analyses of individual struts and a one-bay one-story cross-braced steel frame.

At the Ecole Polytechnique de Montreal in Canada, Archambault, Tremblay, and Filiatrault, (1995, 2003) conducted experimental and analytical studies on the seismic performance of concentrically braced steel frames made with cold-formed rectangular tubular bracing members. They tested a total of 24 quasi-static cyclic tests on full scale X bracing and single diagonal bracing systems. They developed simplified models to predict the out-of-plane deformation of the braces as a function of the ductility level. They then used their models to develop an empirical expression to assess the inelastic deformation capacity before fracture of bracing members made of rectangular hollow sections.

More recently, Jin and El-Tawil, (2003) developed a beam-column element to model the inelastic cyclic behavior of steel braces. In their model a bounding surface plasticity model in stress-resultant space coupled with a backward Euler algorithm is used to keep track of the spread of plasticity through the cross-section. Deterioration of the cross-section stiffness due to local buckling is accounted for through a damage model.

Further information on the experimental and analytical response of steel bracing is available in Tremblay (2002), Jin and El-Tawil, (2003) and Uriz (2005), which provide summaries of experimental and analytical work. In particular Tremblay (2002) conducted a survey of past experimental studies on the inelastic response of diagonal steel bracing members subjected to cyclic inelastic loading to collect data for the seismic design of concentrically braced steel frames for which a ductile response is required during earthquakes. He examined the buckling strength of the bracing members, the brace post-buckling compressive resistance at various ductility levels, the brace maximum tensile strength including strain hardening effects, and the lateral deformations of the braces upon buckling. Additionally he proposed equations for each of these parameters and examined the maximum ductility that can be achieved by rectangular hollow bracing members.

Nakashima and Wakabayashi (1992) provide an overview of Japanese experimental and analytical research on steel braces and braced frames. Current Japanese practice is also briefly summarized.

A.2   Detailed Summaries of Relevant Publications

This section presents summaries of publications that were judged to be particularly relevant to the subject of nonlinear degrading response, and were, therefore, reviewed in detail. Each summary includes the list of authors, an abstract, a narrative summary of the work, relevant figures, a summary of important findings, and a listing of relevant publications included in the list of references. The following publications were selected for detailed review:

- Bernal, D., 1998, "Instability of buildings during seismic response," *Engineering Structures*, Vol. 20, No. 4-6, pp. 496-502.

- Pincheira, J.A, Dotiwala, F.S., and D' Souza J.T., 1999, "Spectral displacement demands of stiffness- and strength-degrading systems," *Earthquake Spectra*, 15(2), 245–272.

- Song, J.-K., and Pincheira, J.A, 2000, "Seismic analysis of older reinforced concrete columns," *Earthquake Spectra*, 16(4), 817–851.

- Miranda, E. and Akkar, S.D., 2003 "Dynamic instability of simple structural systems," *Journal of Structural Engineering*, ASCE, 129(12), pp 1722-1727.

- Vian, D. and Bruneau, M., 2003, "Tests to structural collapse of single-degree-of-freedom frames subjected to earthquake excitations," *Journal of Structural Engineering*, ASCE, 129(12), 1676-1685.

- Kanvinde, A.M., 2003, "Methods to evaluate the dynamic stability of structures – shake table tests and nonlinear dynamic analyses," EERI Annual Student Paper Competition, *Proceedings of 2003 EERI Annual Meeting*, Portland, Oregon.

- Vamvatsikos, D. and Cornell, C.A., 2005, *Seismic performance, capacity and reliability of structures as seen through incremental dynamic analysis*, John A. Blume Earthquake Engineering Research Center, Report No. 151, Department of Civil and Environmental Engineering, Stanford University, Stanford, California.

- Ibarra, L., Medina, R., and Krawinkler, H., 2005, "Hysteretic models that incorporate strength and stiffness deterioration, *Earthquake Engineering and Structural Dynamics*, Vol. 34, no. 12, pp. 1489-1511.

- Ibarra, L.F., and Krawinkler, H., 2005, *Global collapse of frame structures under seismic excitations*, John A. Blume Earthquake Engineering Research Center, Report No. 152, Department of Civil and Environmental Engineering, Stanford University, Stanford, California.

- Kaul, R., 2004, *Object-oriented development of strength and stiffness degrading models for reinforced concrete structures*, Ph.D. Thesis, Department of Civil and Environmental Engineering, Stanford University, Stanford, California.

- Elwood, K.J., 2002, *Shake table tests and analytical studies on the gravity load collapse of reinforced concrete frames*, Ph.D. Dissertation, University of California, Berkeley, California.

- Lee, L.H., Han, S.W., and Oh, Y.H., 1999, "Determination of ductility factor considering different hysteretic models," *Earthquake Engineering and Structural Dynamics*, Vol. 28, 957–977.

- Foutch, D.A. and Shi, S., 1998, "Effects of hysteresis type on the seismic response of buildings," *Proc. 6th U.S. National Conference on Earthquake Engineering*, Seattle, Washington, Earthquake Engineering Research Institute, Oakland, California.

- Ruiz-Garcia, J. and Miranda, E., 2003, "Inelastic displacement ratio for evaluation of existing structures," *Earthquake Engineering and Structural Dynamics*. 32(8), 1237-1258.

- Dolsek, M. and Fajfar, P., 2004, "Inelastic spectra for infilled reinforced concrete frames," *Earthquake Engineering and Structural Dynamics*, Vol. 33, 1395–1416.

A.2.1 Instability of Buildings During Seismic Response

Authors:

Bernal, D. (1998)

Abstract:

The issue of gravity-induced instability during response to severe seismic excitation is examined. While static instability is fully determined by the existence of at least one negative eigenvalue in the second-order tangent stiffness, this condition is necessary but not sufficient for instability during dynamic response. The likelihood of collapse is strongly dependent on the shape of the mechanism that controls during the critical displacement cycle and this shape can be reasonably identified using a pushover analysis with an appropriately selected lateral load distribution. A characterization of the instability limit state based on the reduction of a multistory building to an equivalent SDOF system is presented

Summary:

Dynamic instability is a phenomenon whereby the response changes from vibration to drift in a single direction. In this study a structure is defined as stable if small increases in the ground motion intensity result in small changes in the response. The study shows that the distribution of inelastic action along the height of the building plays a critical role in the likelihood of instability. The study emphasizes that gravity generally has little effect on the dynamic response of structures, except when failure from instability is near. In particular, the study shows that the static based approach of accounting for second-order effects through amplifications of the first-order solution is not appropriate in a dynamic setting. Specifically, design in a region where amplifications from P-Δ are significant implies unacceptably low safety factors.

Representative Figures:

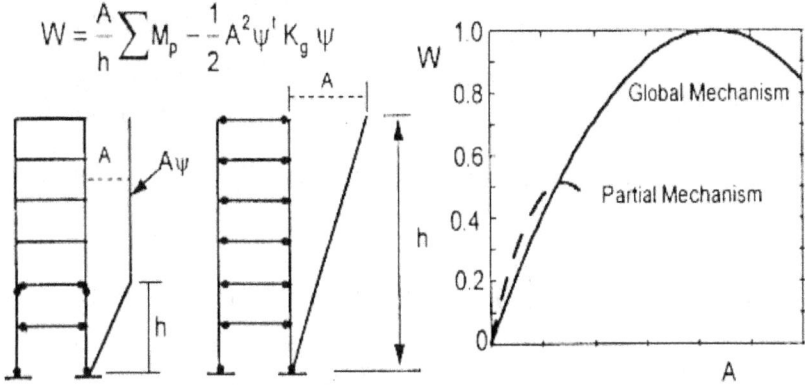

Figure A-1   Effect of mechanism shape on the monotonic work *vs.* amplitude relationship.

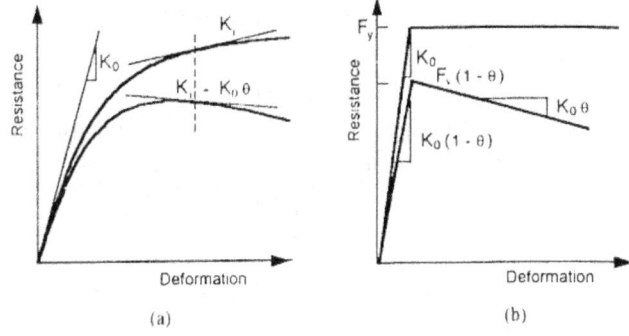

Figure A-2   Illustration of the definition of stability coefficient: (a) general load deformation relationship, (b) elasto-plastic system.

Summary of Findings:

Dynamic instability takes place when the strength of the structure is below a certain threshold and is strongly dependent on the shape of the failure (or collapse) mechanism that controls. Safety against dynamic instability cannot be guaranteed by placing controls on initial elastic stiffness; a rational check of the safety against collapse must contemplate the strength and shape of the critical mechanism. In particular, a rational approach is to estimate the strength level associated with the instability threshold and to ensure that the strength level provided exceeds the required limit by an appropriate safety margin.

Relevant Publications:

Bernal, D., 1998, "Instability of buildings during seismic response," *Engineering Structures*, Vol. 20, No. 4-6, pp. 496-502.

Bernal, D., 1992, "Instability of buildings subjected to earthquakes." *Journal of Structural Engineering*, ASCE, Vol. 118, No. 8, pp. 2239-2260.

Bernal, D., 1987, "Amplification factors for inelastic dynamic P-Δ effects in earthquake analysis," *Earthquake Eng. Struct. Dyn.*, 15(5), pp. 117-144.

A.2.2 Seismic Analysis of Older Reinforced Concrete Columns

Authors:

Pincheira, J.A., Dotiwala, F.S., and D'Souza J.T. (1999)

Abstract:

A nonlinear model and an analytical procedure for calculating the cyclic response of nonductile reinforced concrete columns are presented. The main characteristics of the model include the ability to represent flexure or shear failure under monotonically increasing or reversed cyclic loading. Stiffness degradation with cyclic loading can also be represented. The model was implemented in a multipurpose analysis program and was used to calculate the response of selected columns representative of older construction. A comparison of the calculated response with experimental results shows that the strength, failure mode and general characteristics of the measured cyclic response can be well represented by the model.

Summary:

A beam-column element was created in order to simulate the behavior of older non-ductile or shear-critical reinforced concrete columns in 2D frames. This is a lumped plasticity element using two flexural springs at the beam ends and a shear spring at the midpoint (Figure A-3). For the flexural springs a Takeda hysteresis law is used together with a quadrilinear backbone curve that incorporates a hardening post-yield segment followed by a negative post-peak slope that stops at a residual plateau; essentially, only in-cycle strength degradation is considered (Figure A-4a). On the other hand, the shear spring uses a similar quadrilinear backbone but a pinching hysteresis together with cyclic degradation of the post-peak strength (Figure A-4b).

The element was incorporated in the Drain-2D analysis program, resulting in important limitations in its implementation and applicability. The solution algorithm of Drain-2D cannot handle negative stiffness, thus necessitating the use of numerical techniques to find an approximate solution. Specifically, on a negative slope (for any of the three springs), the load steps are performed first with an arbitrary positive stiffness, and the load unbalance is then subtracted from the resulting increased load to move down to the actual negative slope (Figure A-5). The results can be considered reliable only under small load-steps and they may indeed lead to gross numerical errors and possible numerical instabilities at the MDOF level. Furthermore, the use of Drain-2D means that only a load-control pushover is possible, thus severely reducing the applicability of this element for anything but time-history analysis with small time steps.

Significant effort has gone into the definition of the spring backbones, using modified compression field theory for the shear backbone, and considering anchorage slip, lap-splice slip, and section degradation for the flexural springs. Calibration and testing of the element were performed with regard to the experimental results.

Representative Figures:

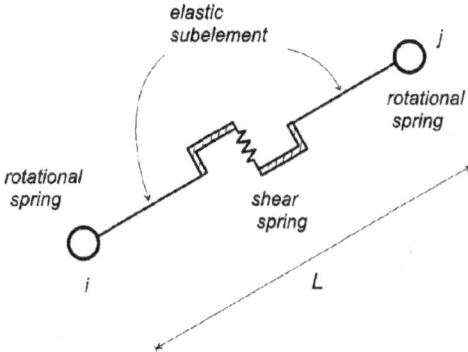

Figure A-3   The RC column element formulation.

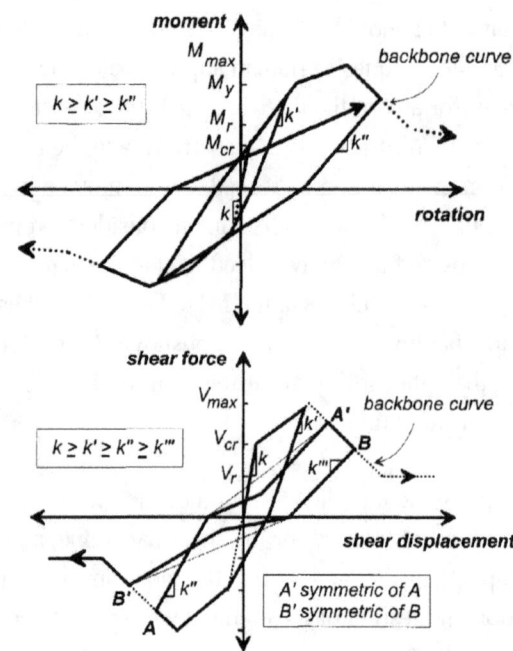

Figure A-4   The hysteretic laws for shear and moment springs.

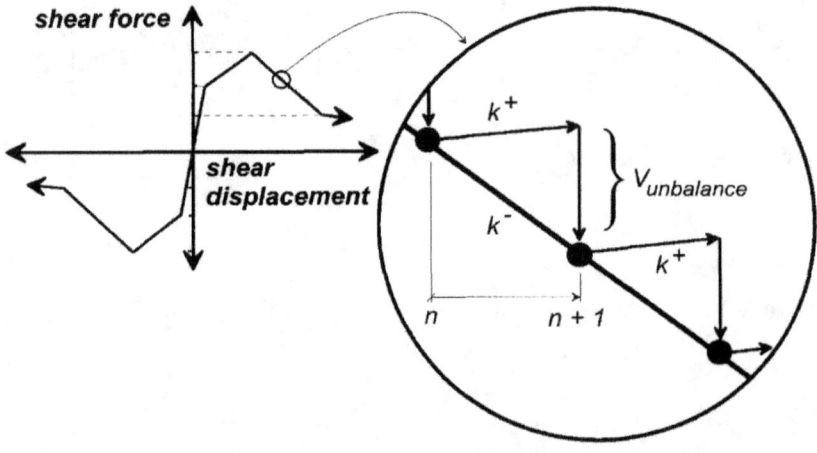

Figure A-5   The force (or moment) unbalance is subtracted after an arbitrary positive stiffness step towards the "correct" displacement. Very small load steps are needed for accuracy, even at the SDOF level.

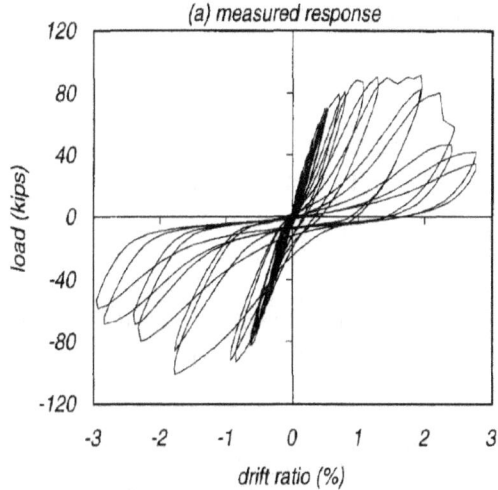
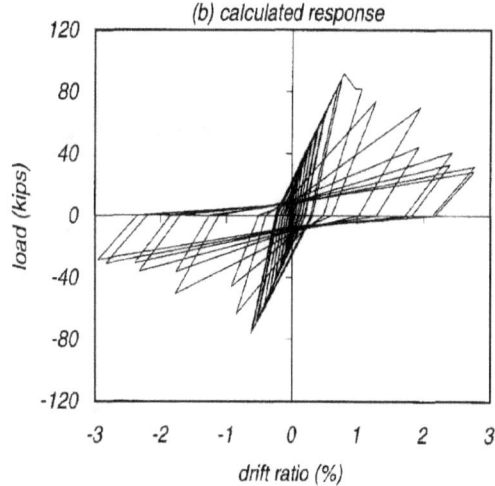

Figure A-6   Observed versus calculated response for a column specimen SC3 (shear critical).

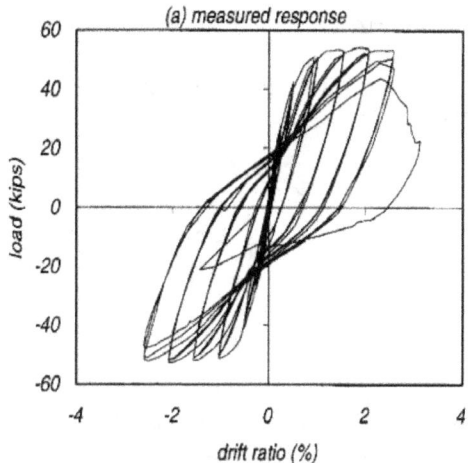
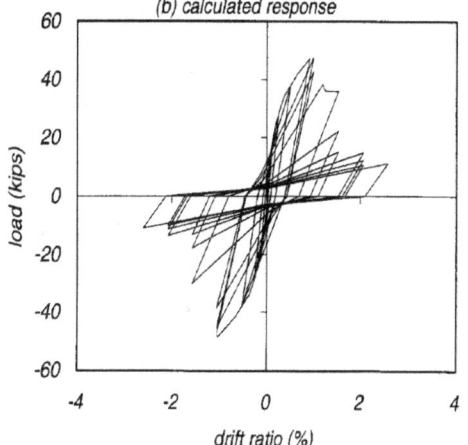

Figure A-7   Observed *versus* calculated response for a column specimen 2CLH18 (fails in shear after considerable flexural deformation).

Summary of Findings:

The comparison with experimental results showed that good correlation of the observed versus the calculated post-peak response was in many cases not possible (Figures A-6, A-7). The cyclic degradation parameters were shown to have a large influence on the post-peak response while significant epistemic uncertainty was identified in the cyclic degradation.

The column failure mode was captured in every test considered, but the estimated failure loads and drifts were generally conservative. Nonetheless, the model was able to capture satisfactorily the overall strength degradation, stiffness degradation, and in-cycle and cyclic degradation properties of the specimens.

Relevant Publications:

Pincheira, J.A, Dotiwala, F.S., and D'Souza J.T., 1999, "Spectral displacement demands of stiffness- and strength-degrading systems," *Earthquake Spectra*, 15(2), 245–272.

Song J.-K., and Pincheira, J.A., 2000, "Seismic analysis of older reinforced concrete columns," *Earthquake Spectra*, 16(4), 817–851.

Dotiwala, F.S., 1996, *A nonlinear flexural-shear model for RC columns subjected to earthquake loads*, MS Thesis, Department of Civil and Environmental Engineering, University of Wisconsin-Madison.

Pincheira, J.A., and Dotiwala, F.S., 1996, "Modeling of nonductile R/C columns subjected to earthquake loading," *Proc. 11th World Conf. on Earthquake Engineering*, Paper No. 316, Acapulco, Mexico.

A.2.3 Spectral Displacement Demands of Stiffness- and Strength-Degrading Systems

Authors:

Song, J.-K., and Pincheira, J.A. (2000)

Abstract:

The effect of stiffness and strength degradation on the maximum inelastic displacement of single-degree-of-freedom (SDOF) systems was investigated. The SDOF model included strength and stiffness degradation with increasing deformation amplitude and upon reversal of loading cycles. Pinching of the hysteresis loops was also considered. Spectral displacements were calculated for oscillators with a range of degrading characteristics subjected to twelve ground motions on rock, firm, and soft soils. The results show that the maximum displacements of degrading oscillators are, on average, larger than those of non-degrading systems. The displacement amplification depends significantly with the period, strength coefficient, degradation rate, and ground motion considered. Nonetheless, the amplification due to the degradation characteristics of the system is more important in the short-period range where average amplification factors of two or three are credible. The amplification factors proposed in the FEMA 273 report by the ATC-33 project provided conservative estimates for oscillators with periods greater than 0.3 seconds subjected to motions on rock or firm soil. On soft soils, a good correlation was found for periods greater than 1.5 seconds. At shorter periods, the ATC 33 factors underestimate the displacement amplification.

Summary:

About 7,600 SDOF dynamic analyses were performed for 12 ground motions including rock, firm soil, soft soil, and near-field and far-field records. The oscillator had a quadrilinear backbone with a hardening, a softening and a residual strength segment. Still, only a limited set of backbones were considered, all having hardening stiffness 5% of the elastic, residual strength 10% of yield strength and reaching 1.25 ductility at peak strength. Some gentle negative slopes were investigated, namely -1% and -3% of the elastic. The oscillator had 5% damping and used a pinching hysteresis with or without cyclic strength degradation. During the investigation an *ad hoc* collapse limit-state was considered when the post-peak strength reached 10% of the yield strength.

Representative Figures:

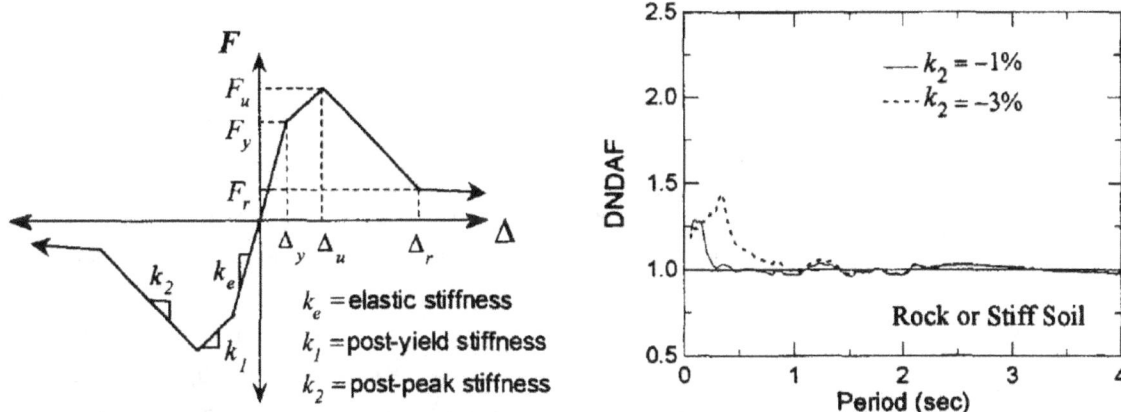

Figure A-8   (a) Hysteresis law used for the SDOF system and (b) ratio of degrading to non-degrading displacement amplification factors for the post-peak stiffness equal to -1% or -3% of the elastic stiffness.

Summary of Findings:

The post-peak stiffness and the unloading stiffness were found to be the most important parameters, while the degree of pinching was important except for soft-soil records. Cyclic degradation generally increases the dynamic response, but mostly for the short periods. Significant differences in the SDOF response and its dependence on cyclic degradation were found between soft soil and firm soil (or rock) and maybe between near-field versus far-field record response, although only one near-source record was used.

Finally, for a given strength reduction factor, collapse (as defined in the paper) was consistently observed below a certain oscillator period. Some duration effects were also reported.

The study seriously suffers from the lack of records. Only twelve were used and they were selected from soft and firm soil sites without differentiating between near-field and far-field. Thus, the statistics on the results are not reliable, although the general observations provided may prove useful.

Relevant Publications:

Song J.-K., and Pincheira, J.A., 2000, "Seismic analysis of older reinforced concrete columns," *Earthquake Spectra*, 16(4), 817–851.

Pincheira, J.A, Dotiwala, F.S., and D'Souza, J.T., 1999, "Spectral displacement demands of stiffness- and strength-degrading systems," *Earthquake Spectra*, 15(2), 245–272.

A.2.4 Dynamic Instability of Simple Structural Systems

Authors:

Miranda, E. and Akkar, S.D., (2003)

Abstract:

Lateral strengths required to avoid dynamic instability of SDOF systems are examined. Oscillators with a bilinear hysteretic behavior with negative post-yield stiffness are considered. Mean lateral strengths, normalized by the lateral strength required to maintain the system elastic, are computed for systems with periods ranging from 0.2 to 3.0 s when subjected to 72 earthquake ground motions recorded on firm soil. The effect of period of vibration and post-yield stiffness is investigated. Results indicate that mean normalized lateral strengths required to avoid dynamic instability decrease as negative post-yield stiffness increases, and that the reductions are much larger for small negative post-yield stiffness than for severe negative post-yield stiffness. It is concluded that there is a significant influence of the period of vibration for short-period systems and for systems with mildly negative post-yield stiffness. Dispersion of normalized lateral strengths required to avoid dynamic instability are found to increase as the negative post-yield stiffness decreases and as the period of vibration increases. Simple equations that capture the effects of period and post-yield stiffness to aid in the evaluation of existing structures are obtained through nonlinear regression analyses.

Summary:

The objective of this study was to assess the minimum lateral strength required to avoid dynamic instability in SDOF systems. The minimum lateral strength is computed as a function of linear elastic spectral ordinates, that is, the lateral strength required to maintain the system elastic. Specific goals of the study were: (a) to study the effect of the post-yield negative stiffness on the minimum strength required to avoid collapse, (b) to study the effect of period of vibration, (c) to compute mean normalized strengths required to avoid dynamic instability, and (d) to develop approximate expressions to assist practicing engineers in evaluating the minimum lateral strengths required in existing structures to avoid dynamic instability.

The study considered SDOF systems with a bilinear force displacement relationship characterized by a linear segment with initial stiffness K followed by a post-yield linear segment with negative stiffness $-\alpha K$. When subjected to earthquake ground motions the likelihood of experiencing dynamic instability in a system with a given negative slope increases as the lateral strength decreases. Lateral strengths required to avoid dynamic instability of bilinear SDOF systems with negative post-yield stiffness were investigated. Mean lateral strengths normalized by the lateral strength required to maintain the system elastic are computed for systems with periods ranging from 0.2 to 3.0 s and post-yield negative stiffness ratios ranging from 0.03 to 2.0 when subjected to 72 earthquake ground motions recorded on firm soil.

Representative Figures:

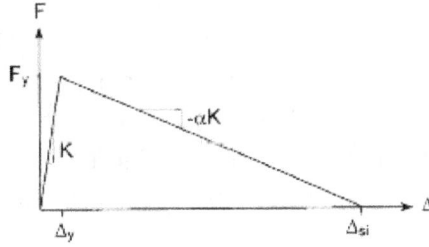

Figure A-9   Force-displacement characteristics of bilinear systems considered

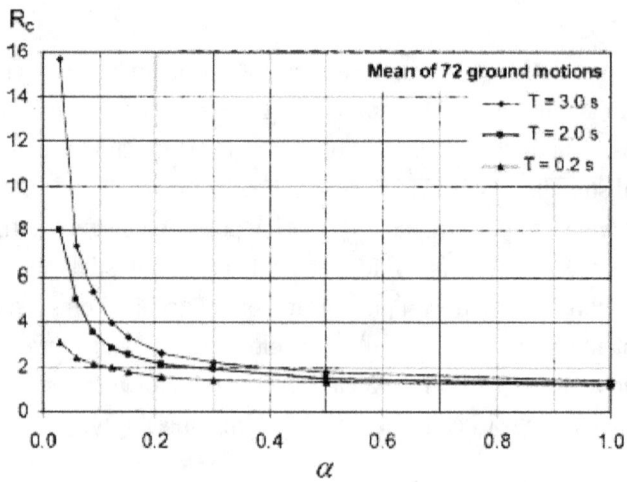

Figure A-10   Effect of period of vibration and post-yield stiffness on the mean strength ratio at which dynamic instability is produced.

Summary of Findings:

Lateral strengths required to avoid dynamic instability of bilinear SDOF systems with negative post-yield stiffness were investigated. Mean lateral strengths, normalized by the lateral strength required to maintain the system elastic, are computed for systems with periods ranging from 0.2 to 3.0 s and post-yield negative stiffness ratios ranging from 0.03 to 2.0 when subjected to 72 earthquake ground motions recorded on firm soil. The following conclusions are drawn from this study.

The strength ratio at which dynamic instability is produced decreases as the post-yield negative stiffness ratio α increases. This means that the lateral strength required to avoid collapse increases as the post-yield descending branch of the force-deformation relationship is steeper. When α is smaller than about 0.2 small increases in α can produce significant increases in required lateral strength to avoid instability. Meanwhile, for values of $\alpha > 1$, the system must remain practically elastic in order to avoid collapse.

The collapse strength ratio increases with increasing period, particularly for post-yielding negative stiffness ratios smaller than 0.3. Mean collapse strength ratios of short period structures are relatively strong, particularly when $\alpha > 0.1$. Dispersion of collapse strength ratios decreases as α increases and as the period of vibration decreases. Coefficients of variation of collapse strength ratios are particularly small for $\alpha > 0.5$. An approximate equation to estimate lateral strengths required to avoid dynamic instability of bilinear SDOF system is proposed.

Relevant Publications:

Miranda, E. and Akkar, S.D., 2003 "Dynamic instability of simple structural systems," *Journal of Structural Engineering*, ASCE, 129(12) , pp 1722-1727.

Vamvatsikos, D., and Cornell, C.A., 2005, *Seismic performance, capacity and reliability of structures as seen through incremental dynamic analysis*, John A. Blume Earthquake Engineering Research Center Report No. 151, Department of Civil and Environmental Engineering, Stanford University, Stanford, California.

A.2.5 Tests to Structural Collapse of Single-Degree-of-Freedom Frames Subjected to Earthquake Excitations

Authors:

Vian, D. and Bruneau, M. (2003)

Abstract:

This paper presents and analyzes experimental results of tests of 15 four-column frame specimens subjected to progressively increasing uniaxial ground shaking until collapse. The specimens were subdivided into groups of three different column slenderness ratios: 100, 150, and 200. Within each group, the column dimensions and supported mass varied. Ground motion of different severity was required to collapse the structures tested. The experimental setup is briefly described and results are presented. Test structure performance is compared with the proposed limits for minimizing $P-\Delta$ effects in highway bridge piers. The stability factor is found to have a strong relation to the relative structural performance in this regard. Performance is also compared with the capacity predicted by currently used design equations dealing with axial and moment interactions for strength and stability by expressing these capacities in terms of acceleration and maximum base shear (represented as a fraction of the system's weight). The experimental results exceeded the maximum spectral accelerations calculated when considering second-order effects, but did not when considering only member strength. Finally, an example of how to use the experimental data for analytical model verification is presented, illustrating the shortcomings and inaccuracies of using a particular simplified model with constant structural damping.

Summary:

Although the first and foremost objective of this project was to provide well-documented data (freely available on the web to be used by others) of tests to collapse, this paper includes results from a preliminary investigation of behavioral trends observed from the shake table results. In particular, peak responses are compared with limits proposed by others to minimize $P-\Delta$ effects in bridge piers. Specimen behavior is also investigated with respect to axial and moment interaction limits considering strength and stability. Finally, to illustrate how the generated experimental data could be used to develop or calibrate analytical models of inelastic behavior to collapse, experimental results are compared with those obtained using a simple analytical model. Progressive bilinear dynamic analyses are performed in two different ways and are compared with the shake table test results.

Representative Figures:

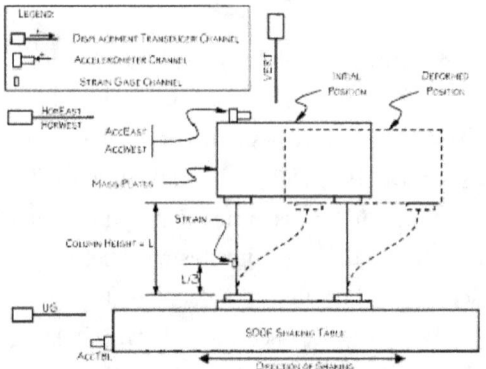

Figure A-11   (a) Schematic of test setup

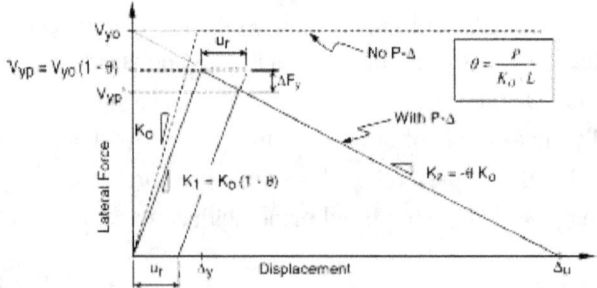

Figure A-12   Simplified bilinear force deformation model

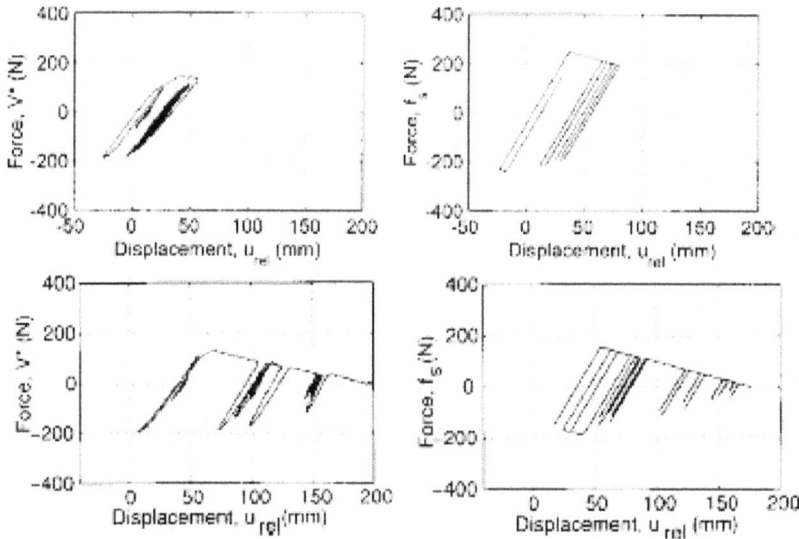

Figure A-13   Comparison of experimental (left) and analytical (right) results.

Summary of Findings:

Specimens showed an approximate bilinear behavior with a tendency to drift to one side and ultimately experience collapse. The stability factor, θ, was observed to have the most significant effect on the structure's propensity to collapse. As θ increases, the maximum attainable ductility, maximum sustainable drift, and maximum spectral acceleration reached before collapse, all decrease. When this factor θ was larger than 0.1, the ultimate values of maximum spectral acceleration, displacement ductility, and drift reached before collapse were all grouped below values of 0.75 g, 5, and 20%, respectively.

Relevant Publications:

Vian, D. and Bruneau, M., 2003, "Tests to structural collapse of single degree of freedom frames subjected to earthquake excitations." *Journal of Structural Engineering*, ASCE, 129(12), 1676-1685.

Bruneau, M. and Vian, D., 2002, "Tests to collapse of simple structures and comparison with existing codified procedures," *Proc. 7th U.S. National Conference on Earthquake Engineering*, Boston, MA.

Bruneau, M. and Vian, D., 2002, "Experimental investigation of P–Δ effects to collapse during earthquakes," *Proc. 12th European Conference on Earthquake Engineering*, London, UK.

Vian, D. and Bruneau, M., 2001, *Experimental investigation of P–Δ effects to collapse during earthquakes,* Report MCEER-01-0001, Multidisciplinary Research for Earthquake Engineering Research Center, Buffalo, N.Y.

A.2.6 Methods to Evaluate the Dynamic Stability of Structures – Shake Table Tests and Nonlinear Dynamic Analyses

Author:

Kanvinde, A.M. (2003)

Abstract:

This paper aims to understand the phenomenon of dynamic instability in structures better, and to suggest and evaluate methods to predict collapse limit states of structures during earthquakes, based on findings of recent shake table tests and nonlinear dynamic analyses conducted at Stanford University. Simple models that collapsed due to a story mechanism were used as test specimens. Data from nineteen experiments suggest that current methods of nonlinear dynamic analysis (using the OpenSees program in this case) are accurate and reliable for predicting collapse and tracing the path of the structure down to the ground during collapse. Moreover, it is found from the experiments that for non-degrading structures, an estimate of collapse drift based on a static pushover analysis can be successfully applied to predict the dynamic collapse or instability due to P-Δ effects. The rationale for this is that the structure has an elongated period at the point of global instability, virtually insulating it from the ground motion and justifying the use of a static-analysis-based drift. Finally, the paper directs the readers to a valuable database of test data from collapse tests of a "clean" structure, which can be used for further verification studies.

Summary:

This was a brief assessment of the collapse performance of two 2D single-story single-bay frames using results from static pushover and incremental dynamic analyses (IDA) and correlating them with experimental tests. The frames tested and simulated had a rigid beam that forced the creation of plastic hinges in the two columns. By using easily replaceable steel plates for the columns it was possible to repeat the shake table tests at various intensities and in effect experimentally reproduce IDA-like results. A total of 19 uniaxial shake table tests was performed using two ground motion records and two different structures (that is, two different column types).

Structure A was ductile and, having a ratio of yield base shear to weight of 1.03, was practically impossible to collapse. Structure B had weaker columns, produced by drilling holes at the bottom and top of the steel plates, and was thus prone to a story-mechanism collapse due to significant $P\text{-}\Delta$ effects (Figure A-14a). The two structures were also simulated in OpenSEES using a Giufre-Menegotto-Pinto hysteresis model for the column hinges (Figure A-14b) without any cyclic deterioration, and an exact corrotational formulation for geometric nonlinearities.

Representative Figures:

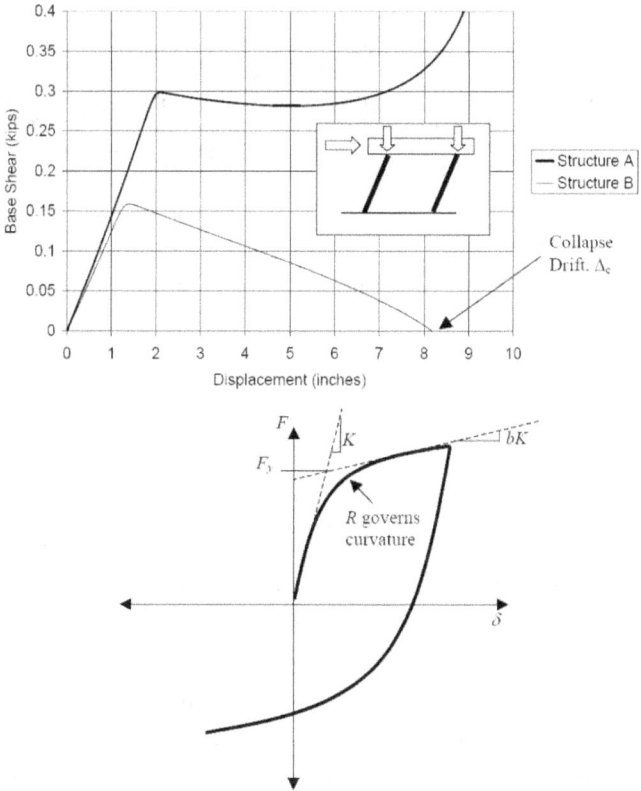

Figure A-14   (a) Static pushover curves for the two frames and (b) modeling of the column plastic hinges in OpenSEES.

Summary of Findings:

The evidence presented shows that nonlinear dynamic analysis is a reliable tool to predict the actual behavior of the two structures. The usefulness of static pushover was also proven, at least when cyclic deterioration is not an issue, as the collapse drift calculated statically was accurately matched by both incremental dynamic analysis and the shake table experiments. Compared to the response at lower intensities, larger scatter was observed close to collapse both in the experiments and in the dynamic analyses, even when using the same earthquake record. This suggests an increased sensitivity of the actual results to the uncertainties in the initial condition of the structure, and an increased difficulty in predicting the collapse drift or intensity level even for such simple specimens.

Relevant Publications:

Kanvinde, A.M., 2003, "Methods to evaluate the dynamic stability of structures – shake table tests and nonlinear dynamic analyses," EERI Annual Student Paper Competition, *Proceedings of 2003 EERI Meeting*, Portland, OR.

Vian, D. and Bruneau, M., 2001, *Experimental investigation of P-Δ effects to collapse during earthquakes*, Report No. MCEER-01-0001, Multidisciplinary Research for Earthquake Engineering Research Center, Buffalo, N.Y.

A.2.7 Seismic Performance, Capacity and Reliability of Structures as Seen Through Incremental Dynamic Analysis

Authors:

Vamvatsikos, D. and Cornell, C.A. (2005)

Abstract:

Incremental Dynamic Analysis (IDA) is an emerging structural analysis method that offers thorough seismic demand and limit-state capacity prediction capability by using a series of nonlinear dynamic analyses under a suite of multiply scaled ground motion records. Realization of its opportunities is enhanced by several innovations, such as choosing suitable ground motion intensity measures and representative structural demand measures. In addition, proper interpolation and summarization techniques for multiple records need to be employed, providing the means for estimating the probability distribution of the structural demand given the seismic intensity. Limit-states, such as the dynamic global system instability, can be naturally defined in the context of IDA. The associated capacities are calculated so that when properly combined with probabilistic seismic hazard analysis, they allow the estimation of the mean annual frequencies of limit-state exceedance.

IDA is resource-intensive. Thus the use of simpler approaches becomes attractive. The IDA can be related to the computationally faster Static

Pushover (SPO), enabling a fast and accurate approximation to be established for SDOF systems. By investigating oscillators with quadrilinear backbones and summarizing the results into a few empirical equations, a new software tool, SPO2IDA, is produced here that allows direct estimation of the summarized IDA results. Interesting observations are made regarding the influence of the period and the backbone shape on the seismic performance of oscillators. Taking advantage of SPO2IDA, existing methodologies for predicting the seismic performance of first-mode-dominated, MDOF systems can be upgraded to provide accurate estimation well beyond the peak of the SPO.

The IDA results may display a large record-to-record variability. By incorporating elastic spectrum information, efficient intensity measures can be created that reduce such dispersions, resulting in significant computational savings. By employing either a single optimal spectral value, a vector of two or a scalar combination of several spectral values, significant efficiency is achieved. As the structure becomes damaged, the evolution of such optimally selected spectral values is observed, providing intuition about the role of spectral shape in the seismic performance of structures.

Summary:

The research presented is entirely based on the concept of incremental dynamic analysis (IDA). The methodology is established and is extensively used to derive (among others) the collapse capacity of MDOF frames. Of particular importance is the exploration of the connection between the fractile IDA curves and the pushover. The authors propose the use of a 5% damped SDOF oscillator with a complex quadrilinear backbone (including a hardening, a softening and a residual plateau segment) with moderately pinching hysteresis to capture the pushover curve shape of actual MDOF frames. No cyclic degradation was considered but in the process millions of nonlinear dynamic SDOF analyses are performed for 30 records and a wide variety of oscillator backbones and periods. The results are fitted and incorporated into a complex R-μ-T relationship, realized in the form of the SPO2IDA Excel tool.

The proposed tool is applied to the MDOF prediction problem using the worst-case pushover concept. This is defined as the pushover that leads to the earliest post-peak collapse, stipulating that it will also help find the collapse mechanism that a dynamic analysis would predict. By applying the SPO2IDA tool on the worst-case pushover the complete IDA curves of

MDOF frames are generated for a 5-story, a 9-story and a 20-story steel frame.

Representative Figures:

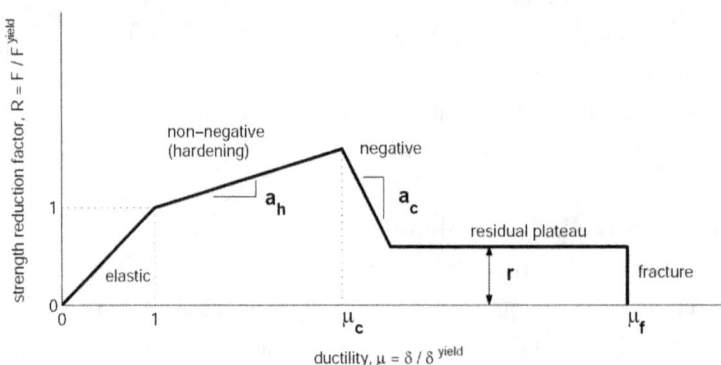

Figure A-15   The backbone of the studied oscillator.

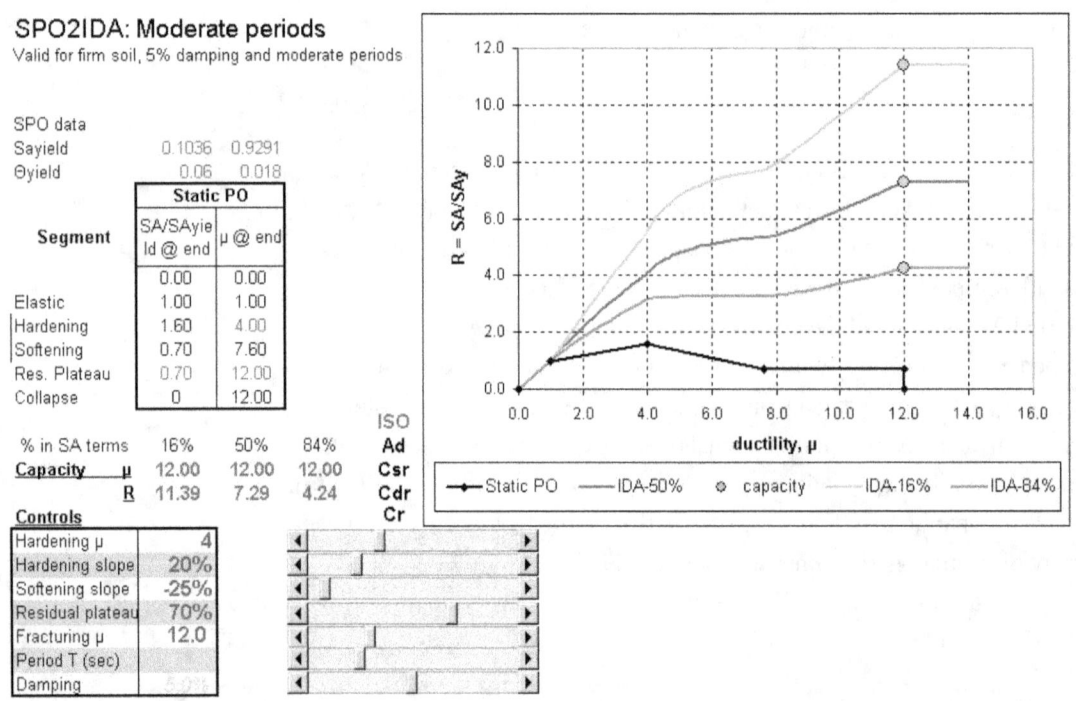

Figure A-16   The interface of the SPO2IDA tool for moderate periods.

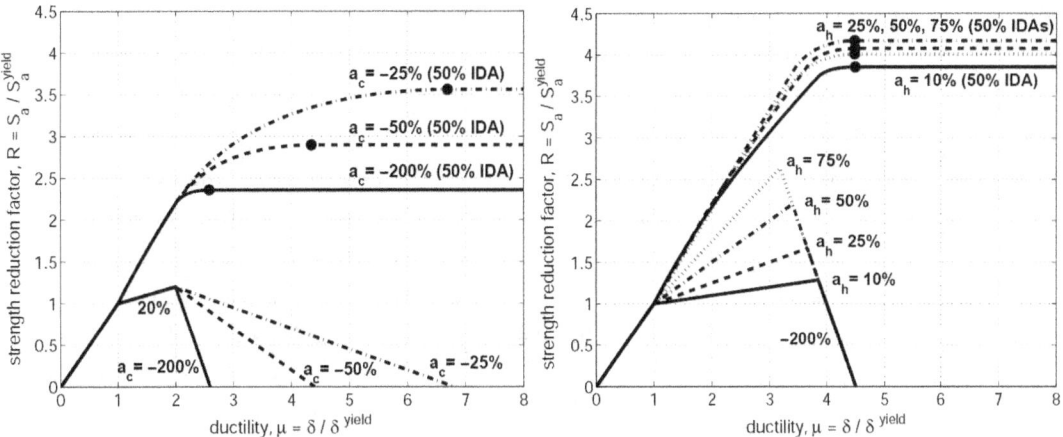

Figure A-17   Influence of (a) the post-peak and (b) post-yield stiffness on the median dynamic response of the oscillator. When the negative segment is the same then the hardening slope has a negligible effect.

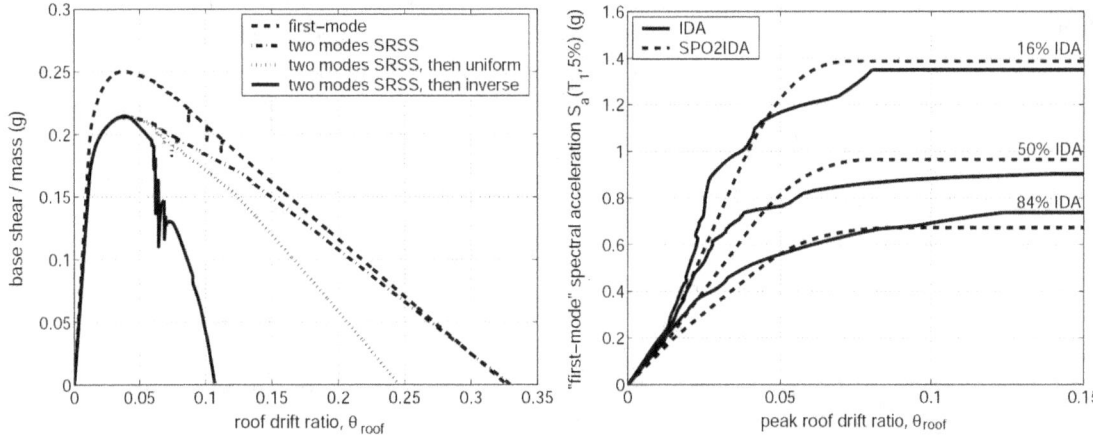

Figure A-18   (a) Influence of the load pattern on the pushover curve shape and (b) the predicted versus actual dynamic response for various intensity levels using SPO2IDA and the worst-case pushover for a 9-story steel moment frame.

Summary of Findings:

Regarding SDOF oscillators, it was found that the shape of the backbone curve has a very complex effect on the dynamic response. For example, the negative slope, the hardening deformation and the residual strength level are the three parameters that dominate (e.g. Figure A-17a). On the other hand, the hardening slope is not as important, while the residual plateau is significant only when long or high enough. Surprisingly, the peak strength of the oscillator is found to be relatively unimportant when the subsequent negative-slope segment is fixed. Thus some very different backbones exist that have almost the same performance (Figure A-17b).

For MDOF application, it was found that the worst-case pushover is not always easy to estimate. For a 9-story steel frame building, in order to find the dominant collapse mechanism it was necessary to change the load pattern after the peak of the pushover and actually try several combinations before getting an acceptable shape (Figure A-18a). Under this condition, the use of SPO2IDA was found to provide accurate results for first-mode-dominated frames. A test conducted on a 20-story frame, where higher modes are a significant issue, showed that it was impossible to get good agreement in the early inelastic range. Curiously, when close to collapse, this approach still managed to provide an accurate answer, leading to the observation that an SDOF can predict reliably the collapse capacity of complex buildings.

Relevant Publications:

Vamvatsikos, D., and Cornell, C.A., 2005, *Seismic performance, capacity and reliability of structures as seen through incremental dynamic analysis*, John A. Blume Earthquake Engineering Research Center, Report No. 151, Department of Civil and Environmental Engineering, Stanford University, Stanford, California.

Ibarra, L.F., and Krawinkler, H., 2005, *Global collapse of frame structures under seismic excitations*, John A. Blume Earthquake Engineering Research Center, Report No. 152, Department of Civil and Environmental Engineering, Stanford University, Stanford, California.

A.2.8 Hysteretic Models that Incorporate Strength and Stiffness Deterioration

Authors:

Ibarra, L., Medina, R.A., and Krawinkler, H., (2005)

Abstract:

This paper presents the description, calibration and application of relatively simple hysteretic models that include strength and stiffness deterioration properties, features that are critical for demand predictions as a structural system approaches collapse. Three of the basic hysteretic models used in seismic demand evaluation are modified to include deterioration properties: bilinear, peak-oriented, and pinching models. The modified models include most of the sources of deterioration, namely, various modes of cyclic deterioration and softening of the post-yielding stiffness, and they also account for a residual strength after deterioration. The models incorporate an energy-based deterioration parameter that controls four cyclic deterioration modes: basic strength, post-capping strength, unloading stiffness, and accelerated reloading stiffness deterioration modes. Calibration of the hysteretic models on steel, plywood, and reinforced-concrete components demonstrates that the proposed models are capable of simulating the main characteristics that influence deterioration. An application of a peak-oriented deterioration model in the seismic evaluation of SDOF systems is illustrated. The advantages of using deteriorating hysteretic models for obtaining the response of highly inelastic systems are discussed.

Summary:

This study presents an improved piece-wise linear hysteretic model that is capable of considering stiffness degradation, pinching cyclic strength degradation as well as in-cycle strength degradation. The paper has a threefold objective: (a) to describe the properties of proposed hysteretic models that incorporate both monotonic and cyclic deterioration; (b) to illustrate the calibration of these hysteretic models on component tests of steel, plywood, and reinforced-concrete specimens; and (c) to exemplify the utilization of the hysteretic models in the seismic response evaluation of SDOF systems. In this study the term deteriorating hysteretic models refers to models that include strength deterioration of the backbone curve or cyclic deterioration or both.

As shown in Figure A-19, the model considers a backbone curve consisting of four linear segments: an elastic segment until a yield displacement, post-yield strain-hardening segment until the 'capping' displacement is reached, a post-capping segment with negative stiffness (that is, in-cycle degradation), and a final residual horizontal segment.

Representative Figures:

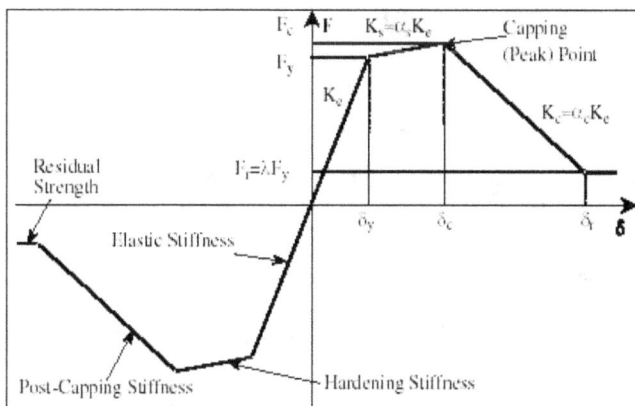

Figure A-19   The backbone of the proposed hysteretic model.

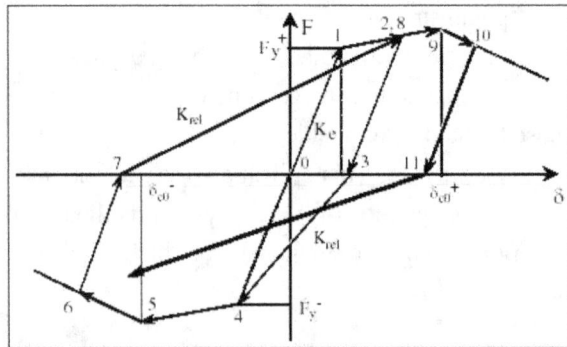

Figure A-20   Basic rules for peak-oriented hysteretic model.

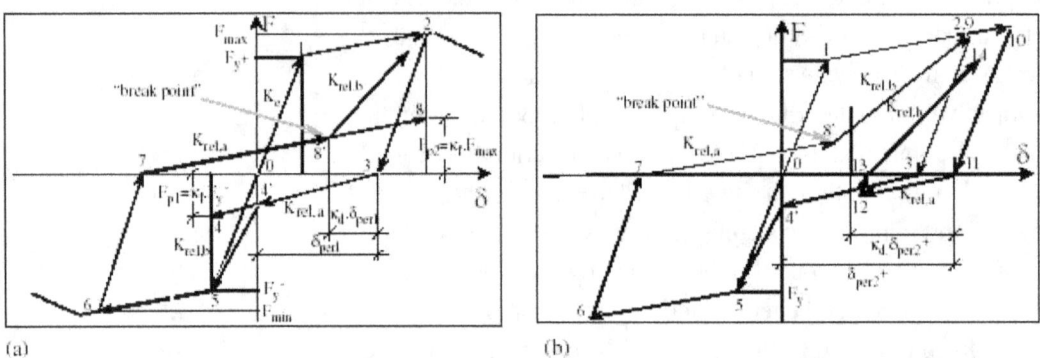

(a)   (b)

Figure A-21   Pinching hysteretic model: (a) basic model rules; and (b) modification if reloading deformation is to the right of break point.

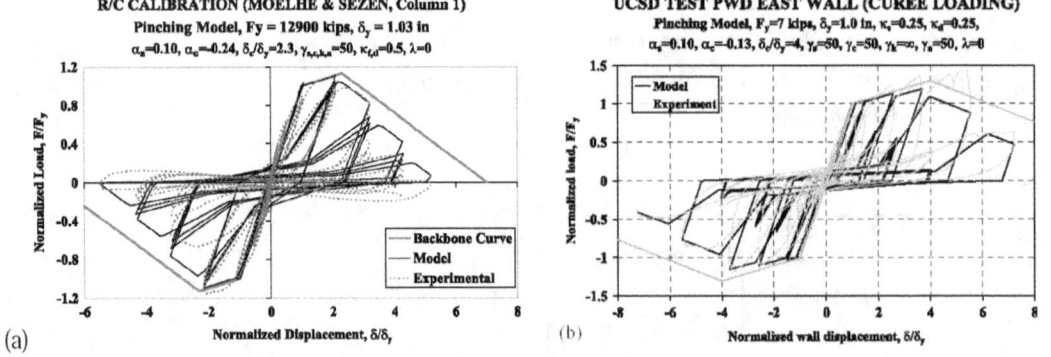

(a)   (b)

Figure A-22   Examples of comparisons between experimental and analytical results for (a) non-ductile reinforced concrete column; and (b) plywood shear wall.

Summary of Findings:

The hysteretic models include a post-capping softening branch, residual strength, and cyclic deterioration. Cyclic deterioration permits deterioration to be traced as a function of past loading history, the rate of deterioration (which depends on the hysteretic energy dissipated in past cycles), and on a reference energy dissipation capacity. Four modes of cyclic deterioration can be simulated: basic strength, postcapping strength, unloading stiffness, and accelerated reloading stiffness deterioration. Based on calibrations performed with experimental data from component tests of steel, wood, and reinforced-concrete specimens, they concluded that it appears that, for a given component, the backbone characteristics and a single parameter that controls all four modes of cyclic deterioration are adequate to represent component behavior regardless of the loading history.

Results from the seismic evaluation of various SDOF systems demonstrate that strength deterioration becomes a dominant factor when the response of a structure approaches the limit state of collapse. At early stages of inelastic behavior, both deteriorating and nondeteriorating systems exhibit similar responses. The differences become important when the post-capping stiffness is attained in the response.

Relevant Publications:

Ibarra, L., Medina, R., Krawinkler, H., 2005, "Hysteretic models that incorporate strength and stiffness deterioration, *Earthquake Engineering and Structural Dynamics*, Vol. 34, no. 12, pp. 1489-1511.

Ibarra, L.F., and Krawinkler, H., 2005, *Global collapse of frame structures under seismic excitations*, John A. Blume Earthquake Engineering Research Center, Report No. 152, Department of Civil and Environmental Engineering, Stanford University, Stanford, California.

Ibarra, L., Medina, R., Krawinkler, H., 2002, "Collapse assessment of deteriorating SDOF systems," *Proc. 12th European Conference on Earthquake Engineering*, London, UK, Paper 665, Elsevier Science Ltd.

A.2.9 *Global Collapse of Frame Structures Under Seismic Excitations*

Authors:

Ibarra, L.F. and Krawinkler, H. (2005)

Abstract:

Global collapse in earthquake engineering refers to the inability of a structural system to sustain gravity loads in the presence of seismic effects. This research proposes a methodology for evaluating global incremental (sidesway) collapse based on a relative intensity measure instead of an Engineering Demand Parameter (EDP). The relative intensity is the ratio of ground motion intensity to a structure strength parameter, which is increased until the response of the system becomes unstable, which means that the relative intensity - EDP curve becomes flat (that is, with zero slope). The largest relative intensity is referred to as "collapse capacity."

In order to implement the methodology, deteriorating hysteretic models are developed to represent the monotonic and cyclic behavior of structural components. Parameter studies that utilize these deteriorating models are performed to obtain collapse capacities and quantify the effects of system parameters that most influence the collapse for SDOF and MDOF structural systems. The range of collapse capacity due to record-to-record variability and uncertainty in the system parameters is evaluated. The latter source of dispersion is quantified by means of the first order second moment (FOSM) method. The studies reveal that softening of the post-yield stiffness in the backbone curve (postcapping stiffness) and the displacement at which this softening commences (defined by the ductility capacity) are the two system parameters that most influence the collapse capacity of a system. Cyclic deterioration appears to be an important but not the dominant issue for collapse evaluation. P-Δ effects greatly accelerate collapse of deteriorating systems and may be the primary source of collapse for flexible, but very ductile, structural systems.

The dissertation presents applications of the proposed collapse methodology to the development of collapse fragility curves and the evaluation of the mean annual frequency of collapse.

An important contribution is the development of a transparent methodology for the evaluation of incremental collapse, in which the assessment of collapse is closely related with the physical phenomena that lead to this limit state. The methodology addresses the fact that collapse is caused by deterioration in complex assemblies of structural components that should be modeled explicitly.

Summary:

The authors used an oscillator with a quadrilinear backbone curve with hardening, softening and residual segment to conduct an extensive parametric study. Pinching, peak-oriented and (bilinear-like) kinematic hysteresis rules were considered, while the cyclic degradation of the backbone stiffness and strength and of the unloading/reloading stiffness were also included. The effects of P-Δ were added separately, as a rotation of the backbone around the center of the axes. The investigation used 40 "ordinary" ground motion records and it was focused on determining the influence of all the parameters on the collapse capacity, which was considered to occur in an IDA (incremental dynamic analysis) fashion, when numerical instability occurred or when the IDA curve becomes horizontal.

Additionally a number of 2D single-bay frames with 3, 6, 9, 12, 15 and 18 stories was considered; they were designed according to a strong column, weak beam, concept, with the beam hinges having a hysteretic model of the same type as the one used for the SDOF studies. By maintaining a uniform hysteretic model for all beam hinges and globally varying its parameters, another parametric study was performed, focused now on the effect of the hysteretic parameters on the MDOF response.

Representative Figures:

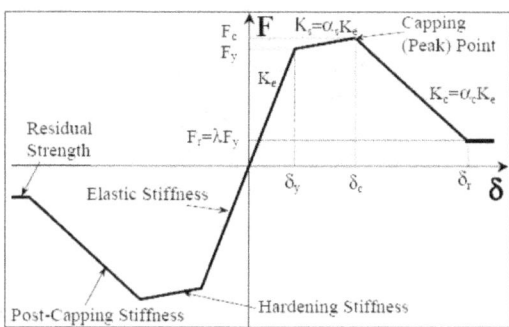 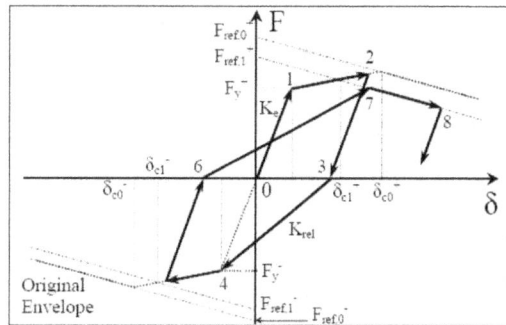

Figure A-23   (a) Backbone curve used for the investigations and (b) post-peak stiffness cyclic deterioration considered.

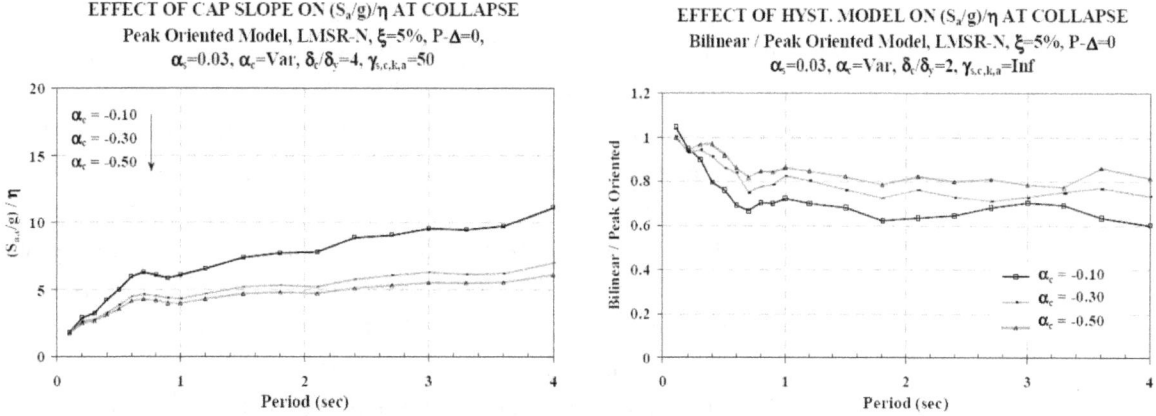

Figure A-24   (a) Effect of the post-peak stiffness to the median collapse capacity spectra for a peak-oriented model and (b) the ratio of collapse capacities for different hysteretic models.

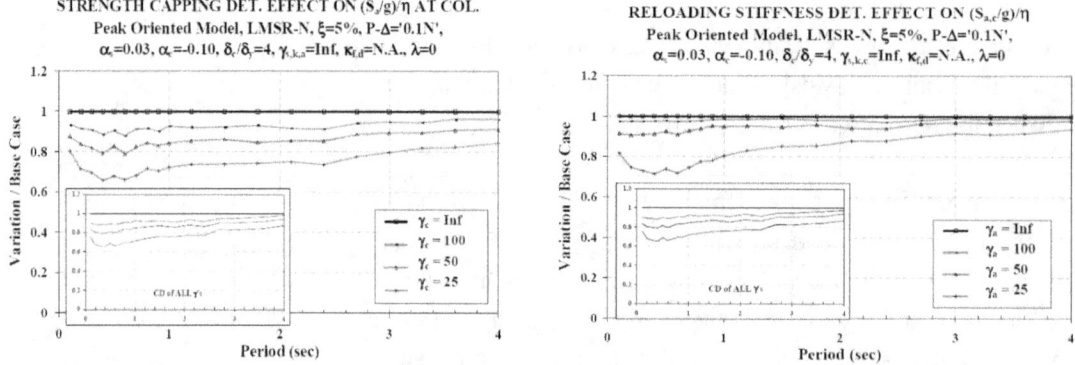

Figure A-25   Effect of (a) post-yield slope and (b) reloading stiffness cyclic deterioration on the collapse capacity.

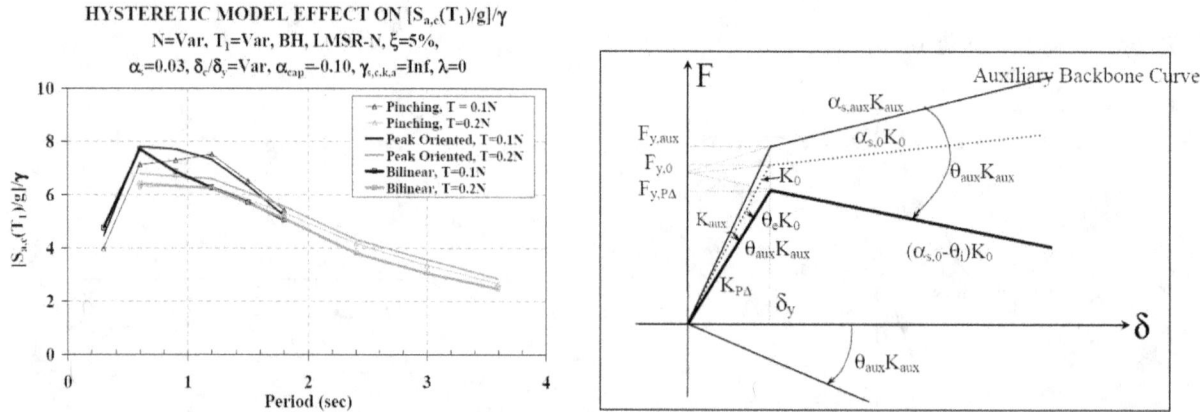

Figure A-26   (a) Effect of the beam-hinge hysteretic model on the median MDOF collapse capacity and (b) the generation of an equivalent SDOF system by using an auxiliary backbone curve to incorporate *P-Δ*.

Summary of Findings:

For the SDOF oscillator it was found that there is a complex interplay between the parameters that defines the dynamic response. Their relative values are significant. The cyclic deterioration was found to be unimportant in the pre-peak region and only mildly important post-peak. Furthermore, its influence does not depend on the type of ground motion, i.e. whether it is near or far field or long in duration. The peak-strength ductility and the post-peak slope are the most dominant parameters. Regarding the hysteresis laws, kinematic hysteresis produces lower capacities than the pinching or the peak-oriented, which are roughly similar. The residual strength becomes important only when it is large enough. Finally, the application of damping in a mass-proportional or stiffness-proportional formulation becomes an important issue after yielding, as the tangent stiffness is constantly changing.

The study of the MDOF frames concluded that these higher-mode-influenced frames fail mostly due to a lower-story mechanism. There is a large dependence of the collapse capacity on the first-mode period. In general the effects of the hinge hysteresis and backbone parameters were similar to the SDOF results. It was also observed that the inelastic instability coefficient (i.e. the difference between the post-peak slope with and without P-Δ) was often much larger than the elastic stability coefficient. Thus, surrogate SDOFs need a separate inclusion of the P-Δ effects in the pre- and post-peak regions. Such equivalent SDOFs were shown to have good accuracy in predicting the collapse capacity.

Relevant Publications:

Ibarra, L.F., and Krawinkler, H., 2005, *Global collapse of frame structures under seismic excitations,* John A. Blume Earthquake Engineering Research Center, Report No. 152, Department of Civil and Environmental Engineering, Stanford University, Stanford, California.

Rahnama, M. and Krawinkler, H., 1993, *Effect of soft soils and hysteresis models on seismic design spectra,* John A. Blume Earthquake Engineering Research Center, Report No. 108, Department of Civil Engineering, Stanford University, Stanford, California.

Medina, R., 200), *Seismic demands for nondeteriorating frame structures and their dependence on ground motions,* Ph.D.. dissertation submitted to the Department of Civil and Environmental Engineering, Stanford University, Stanford, California.

Ibarra, L., Medina, R., and Krawinkler, H., 2002, "Collapse assessment of deteriorating SDOF systems," *Proc. 12th European Conference on Earthquake Engineering,* London, UK, Paper 665, Elsevier Science Ltd.

A.2.10 Object-Oriented Development of Strength and Stiffness Degrading Models for Reinforced Concrete Structures

Author:

Kaul, R. (2004)

Abstract:

The aim of this research is to develop structural simulation models that can capture the strength and stiffness degradation of reinforced concrete frames up to collapse under earthquake-induced motions. The key modeling aspects of the element formulations include: (1) rigorous modeling of large deformation response, (2) flexural yielding and inelastic interaction between axial force and moment (3) degradation of the element stiffness under cyclic loading and (4) axial force-moment-shear interaction for shear-critical reinforced concrete columns. Beam-column models are developed and implemented in an object-oriented analysis framework called OpenSees (Open System for Earthquake Engineering Simulation). The large deformation element formulations employ an updated Lagrangian approach. Inelastic models are based on stress-resultant plasticity to simulate inelastic hardening and softening response under combined axial loads and bending. A two-surface evolution model is proposed for combined nonuniform expansion or contraction and kinematic motion of the yield surface. The yield surface can be used to simulate inelastic section response at integration points along a beam-column element (distributed plasticity) or inelastic hinging at the ends of a beam-column element (concentrated plasticity). In the concentrated plasticity approach, the element between the hinges is quasi-elastic, in which hysteretic models are developed to model the cyclic degradation. This concentrated plasticity model is extended to simulate shear-critical column behavior, including shear strength degradation and failure, interaction between axial and shear forces, and pinched cyclic response. Implementation of the models in OpenSees is planned and structured using object-oriented programming concepts. Individual components of the inelastic modeling problem are identified and the interactions between the governing classes are established. The models are implemented in a hierarchal structure, which provides a modular and extensible software design. The accuracy and the capabilities of the proposed models are verified by comparing the analytical results with the experimental data. The models developed as part of this research provide ideal tools for conducting extensive application studies. An extensible framework is provided to facilitate tool development for nonlinear or inelastic analysis.

Summary:

The objective was the creation, and incorporation into OpenSEES, of a beam-column element with concentrated plasticity, that is appropriate for multiaxial loading of older, shear-critical RC columns. The element has been based on a yield-surface formulation and the focus was on modeling the multiaxial response of a complete RC section. The model incorporates in-cycle strength degradation, allowing for a quadrilinear backbone with a negative stiffness segment (Figure A-27). Inelastic hardening and softening were formulated according to a combined kinematic and isotropic hardening rule with either peak-oriented (for moment) or pinching (for shear) hysteretic rules (Figure A-28). There is little provision for cyclic degradation; the formulation is entirely based on the peak plastic strains and rotations, and the direction of evolution, so depending on the details there may be no cyclic degradation.

The significant advantage of the models is their apparent extensibility and the possibilities for easy modification and incorporation into a variety of elements, an inherent feature of the object-oriented programming upon which OpenSEES has been built.

The model behavior has been calibrated and tested against a variety of RC beam-column experiments, including both shear and moment-critical columns, as well as a set of theoretical solutions for large deformation response (Figure A-29).

Representative Figures:

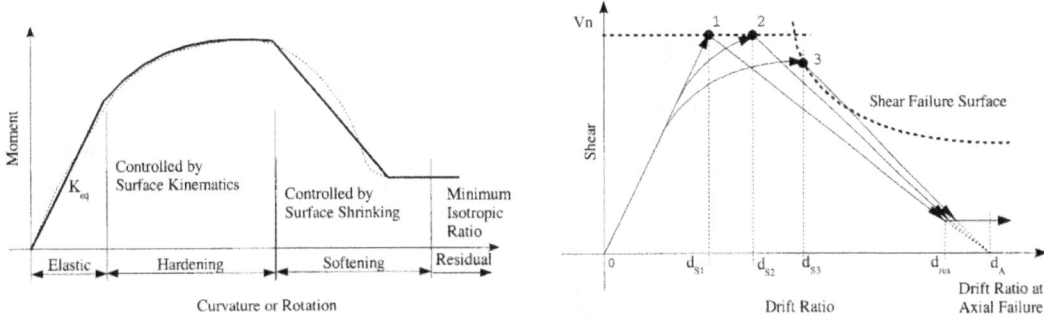

Figure A-27   Idealization of the (a) flexure spring and (b) shear spring backbones.

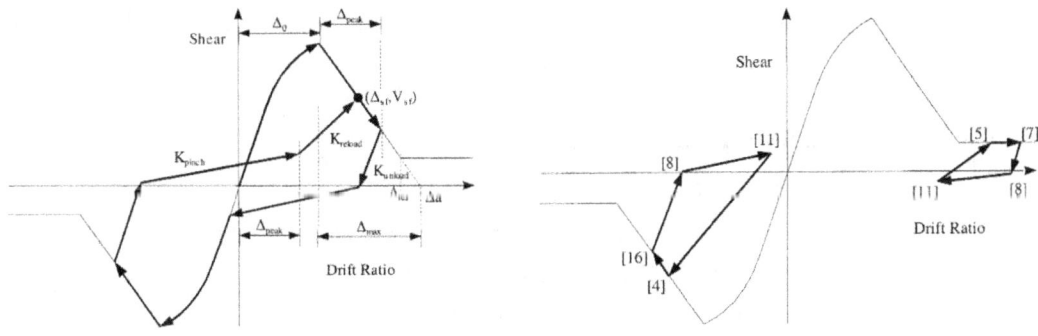

Figure A-28   (a) Full and (b) half cycle pinching hysteresis for the shear spring.

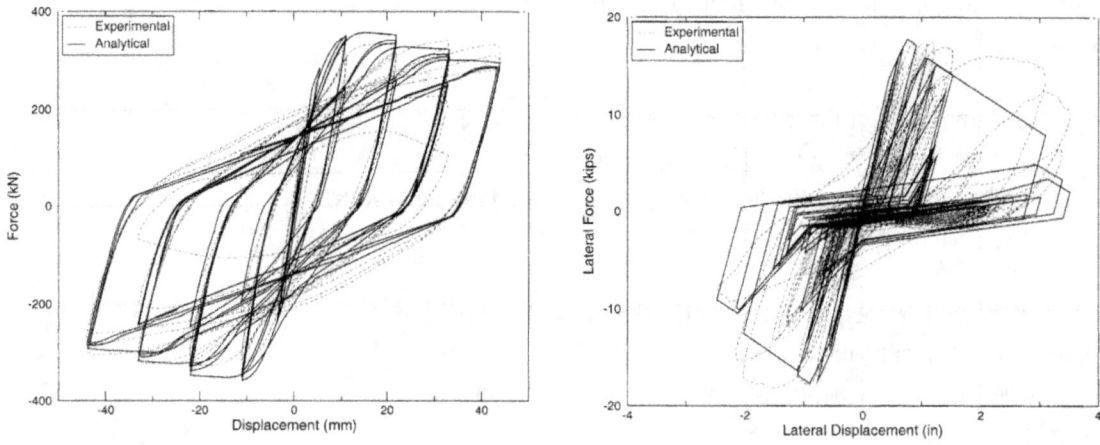

Figure A-29   Comparison of calculated versus experimental results for (a) a moment-critical column and (b) a shear-critical column.

Summary of Findings:

A reliable and extensible concentrated-plasticity beam-column element was created for RC members. Extensive testing and calibration has shown good agreement for a variety of experimental results, including shear-critical columns, large deformations, and planar moment and axial-force interaction. The only serious limitation is the limited formulation of cyclic degradation, an issue that can be potentially solved with the incorporation of damage models. The extension to 3D beam-column elements is somewhat hampered. Appropriate yield surface and evolution rules have not been incorporated, although the hysteretic material models presented are directly usable. The absence of the bond-slip effect and longitudinal reinforcement development in the element springs, are issues that still remain to be addressed.

Relevant Publications:

Kaul, R., 2004, *Object oriented development of strength and stiffness degrading models for reinforced concrete structures*, Ph.D. Thesis, Department of Civil and Environmental Engineering, Stanford University, Stanford, California.

McKenna, F.T. (1997), *Object-oriented finite element programming: framework for analysis, algorithms, parallel computing*, Ph.D. Dissertation, University of California, Berkeley, California.

Mehanny, S.S.F., and Deierlein, G.G., 2001, "Seismic collapse assessment of composite RCS moment frames," *Journal of Structural Engineering*, ASCE, Vol. 127(9).

El-Tawil, S., 1996, *Inelastic dynamic analysis of mixed steel-concrete space frames*, Ph.D. Dissertation, Cornell University, Ithaca, NY.

Elwood, K.J., 2002, *Shake table tests and analytical studies on the gravity load collapse of reinforced concrete frames*, Ph.D. Dissertation, University of California, Berkeley, California.

A.2.11 Shake Table Tests and Analytical Studies on the Gravity Load Collapse of Reinforced Concrete Frames

Author:

Elwood, K.J. (2002).

Abstract:

An empirical model, based on the evaluation of results from an experimental database, is developed to estimate the drift at shear failure of existing reinforced concrete building columns. A shear-friction model is also developed to represent the general observation from experimental tests that the drift at axial failure of a shear-damaged column is directly proportional to the amount of transverse reinforcement and is inversely proportional to the magnitude of the axial load. The two drift-capacity models are incorporated in a nonlinear uniaxial constitutive model implemented in a structural analysis platform to allow for the evaluation of the influence of shear and axial load column failures on the response of a building. Shake table tests were designed to observe the process of dynamic shear and axial load failures in reinforced concrete columns when an alternative load path is provided for load redistribution. The results from these tests provide data on the dynamic shear strength and the hysteretic behavior of columns failing in shear, the loss of axial load capacity after shear failure, the redistribution of loads in a frame after shear and axial failures of a single column, and the influence of axial load on each of the above-mentioned variables. An analytical model of the shake table specimens, incorporating the proposed drift-capacity models to capture the observed shear and axial load failures, provides a good estimate of the measured response of the specimens.

Summary:

The objective was the creation and incorporation into OpenSEES of a beam-column element with concentrated plasticity that is appropriate for multiaxial loading of older, shear-critical RC columns. The element has been based on a yield-surface formulation and the focus was on modeling the multiaxial response of a complete RC section. The model incorporates in-cycle strength degradation, allowing for a quadrilinear backbone with a negative stiffness segment (Figure A-30). Inelastic hardening and softening were formulated according to a combined kinematic and isotropic hardening rule with either peak-oriented (for moment) or pinching (for shear) hysteretic rules (Figure A-31). There is little provision for cyclic degradation; the formulation is entirely based on the peak plastic strains and rotations, and the direction of evolution, so depending on the details there may be no cyclic degradation.

The significant advantage of the models is their apparent extensibility and the possibilities for easy modification and incorporation into a variety of elements, an inherent feature of the object-oriented programming upon which OpenSEES has been built.

Representative Figures:

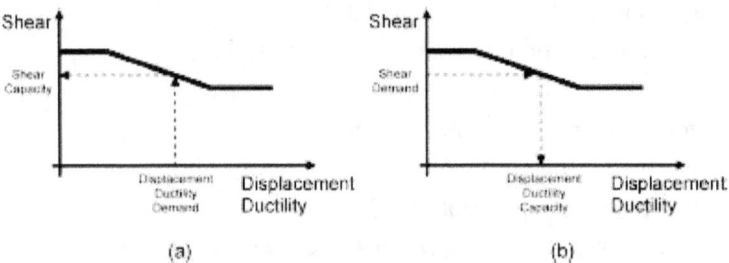

Figure A-30   Use of Sezen model to estimate (a) shear capacity and (b) displacement ductility capacity.

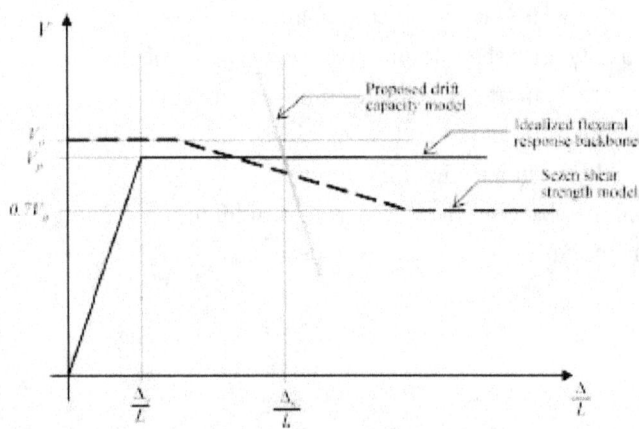

Figure A-31   Comparison of the Sezen shear strength model and the proposed drift capacity model.

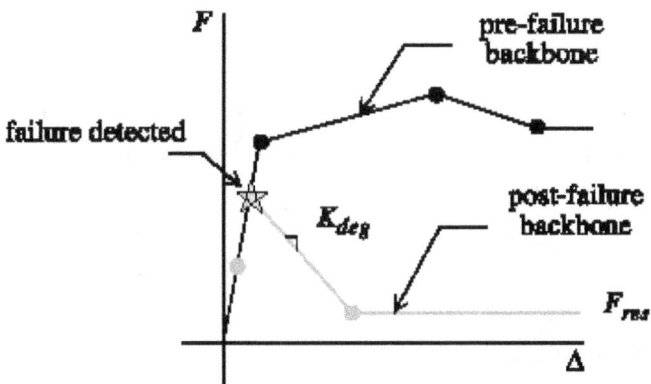

Figure A-32   Redefinition of backbone in Elwood's model after shear failure
is detected.

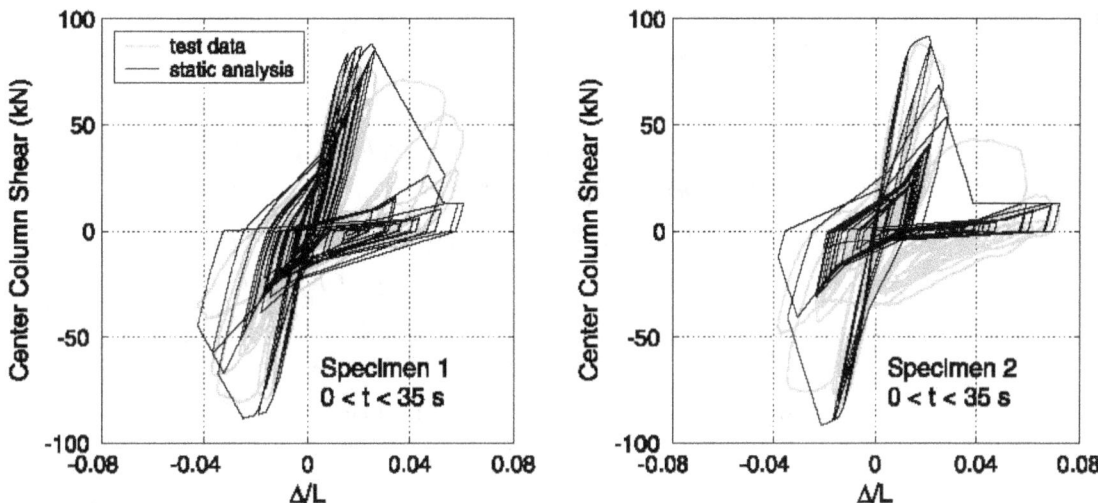

Figure A-33   Comparison of calculated versus experimental results for two shear-critical columns.

Summary of Findings:

Given the lack of agreement between existing models for the drift at shear
failure and results from an experimental database of shear-critical building
columns, two empirical models were developed to provide a more reliable
estimate of the drift at shear failure for existing reinforced concrete columns:

$$\frac{\Delta_s}{L} = \frac{3}{100} + 4\rho'' - \frac{1}{500}\frac{v}{\sqrt{f_c'}} - \frac{1}{40}\frac{P}{A_g f_c'} \geq \frac{1}{100} \text{ (psi units)}$$

Based on shear-friction concepts and the results from 12 columns tested to
axial failure, a model was also developed to estimate the drift at axial failure
for a shear-damaged column:

$$\frac{\Delta_a}{L} = \frac{4}{100} \frac{1 + (\tan\theta)^2}{\tan\theta + P\left(\frac{s}{A_{st}f_{yt}d_c\tan\theta}\right)} \quad \text{where } \theta = 65°$$

The capacity models for the drift at shear and axial load failure were used to initiate the strength degradation of a uniaxial material model implemented in the OpenSees analytical platform (OpenSees, 2002). When attached in series with a beam-column element, the material model can be used to model either shear or axial failure, or both if two materials are used in series. Based on experimental evidence suggesting that an increase in lateral shear deformations may lead to an increase in axial deformations and a loss of axial load, shear-to-axial coupling was incorporated in the material model to approximate the response of a column after the onset of axial failure.

Relevant Publications:

Elwood, K.J., 2002, *Shake table tests and analytical studies on the gravity load collapse of reinforced concrete frames*, Ph.D. Dissertation, University of California, Berkeley.

Elwood, K.J., and Moehle, J.P., 2003, *Shake table tests and analytical studies on the gravity load collapse of reinforced concrete frames*. Pacific Earthquake Engineering Research Center, PEER Report 2003/01, University of California, Berkeley, Calif.

Elwood, K.J., 2004, "Modeling failures in existing reinforced concrete columns," *Can. J. Civ. Eng.*, 31: 846–859 (2004)

Elwood, K.J., and Moehle, J.P., 2005, "Drift capacity of reinforced concrete columns with light transverse reinforcement." *Earthquake Spectra*, Volume 21, No. 1, pp. 71–89,

A.2.12 Determination of Ductility Factor Considering Different Hysteretic Models

Authors:

Lee, L.H., Han, S.W., and Oh, Y.H. (2003)

Abstract:

In current seismic design procedures, base shear is calculated by the elastic strength demand divided by the strength reduction factor. This factor is well known as the response modification factor, R, which accounts for ductility, overstrength, redundancy, and damping of a structural system. In this study, the R factor accounting for ductility is called the ductility factor, R_μ. The R_μ

factor is defined as the ratio of elastic strength demand imposed on the SDOF system to inelastic strength demand for a given ductility ratio. The R_μ factor allows a system to behave inelastically within the target ductility ratio during the design level earthquake ground motion. The objective of this study is to determine the ductility factor considering different hysteretic models. It usually requires large computational efforts to determine the R_μ factor. In order to reduce the computational efforts, the R_μ factor is prepared as a functional form in this study. For this purpose, statistical studies are carried out using forty different earthquake ground motions recorded at a stiff soil site. The R_μ factor is assumed to be a function of the characteristic parameters of each hysteretic model, target ductility ratio and structural period. The effects of each hysteretic model on the R_μ factor are also discussed.

Summary:

The focus of the research was the creation of an R-μ-T relationship that would include the effect of several backbone and hysteretic characteristics. The authors' approach was to consider such effects as completely independent from each other by considering the following models: (a) kinematic hysteresis with a bilinear backbone having positive post-yield stiffness; (b) bilinear backbone and kinematic hysteresis with cyclic degradation of the reloading/unloading stiffness; (c) bilinear backbone with peak-oriented (Clough-like) hysteresis with cyclic strength degradation; and (d) bilinear backbone with pinching hysteresis (Figure A-34). Details of the models are provided by Kunnath *et al.*, (1990).

Using the elastic-perfectly-plastic model with kinematic hardening as a basis, the researchers investigated the influence of each of the four parameters and provided correction factors, in a "coefficient-like" method. These can be applied to an R-μ-T relationship based on the elastic-plastic system to account for the effect of each of the parameters.

Representative Figures:

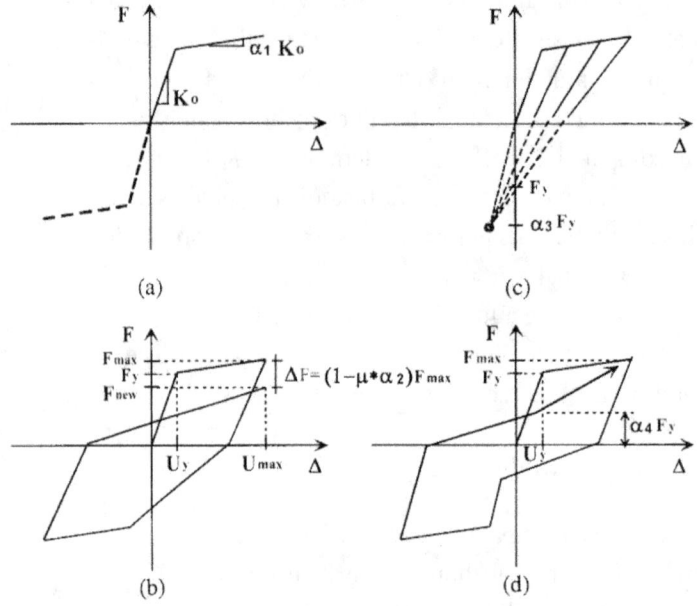

Figure A-34   The parameters investigated: (a) backbone hardening ratio;
(b) unloading/reloading cyclic stiffness degradation; (c) strength degradation;
and (d) degree of pinching.

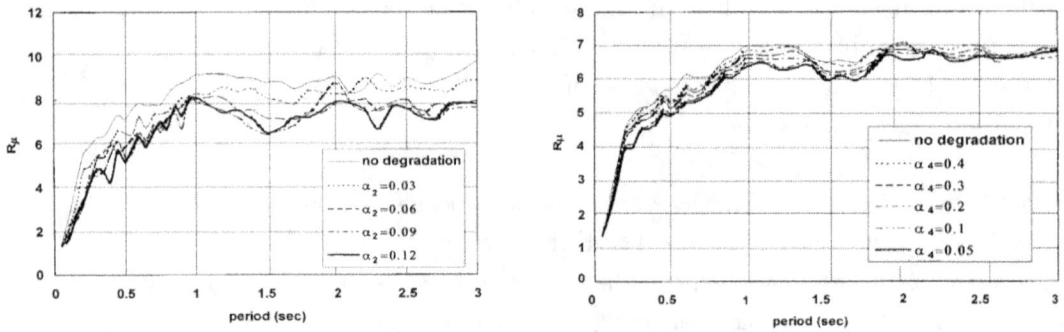

Figure A-35   The effect of (a) cyclic strength degradation and (b) degree of pinching on the
mean R-factor for a given ductility.

Summary of Findings:

The results presented show the effect of the cyclic degradation of strength or
of reloading and unloading stiffness, and the degree of pinching on the mean
R-factor observed for a given ductility for SDOF systems (Figure A-35). In
general the effects are relatively small. Unfortunately the influence of such
parameters is evaluated separately for each parameter, only for a relatively
limited range of values, and always in relation to the pure elasto-plastic
system. Still, the authors do perform verifications for systems having a
combination of all such characteristics, thus providing evidence that the

proposed formulas can approximate more complex systems. Perhaps the greatest limitation of this research is that it does not apply to systems with in-cycle strength degradation. Only positive post-yield stiffnesses are considered. Therefore, the influence of several investigated parameters is small, and the results cannot be applied when negative backbone slopes are present.

Relevant Publications:

Lee, L.H, Han, S.W. and Oh, Y.H., 1999, "Determination of ductility factor considering different hysteretic models," *Earthquake Engineering and Structural Dynamics*, Vol. 28, 957–977.

Kunnath, S.K., Reinhorn, A.M. and Park, Y.J., 1990, "Analytical modeling of inelastic seismic response of RC structures," *Journal of Structural Engineering*, ASCE, 116, 996–1017.

A.2.13 Effects of Hysteresis Type on the Seismic Response of Buildings

Authors:

Foutch, D.A. and Shi, S. (1998)

Abstract:

Current design procedures account for inelastic behavior in a crude manner using the R factor. Although different R values are used for different building types, the determination of a specific R value was not done in a very consistent or scientific manner. The hysteresis behavior of members can be different depending on the material and member type. Buildings with members that dissipate energy through full hysteresis loops (for example, steel moment frames with compact members and no joint fracture) will respond differently from buildings with members that demonstrate strength-degrading hysteresis behavior by having either non-compact steel members, concentric braces, or members with fractured joints. This paper will present results of a study that has closely examined these effects using both SDOF and MDOF systems. A procedure for developing reliability-based design methods which incorporates these effects will also be presented.

Summary:

A total of nine moment-resisting frames with three different configurations (3-story, 6-story and 9-story) were used in this study to examine the effect of the beam-hinge model on the seismic behavior of MDOF structures. Using a

---

suite of 12 ground motion records, all structures (numbering 3x9x8) were analyzed for several *R*-factor levels (or approximately an equivalent number of earthquake intensity levels) and the results were summarized and compared with the buildings having a basic bilinear hinge with kinematic hardening (Figure A-36).

Eight different hinge models were considered: (1) kinematic hardening with bilinear backbone (positive post-yield stiffness), (2) same as 1 but with cyclic strength degradation, (3) same as 1 but having peak-oriented (Clough-like) hysteresis, (4) same as 2 but with peak-oriented hysteresis, (5) same as 1 but with pinching hysteresis, (6) same as 2 but with pinching hysteresis, (7) fracturing connection model with pinching hysteresis and asymmetric backbones including a negative slope and a residual plateau at one direction, and (8) a purely elastic bilinear backbone that dissipates no energy (Figure A-37).

Representative Figures:

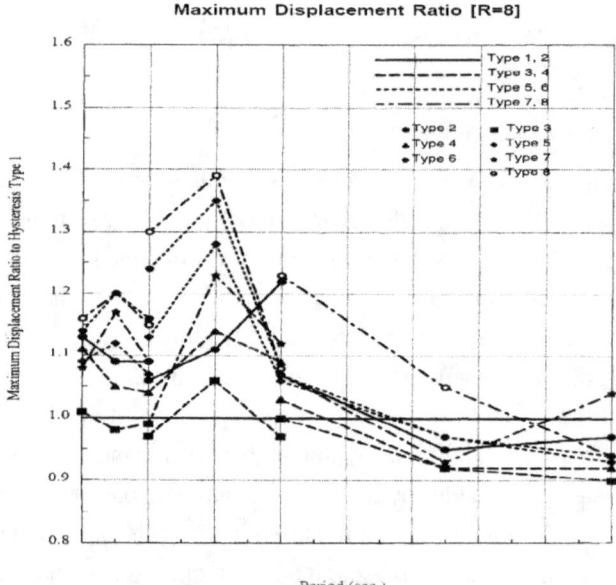

Figure A-36   Ratio of maximum displacement for all buildings and hinge types versus the hinge type 1 (kinematic hardening, no degradation).

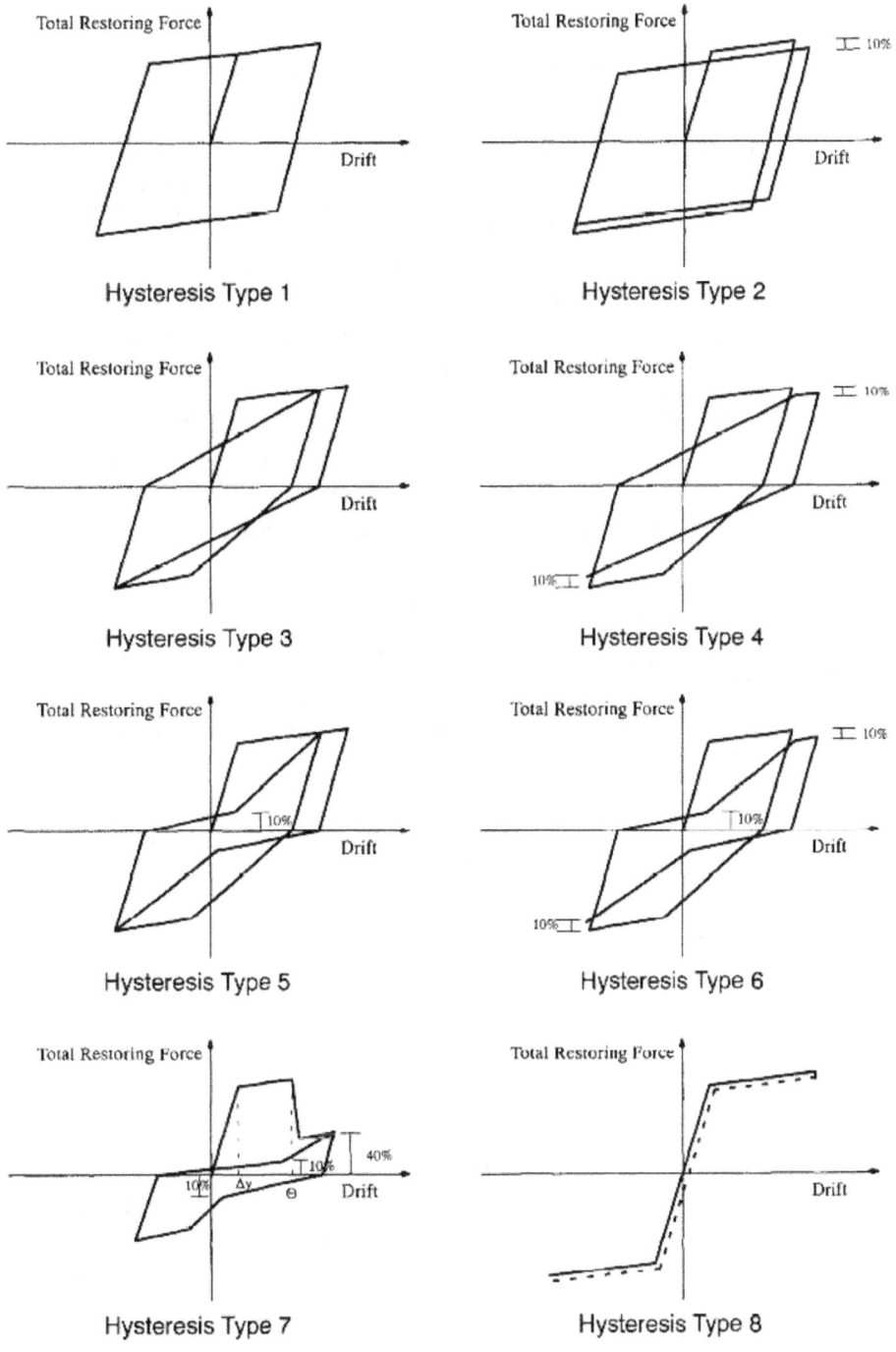

Figure A-37   The hysteresis types considered for the beam-hinges.

Summary of Findings:

The results show that the investigated plastic hinge models have a relatively similar effect on the seismic response of the structures, regardless of the structural period. Even hinge models with no energy dissipation do not produce excessive demands. At most, for an $R = 8$ reduction factor, the differences that appear are in the order of 30-40% and they only appear for a limited range of periods and models. Nevertheless, only a single backbone with in-cycle strength degradation has been considered, and even then the negative slope exists only in one direction of loading. Therefore the conclusions may be of limited use in such cases. This is an interesting exercise and one of the very few investigations that has produced data on the actual impact of local, hinge-level models on the global MDOF response.

Relevant Publications:

Foutch, D.A. and Shi, S., 1998, "Effects of hysteresis type on the seismic response of buildings," *Proc. 6th U.S. National Conference on Earthquake Engineering*, EERI, Seattle, WA.

Shi, S. and Foutch, D.A., 1997, *Evaluation of connection fracture and hysteresis type on the seismic response of steel buildings*, Report No. 617, Structural Research Series, Civil Engineering Studies, University of Illinois at Urbana-Champaign, Urbana, IL.

A.2.14 Performance-Based Assessment of Existing Structures Accounting For Residual Displacements

Authors:

Ruiz-Garcia, J. and Miranda, E. (2005)

Abstract:

The first part of this investigation describes comprehensive statistical studies to quantify residual and maximum displacement demands of inelastic SDOF systems, considering a relatively large earthquake ground motion database, and considering a large number of structural parameters. The second part of this study focuses on the evaluation of permanent (residual) and maximum (transient) drift demands of multi-story framed building models under different levels of ground motion intensity. Both parts include the formulation and implementation of simplified probabilistic approaches to estimate maximum and residual displacement demands accounting for the uncertainty in the structural response and the ground motion hazard. The study provides information towards incorporating explicitly the evaluation of residual displacement demands for assessing the seismic performance of existing structures, or for the preliminary design phase of new structures, where structural damage control is achieved through control of lateral deformation demands.

Summary:

This study examined the effect of hysteretic behavior of maximum deformations of SDOF systems subjected to a large ensemble of 240 ground motions recorded on firm sites in California. They considered seven different types of hysteretic behavior: elastoplastic, bilinear, modified Clough, Takeda, origin-oriented, moderate degrading and severely degrading models. The modified Clough, the Takeda and origin-oriented models only exhibit stiffness degradation while the moderate degrading and severely degrading systems exhibit both stiffness and cyclic strength degradation. This study computed mean ratios of maximum deformation of degrading hysteretic models to non-degrading ones. They also studied the effect of hysteretic behavior for systems subjected to ground motions recorded in very soft soil sites and near-fault ground motions influence by forward directivity.

Representative Figures:

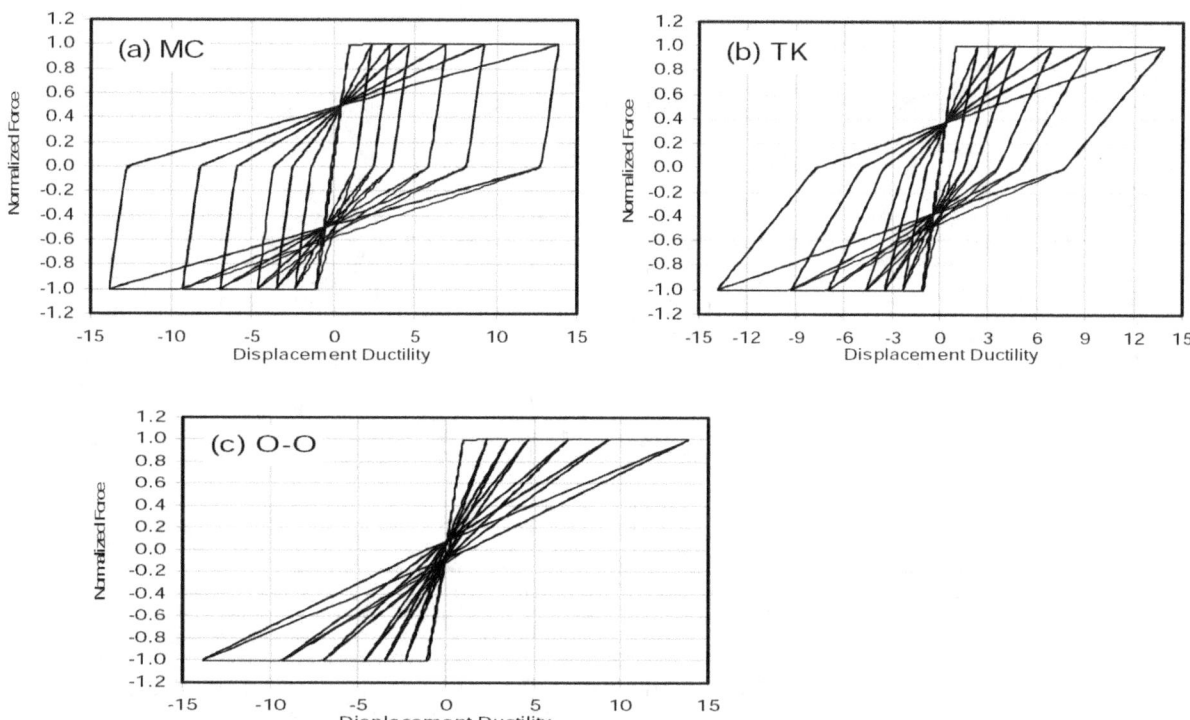

Figure A-38   Hysteretic models used in this investigation that only have stiffness degradation. (a) Modified-Clough (MC); (b) Takeda model (TK); and (c) Origin-Oriented model (O-O).

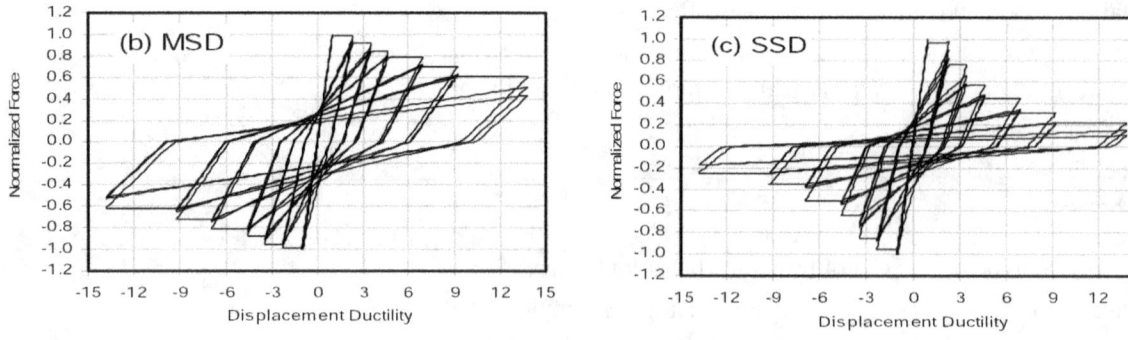

Figure A-39　　Hysteretic models used in this investigation with stiffness and cyclic strength degradation. (b) Moderate Degrading (MSD); and (c) Severely Degrading (SSD).

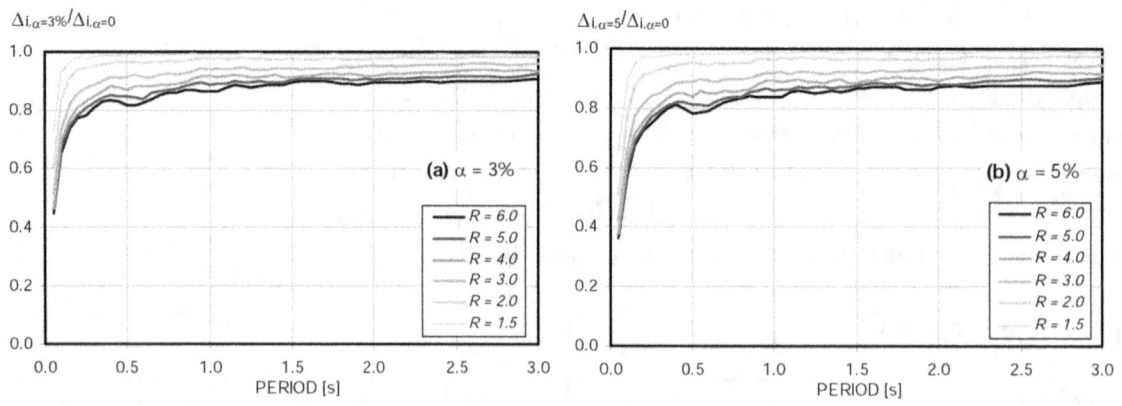

Figure A-40　　Mean ratios of maximum deformation of bilinear to elastoplastic systems: (a) $\alpha = 3\%$; and (b) $\alpha = 5\%$.

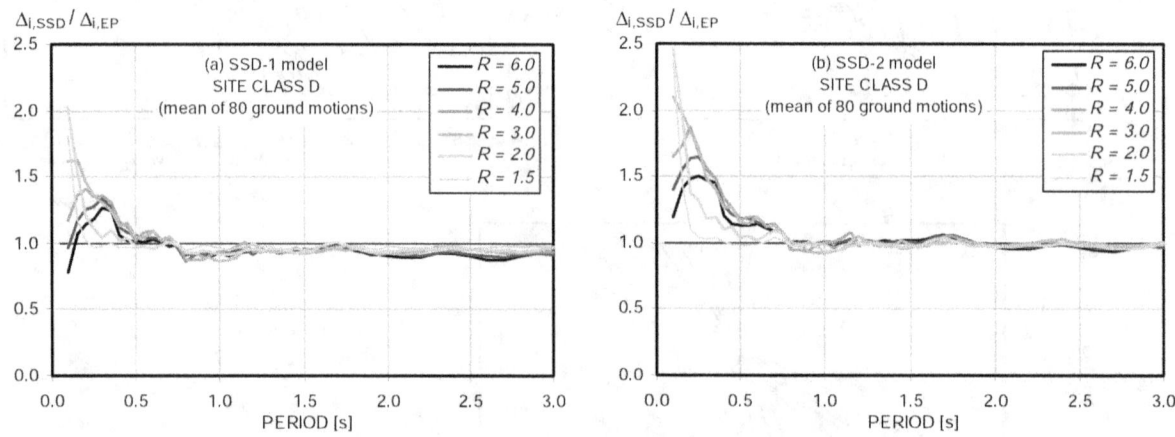

Figure A-41　　Mean ratio of inelastic displacement demands in structural degrading and bilinear systems: (a) SSD-1 model; and (b) SSD-2 model.

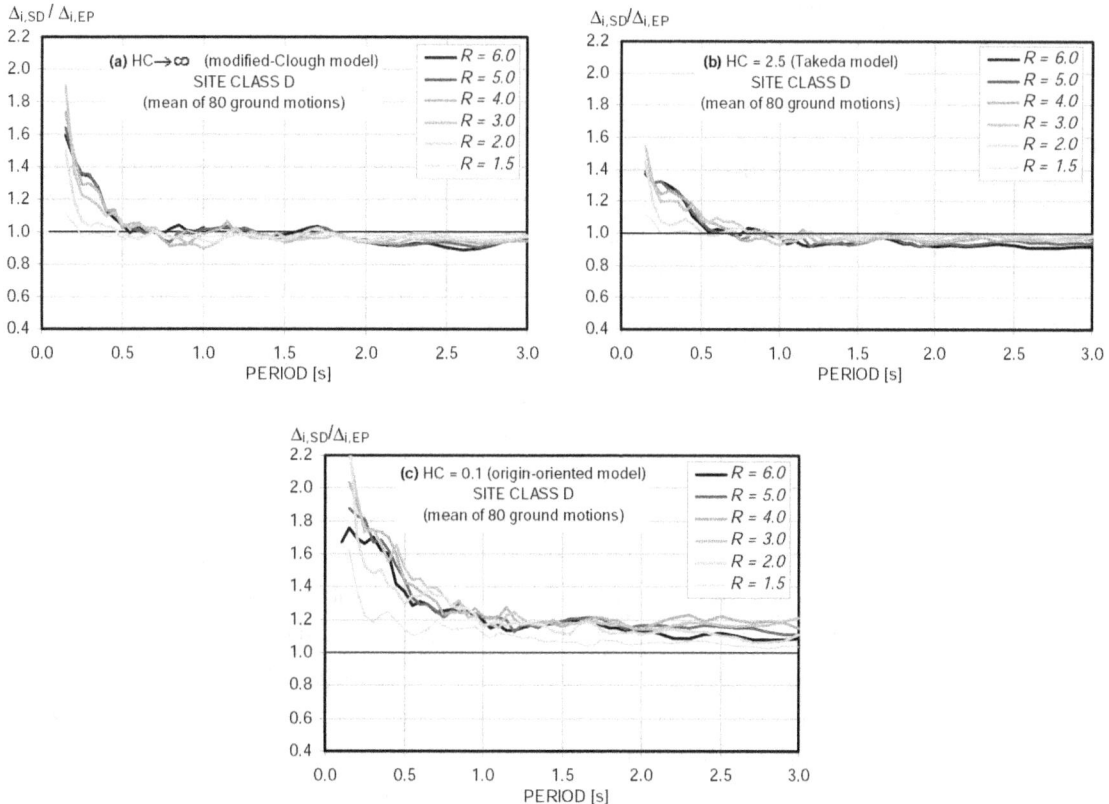

Figure A-42   Influence of hysteretic behavior on maximum deformation for three types of stiffness-degrading systems: (a) Modified-Clough model; (b) Takeda model; and (c) Origin-oriented model.

Summary of Findings:

The effect of positive post-yield stiffness was a function of period of vibration and level of lateral strength relative to the strength required to maintain the system elastic. Positive post-elastic stiffness tended to reduce maximum deformation demands but for realistic values of post-yield stiffness, with exception of systems with very short periods, reductions were small (smaller than 10%).

Maximum deformation demands of short-period degrading structures are, on average, larger than those of non-degrading systems. In general, the increment in displacement produced by degradation effects increases as the strength ratio increases (i.e., as the system becomes weaker relative to the lateral strength required to maintain the system elastic). For structures with periods longer than about 0.7 s, maximum deformation of degrading systems are on average either similar or slightly smaller than those of non-degrading systems.

The effects of stiffness degradation were larger for structures on soft soil sites than those observed for structures on firm sites. For structures with periods of vibration shorter than the predominant period of the ground motion, the lateral displacement demands in stiffness-degrading systems are on average 25% larger than those of non-degrading systems and in order to control lateral deformations to levels comparable to those in non-degrading structures, stiffness-degrading structures in this spectral region need to be designed for higher lateral forces.

Maximum inelastic displacement demands of stiffness-degrading systems are not significantly affected by the unloading stiffness provided that the reduction in unloading stiffness is small or moderate. However, for systems that unload toward the origin (that is, origin-oriented systems), or near the origin, maximum inelastic displacements are on average larger than maximum deformation demands of elastoplastic or bilinear systems and therefore the equal displacement rule should not be used for these systems. Hysteretic behaviors, in particular post-yield stiffness and unloading stiffness, have a large influence on residual displacement demands.

Relevant Publications:

Ruiz-Garcia, J. and Miranda, E., 2003, "Inelastic displacement ratio for evaluation of existing structures," *Earthquake Engineering and Structural Dynamics*. 32(8), 1237-1258.

Ruiz-Garcia, J. and Miranda, E., 2004, "Inelastic displacement ratios for structures built on soft soil sites", *Journal of Structural Engineering*, 130(12), December 2004, pp. 2051-2061

Ruiz-Garcia, J. and Miranda, E., 2005, *Performance-based assessment of existing structures accounting for residual displacements*, John A. Blume Earthquake Engineering Center, Report No. 153, Department of Civil and Environmental Engineering, Stanford University, Stanford, California, 444 p.

Ruiz-Garcia, J. and Miranda, E., 2006a, "Residual displacement ratios for the evaluation of existing structures," *Earthquake Engineering and Structural Dynamics*, Vol. 35, pp. 315-336.

Ruiz-Garcia, J. and Miranda, E., 2006b, "Inelastic displacement ratios for evaluation of structures built on soft soil sites," *Earthquake Engineering and Structural Dynamics,* in press.

A.2.15 Inelastic Spectra for Infilled Reinforced Concrete Frames

Authors:

Dolsek, M. and Fajfar, P. (2004)

Abstract:

In two companion papers a simplified nonlinear analysis procedure for infilled reinforced concrete frames is introduced. In this paper a simple relation between strength reduction factor, ductility and period (R-μ-T relation) is presented. It is intended to be used for the determination of inelastic displacement ratios and of inelastic spectra in conjunction with idealized elastic spectra. The R-μ-T relation was developed from results of an extensive parametric study employing a SDOF mathematical model composed of structural elements representing the frame and infill. The structural parameters used in the proposed R-μ-T relation, in addition to the parameters used in a usual (e.g. elasto-plastic) system, are ductility at the beginning of strength degradation, and the reduction of strength after the failure of the infills. Formulae depend also on the corner periods of the elastic spectrum. The proposed equations were validated by comparing results in terms of the reduction factors, inelastic displacement ratios, and inelastic spectra in the acceleration–displacement format, with those obtained by non-linear dynamic analyses for three sets of recorded and semi-artificial ground motions. A new approach was used for generating semi-artificial ground motions compatible with the target spectrum. This approach preserves the basic characteristics of individual ground motions, whereas the mean spectrum of the complete ground motion set fits the target spectrum excellently. In the parametric study, the R-μ-T relation was determined by assuming a constant reduction factor, while the corresponding ductility was calculated for different ground motions. The mean values proved to be noticeably different from the mean values as determined when based on a constant ductility approach, while the median values determined by the different procedures were between the two means. The approach employed in the study yields an R-μ-T relation which is conservative both for design and performance assessment (compared with a relation based on median values).

Summary:

An R-μ-T relationship was developed that is suitable for use with infilled frames having a quadrilinear elastic-positive-negative-residual backbone. The system used for the analysis contained separate springs, Takeda for the frame and shear-slip for the infill (Figure A-43). These were suitably calibrated to generate a backbone similar to the ones observed in pushovers of infilled frames (Figure A-44). Thus, only in-cycle strength degradation was considered, while any cyclic degradation issues were not investigated.

The system was analyzed using three suites of ground motion records which were spectrum-matched to a target design spectrum. A parametric study of the quadrilinear system was then conducted by varying the period and the backbone parameters within prescribed values.

Representative Figures:

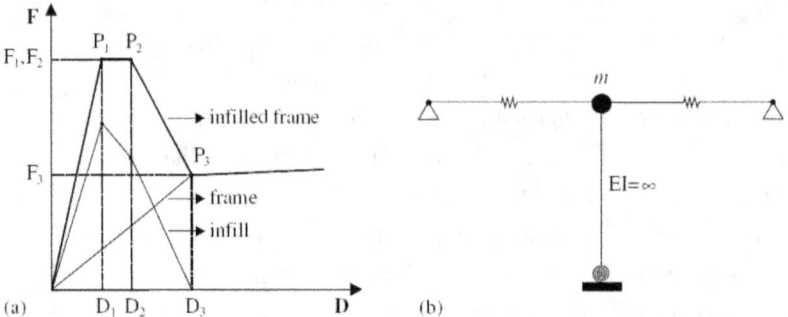

Figure A-43   The SDOF system: (a) force-displacement envelope; and (b) mathematic model.

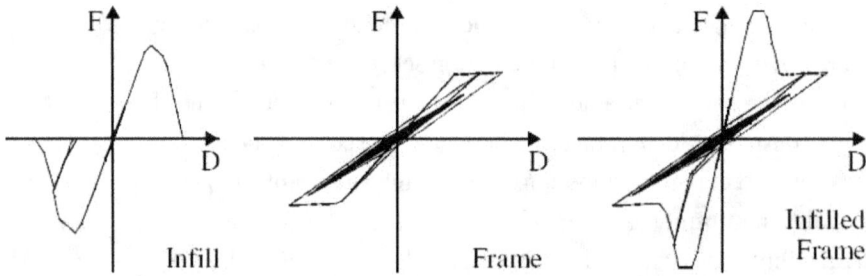

Figure A-44   The hysteretic behavior of the equivalent SDOF system.

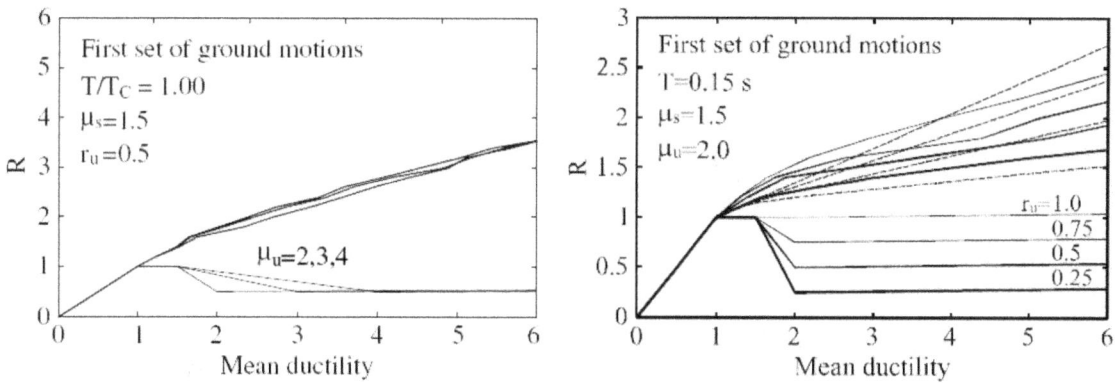

Figure A-45   The influence of negative slope and residual plateau on the mean ductility for given R-factor.

Summary of Findings:

The negative slope was found to have a very small effect on the seismic response of the system when only a short drop in strength to the residual plateau is allowed (Figure A-45a). Actually, the level of the plateau combined with a worsening negative slope were shown to be quite important, lower plateaus increase the ductility demands considerably (Figure A-45b). Approximate values were proposed for the dispersion around the mean provided by the relationship.

Relevant Publications:

Dolsek, M. and Fajfar, P., 2004, "Inelastic spectra for infilled reinforced concrete frames," *Earthquake Engineering and Structural Dynamics*, Vol. 33, 1395–1416.

Dolsek, M. and Fajfar, P., 2000, "Simplified nonlinear seismic analysis of infilled reinforced concrete frames," *Earthquake Engineering and Structural Dynamics*, Vol. 34, 49–66.

Dolsek, M., 2002, *Seismic response of infilled reinforced concrete frames*, Ph.D. thesis, University of Ljubljana, Faculty of Civil and Geodetic Engineering, Ljubljana, Slovenia [in Slovenian].

Appendix B

Quantile IDA Curves for Single-Spring Systems

This appendix presents quantile (16th, 50th and 84th percentile) incremental dynamic analysis (IDA) curves from focused analytical studies on individual spring single-degree-of-freedom (SDOF) systems. These systems consist of spring types 1 through 8, with characteristics described in Chapter 3. This collection of curves is intended to present the range of results for representative short (T=0.5s), moderate (T=1.0s), and long (T=2.0s) period systems, both with and without cyclic degradation. In the figures, the vertical axis is the intensity measure $S_a(T,5\%)$, which is not normalized, and the horizontal axis is the maximum story drift ratio, θ_{max}, in radians. IDA curves with cyclic degradation (black lines) are shown along with IDA curves without cyclic degradation (grey lines). Differences between the black and grey lines in the plots indicate the effect of cyclic degradation given the characteristics of the particular spring and period of vibration.

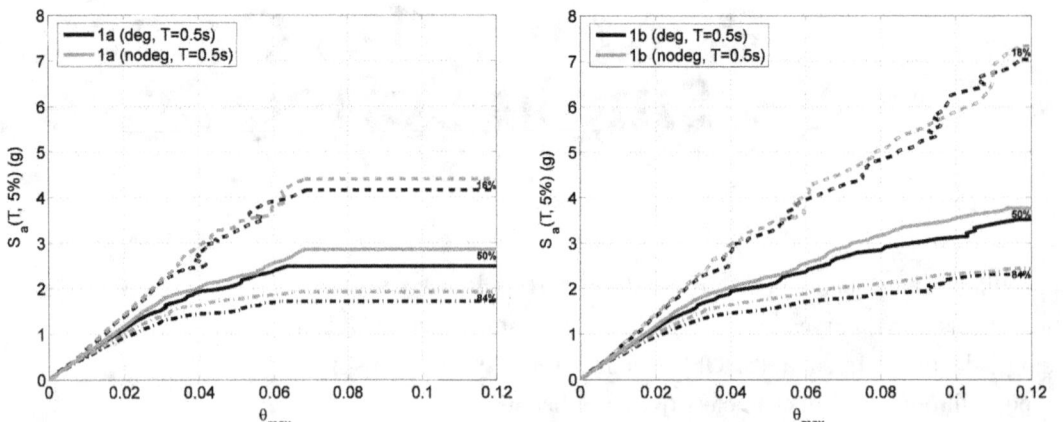

Figure B-1   Quantile IDA curves plotted versus $S_a(T,5\%)$ for Spring 1a and Spring 1b with a period of T = 0.5s

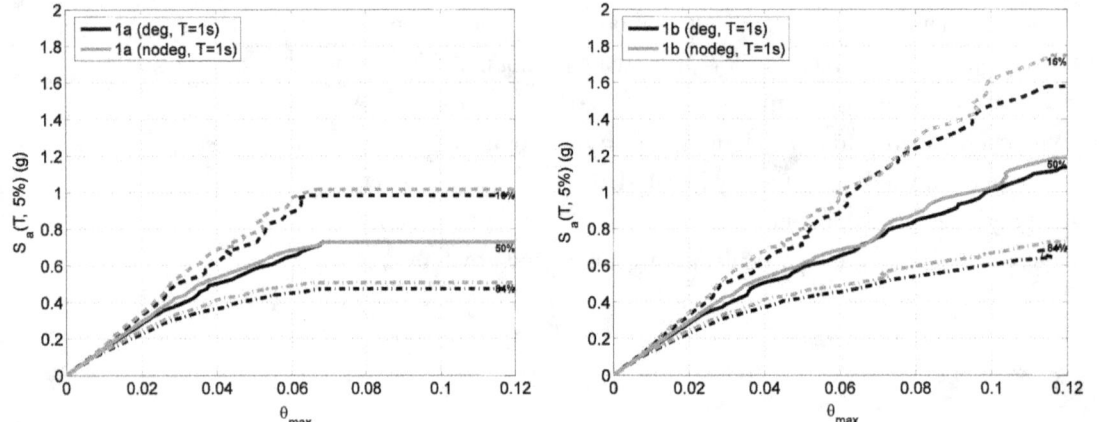

Figure B-2   Quantile IDA curves plotted versus $S_a(T,5\%)$ for Spring 1a and Spring 1b with a period of T = 1.0s.

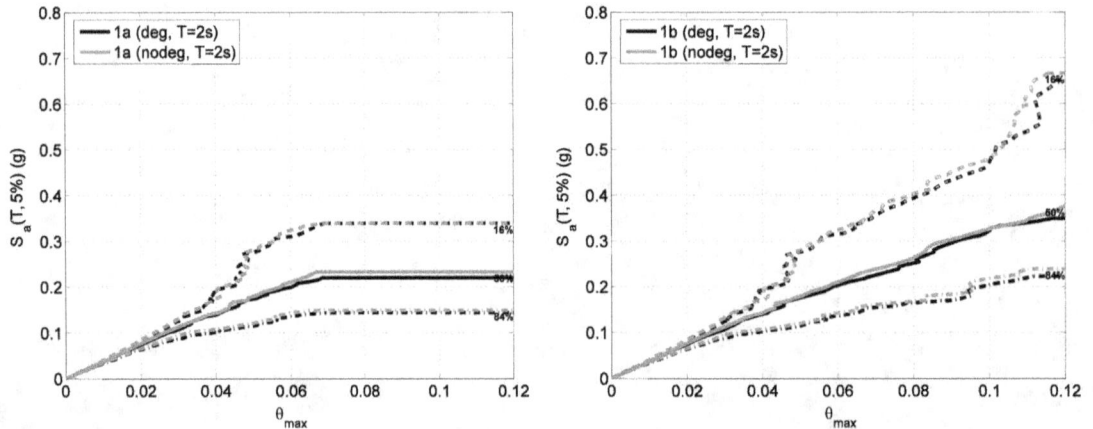

Figure B-3   Quantile IDA curves plotted versus $S_a(T,5\%)$ for Spring 1a and Spring 1b with a period of T = 2.0s.

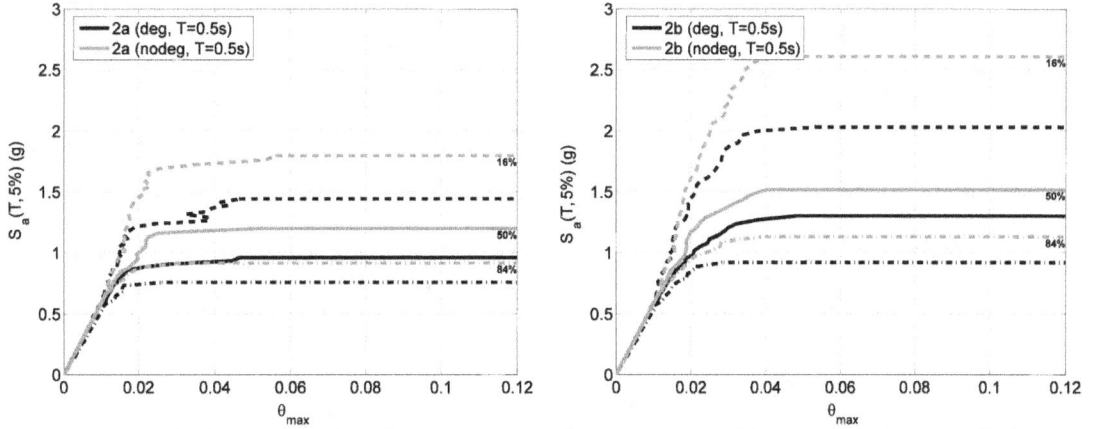

Figure B-4   Quantile IDA curves plotted versus $S_a(T,5\%)$ for Spring 2a and Spring 2b with a period of T = 0.5s.

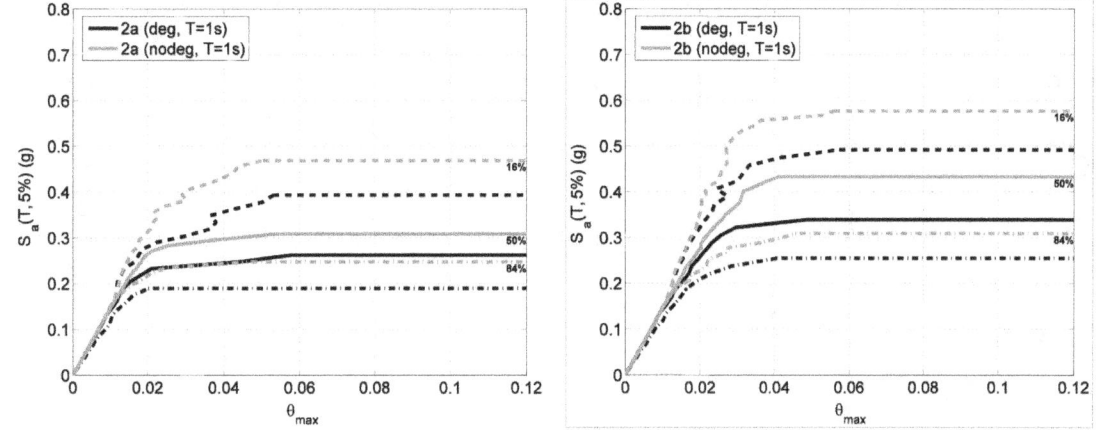

Figure B-5   Quantile IDA curves plotted versus $S_a(T,5\%)$ for Spring 2a and Spring 2b with a period of T = 1.0s.

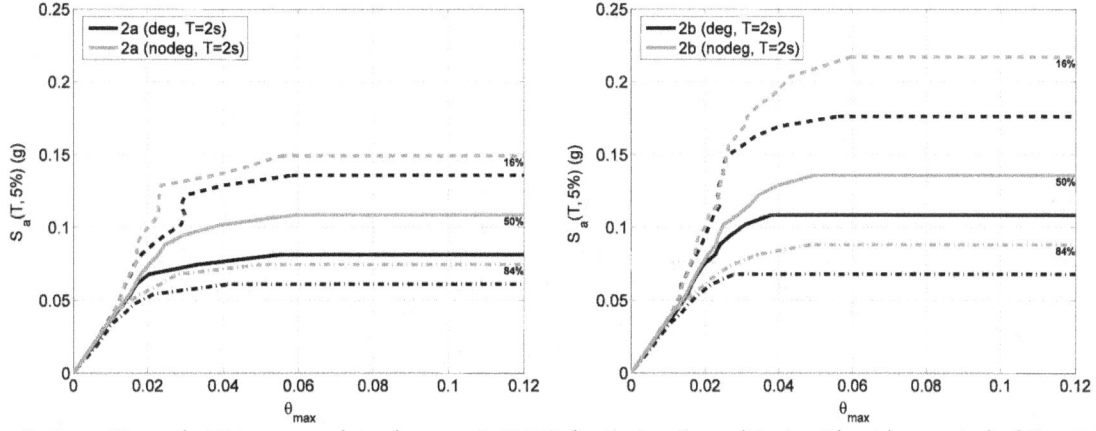

Figure B-6   Quantile IDA curves plotted versus $S_a(T,5\%)$ for Spring 2a and Spring 2b with a period of T = 2.0s.

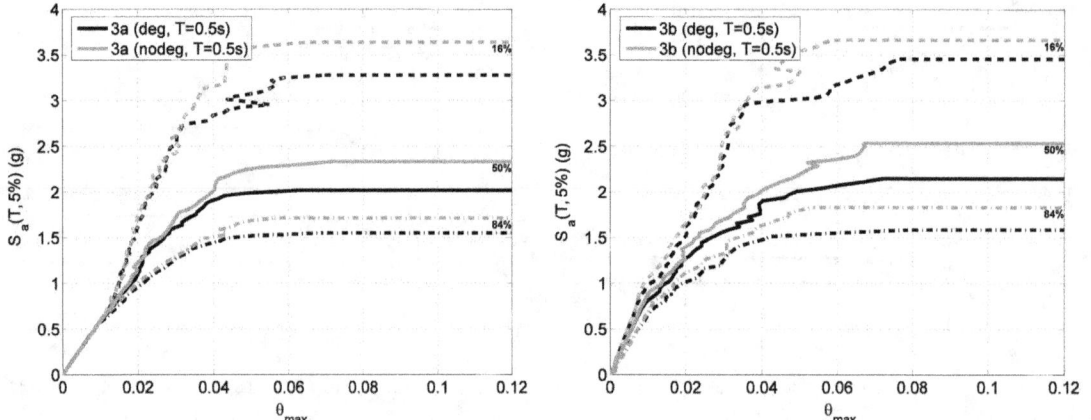

Figure B-7   Quantile IDA curves plotted versus $S_a(T,5\%)$ for Spring 3a and Spring 3b with a period of T = 0.5s.

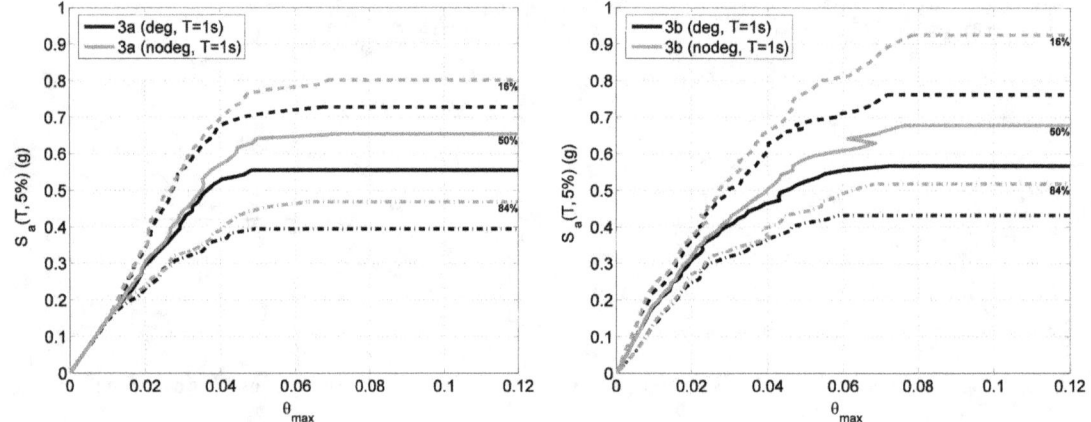

Figure B-8   Quantile IDA curves plotted versus $S_a(T,5\%)$ for Spring 3a and Spring 3b with a period of T = 1.0s.

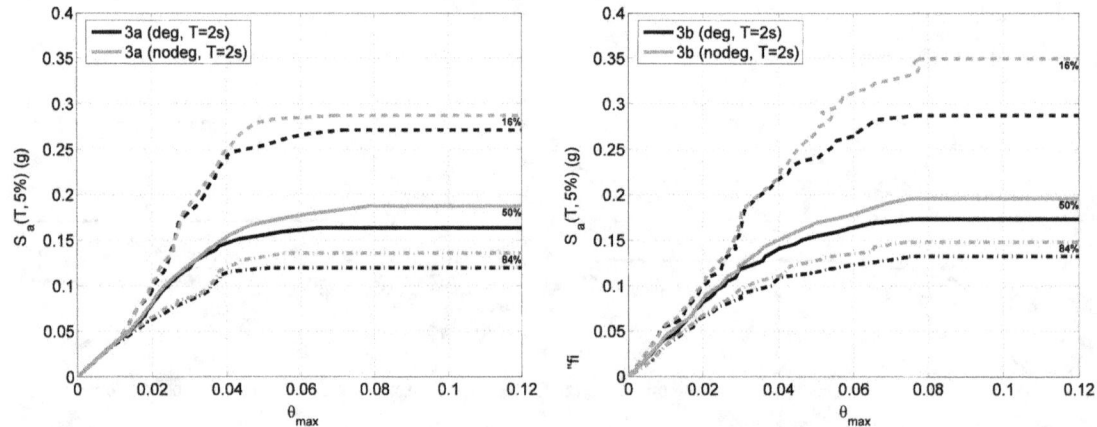

Figure B-9   Quantile IDA curves plotted versus $S_a(T,5\%)$ for Spring 3a and Spring 3b with a period of T = 2.0s.

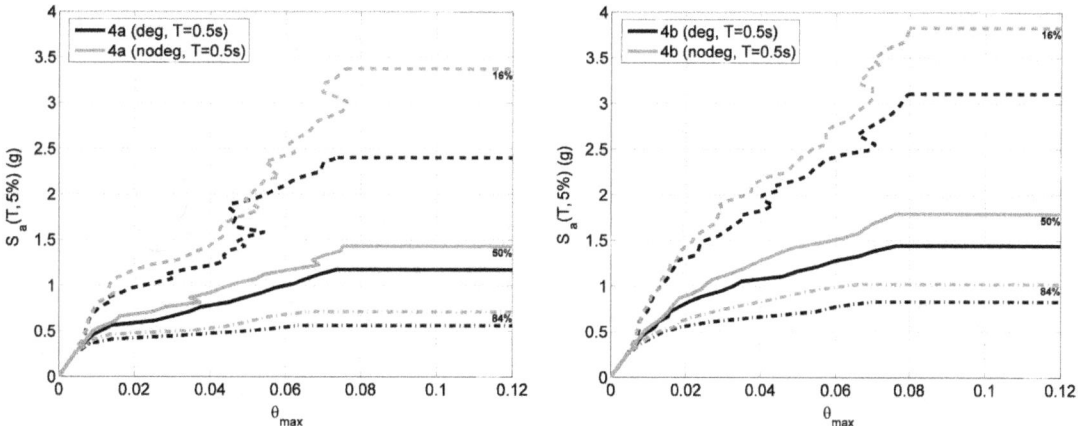

Figure B-10   Quantile IDA curves plotted versus $S_a(T,5\%)$ for Spring 4a and Spring 4b with a period of T = 0.5s.

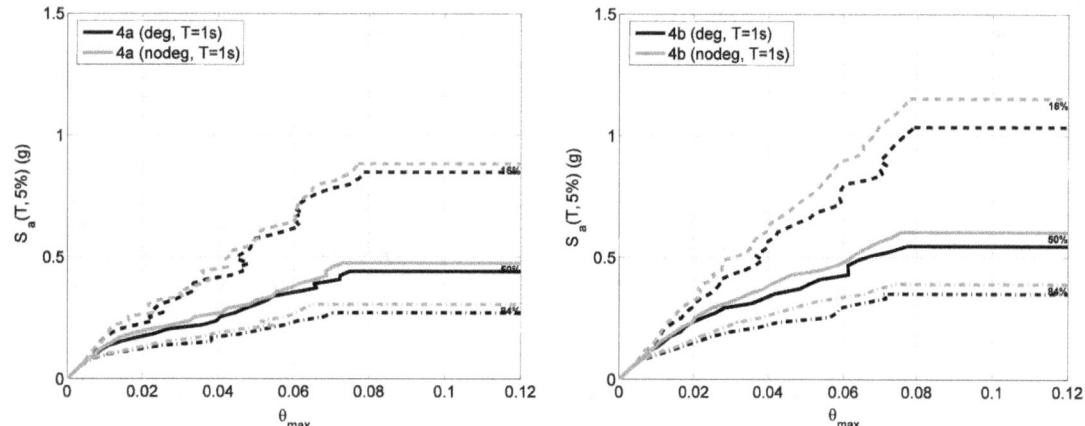

Figure B-11   Quantile IDA curves plotted versus $S_a(T,5\%)$ for Spring 4a and Spring 4b with a period of T = 1.0s.

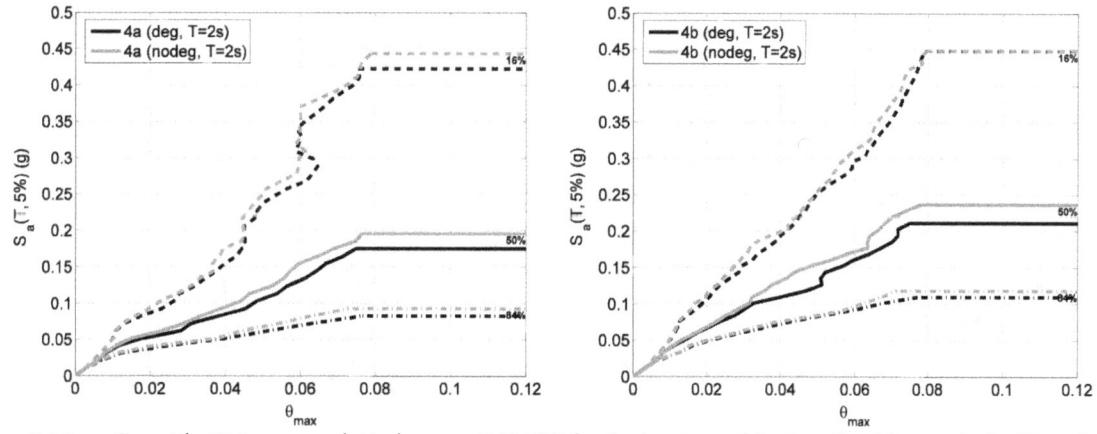

Figure B-12   Quantile IDA curves plotted versus $S_a(T,5\%)$ for Spring 4a and Spring 4b with a period of T = 2.0s.

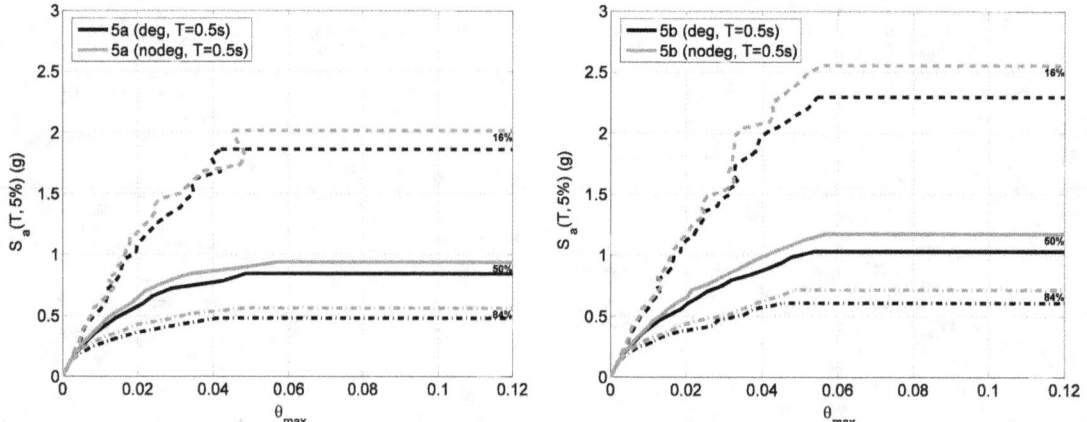

Figure B-13   Quantile IDA curves plotted versus Sa(T,5%) for Spring 5a and Spring 5b with a period of T = 0.5s.

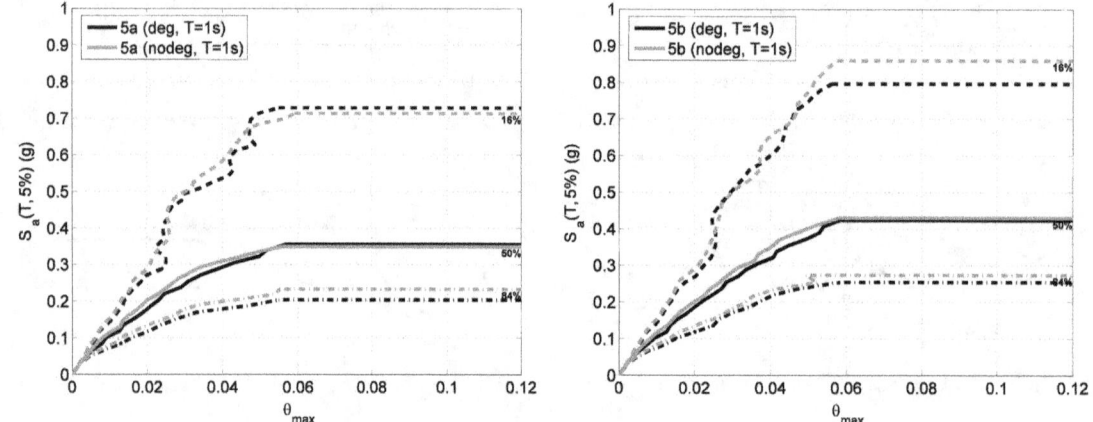

Figure B-14   Quantile IDA curves plotted versus $S_a(T,5\%)$ for Spring 5a and Spring 5b with a period of T = 1.0s.

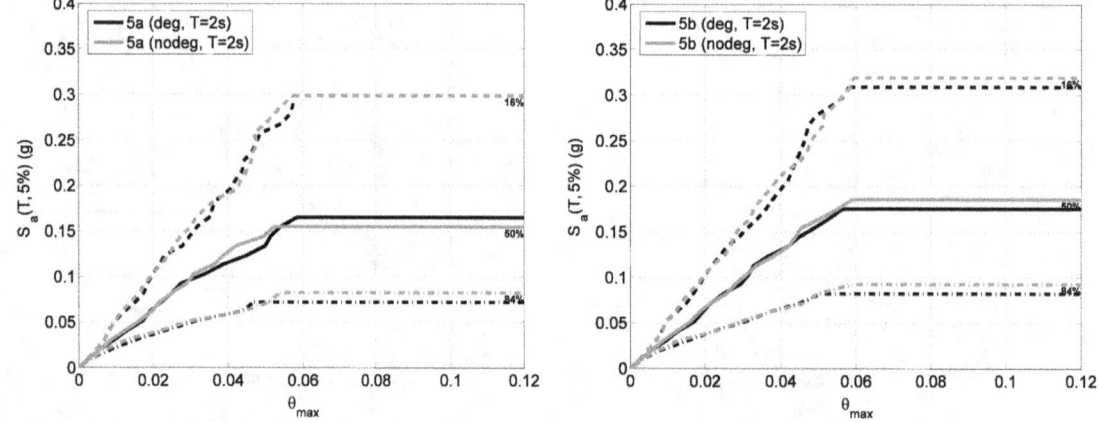

Figure B-15   Quantile IDA curves plotted versus $S_a(T,5\%)$ for Spring 5a and Spring 5b with a period of T = 2.0s.

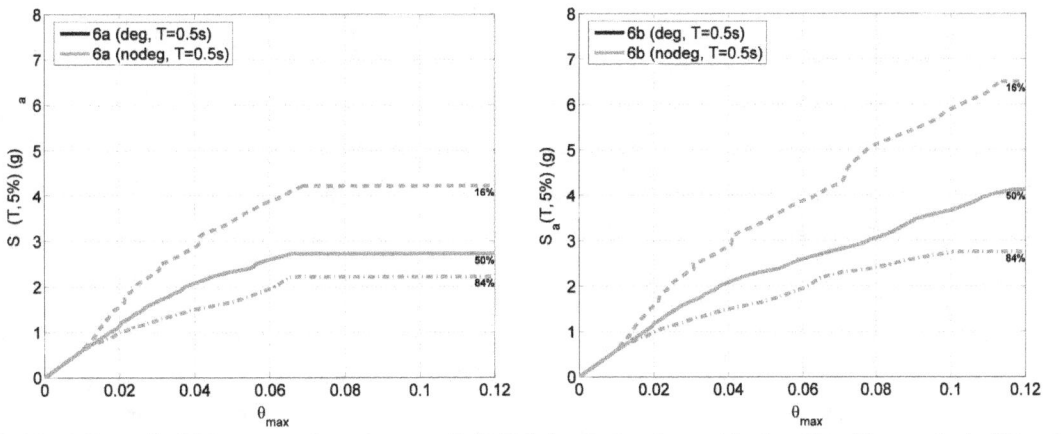

Figure B-16   Quantile IDA curves plotted versus Sa(T,5%) for Spring 6a and Spring 6b with a period of T = 0.5s.

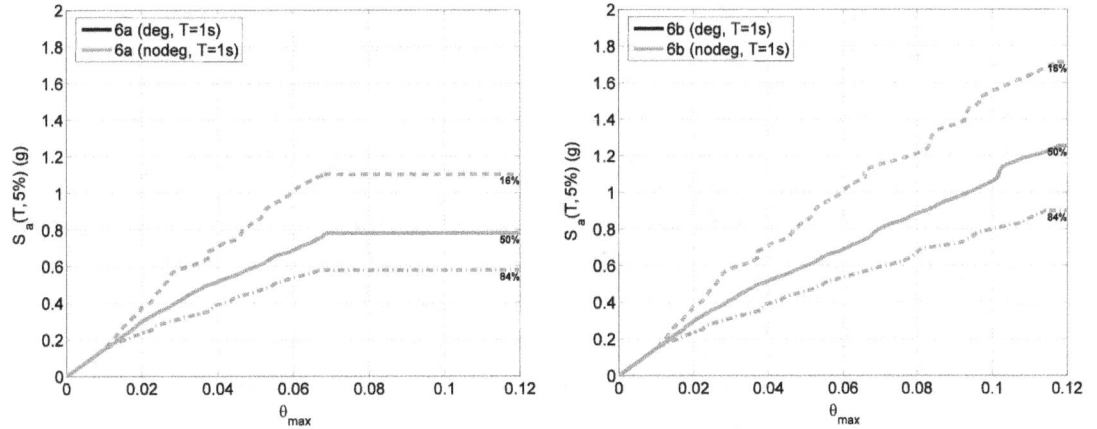

Figure B-17   Quantile IDA curves plotted versus $S_a(T,5\%)$ for Spring 6a and Spring 6b with a period of T = 1.0s.

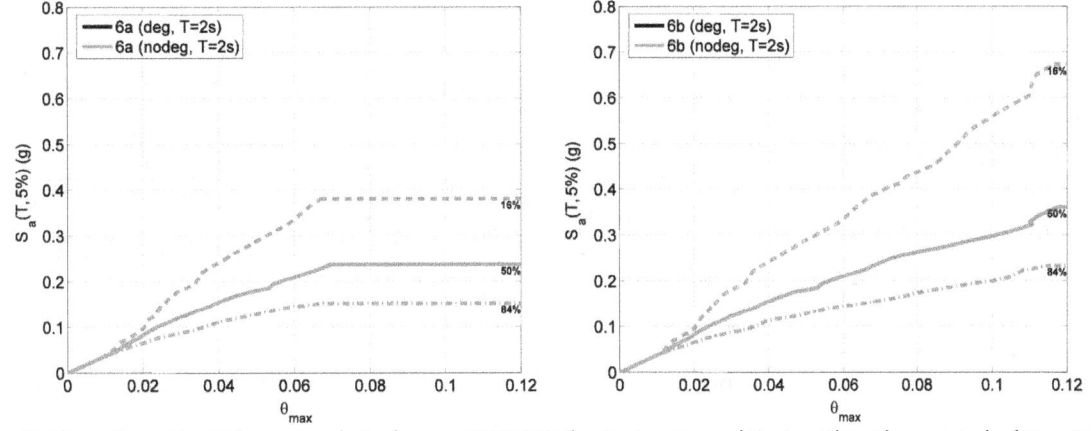

Figure B-18   Quantile IDA curves plotted versus $S_a(T,5\%)$ for Spring 6a and Spring 6b with a period of T = 2.0s.

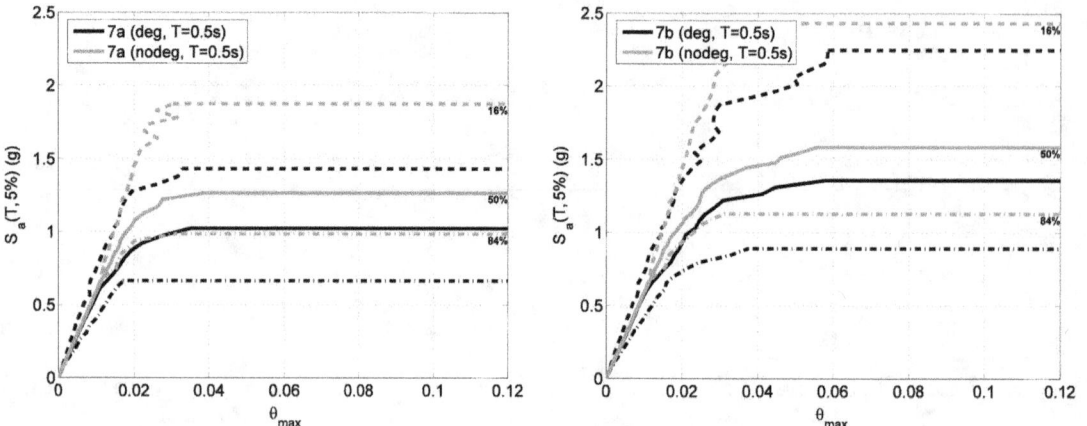

Figure B-19   Quantile IDA curves plotted versus Sa(T,5%) for Spring 7a and Spring 7b with a period of T = 0.5s.

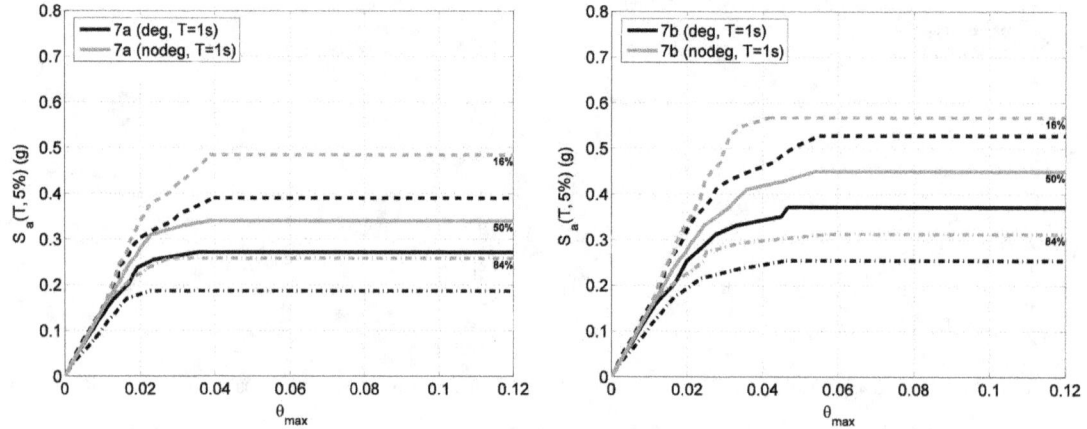

Figure B-20   Quantile IDA curves plotted versus $S_a(T,5\%)$ for Spring 7a and Spring 7b with a period of T = 1.0s.

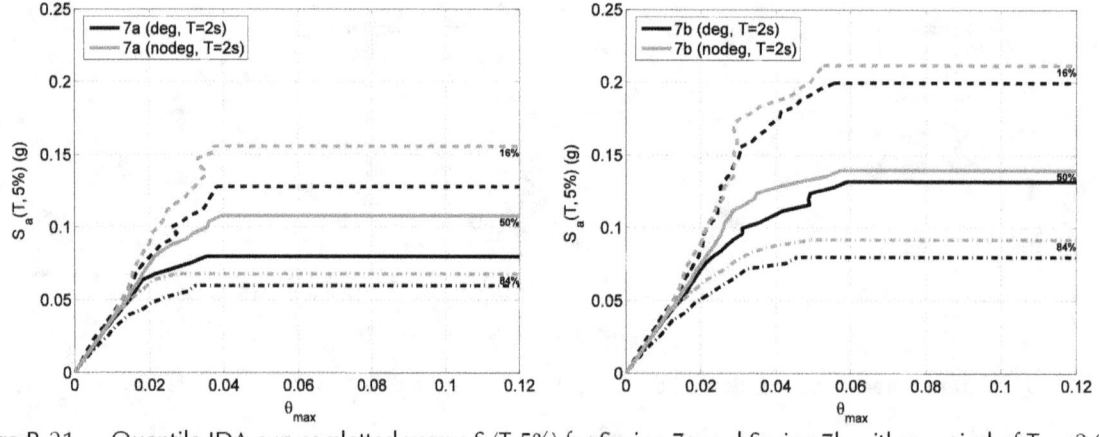

Figure B-21   Quantile IDA curves plotted versus $S_a(T,5\%)$ for Spring 7a and Spring 7b with a period of T = 2.0s.

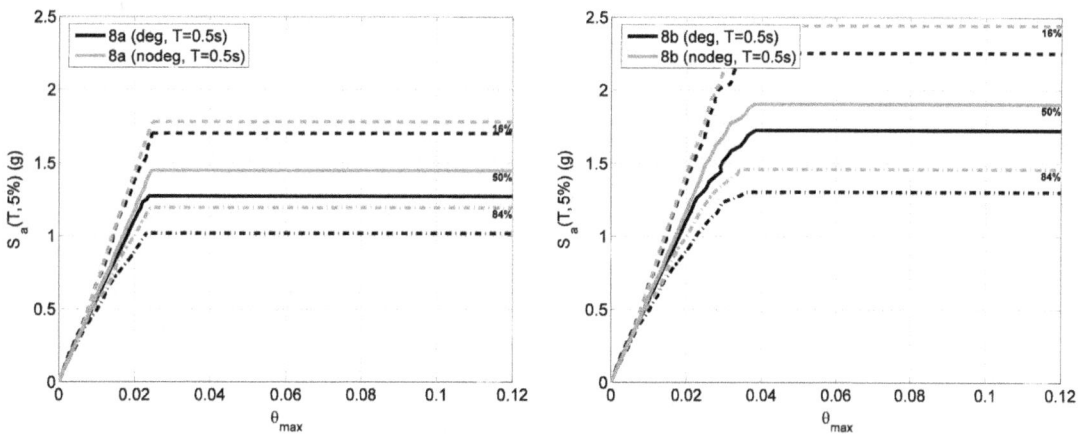

Figure B-22   Quantile IDA curves plotted versus Sa(T,5%) for Spring 8a and Spring 8b with a period of T = 0.5s.

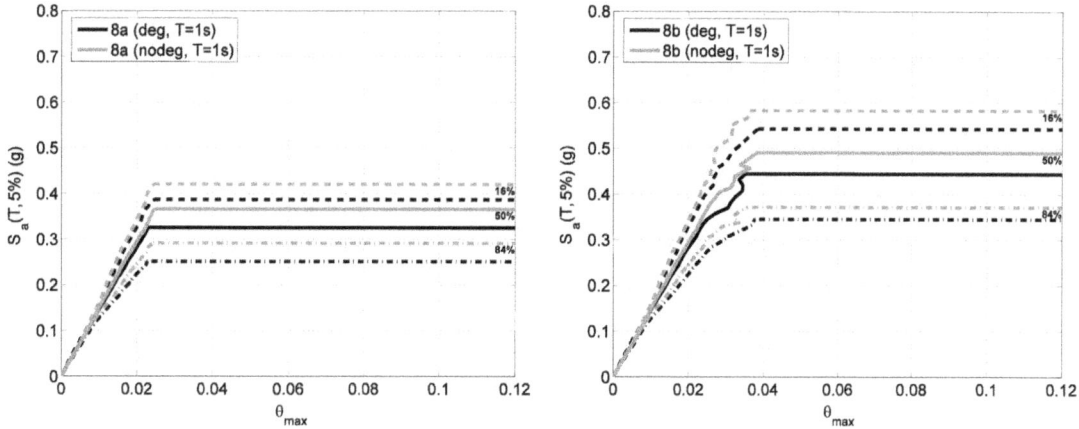

Figure B-23   Quantile IDA curves plotted versus $S_a(T,5\%)$ for Spring 8a and Spring 8b with a period of T = 1.0s.

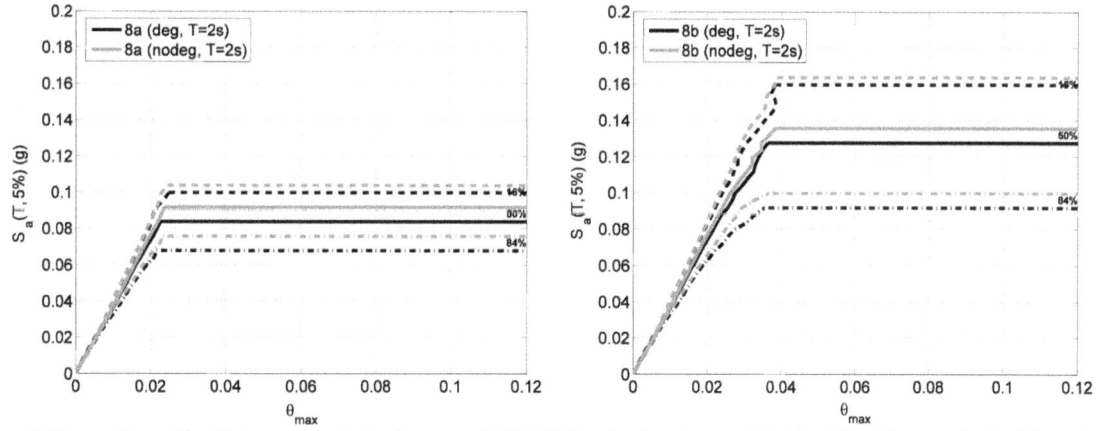

Figure B-24   Quantile IDA curves plotted versus $S_a(T,5\%)$ for Spring 8a and Spring 8b with a period of T = 2.0s.

Appendix C

Median IDA Curves for Multi-Spring Systems versus Normalized Intensity Measures

This appendix contains normalized plots of median incremental dynamic analysis (IDA) curves from focused analytical studies on multi-spring single-degree-of-freedom (SDOF) systems. All systems are composed of two springs representing a primary lateral-force-resisting system and a secondary gravity system with the characteristics described in Chapter 3. Multi-spring systems carry a designation of "NxJa+1a" or "NxJa+1b" where "N" is the peak strength multiplier (N = 1, 2, 3, 5, or 9), "J" is the lateral-force-resisting spring number (J = 2, 3, 4, 5, 6, or 7), and 1a or 1b is the gravity system identifier. In all figures, the vertical axis is the normalized intensity measure $R = S_a(T_1, 5\%)/S_{ay}(T_1, 5\%)$, and the horizontal axis is the maximum story drift ratio, θ_{max}, in radians. The period of vibration for each system is indicated in parentheses.

C.1 Visualization Tool

Given the large volume of analytical data, customized algorithms were developed for post-processing, statistical analysis, and visualization of results. The accompanying CD includes an electronic visualization tool that was developed to view results of multi-spring studies. The tool is a Microsoft Excel based application with a user-interface that accesses a database of all available multi-spring data. By selecting a desired spring combination ("NxJa+1a" or "NxJa+1b"), stiffness level (stiff or flexible), and intensity measure (normalized or non-normalized), users can view the resulting quantile (median, 16th, and 84th percentile) IDA curves for the combination of interest.

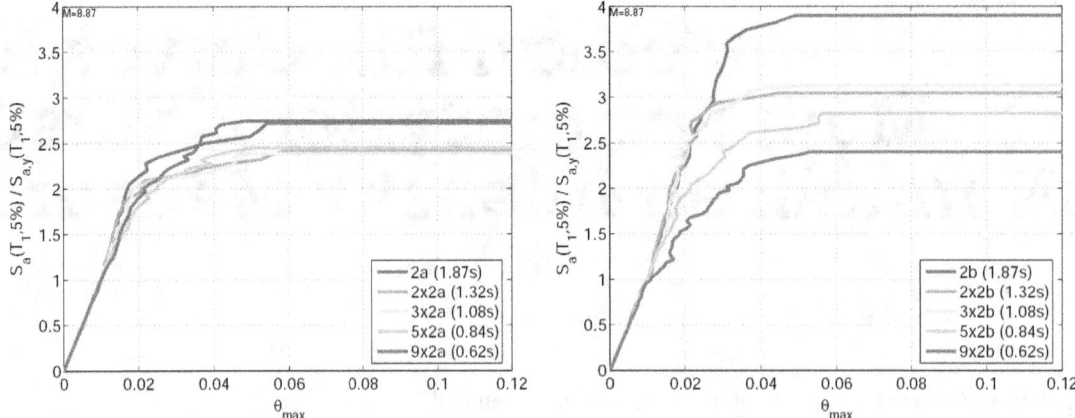

Figure C-1   Median IDA curves plotted versus the normalized intensity measure $R = S_a(T_1,5\%)/S_{ay}(T_1,5\%)$ for systems Nx2a and Nx2b with mass M=8.87ton.

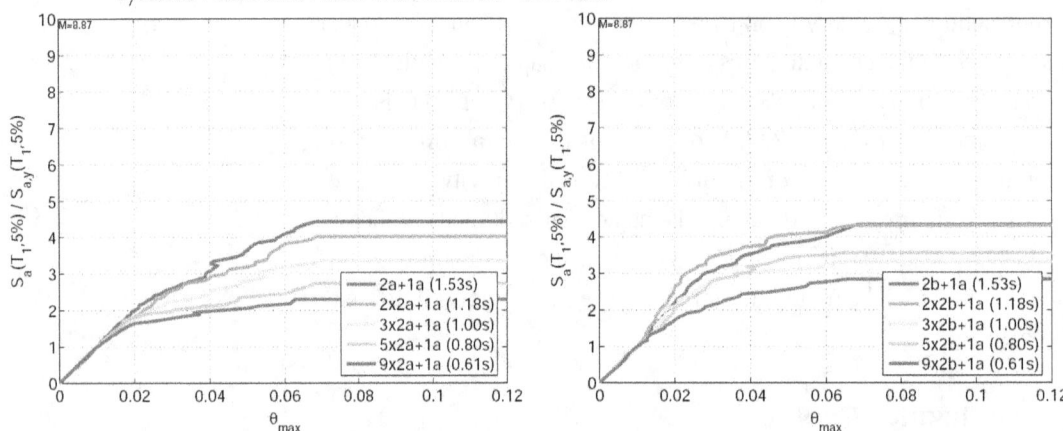

Figure C-2   Median IDA curves plotted versus the normalized intensity measure $R = S_a(T_1,5\%)/S_{ay}(T_1,5\%)$ for systems Nx2a+1a and Nx2b+1a with mass M=8.87ton.

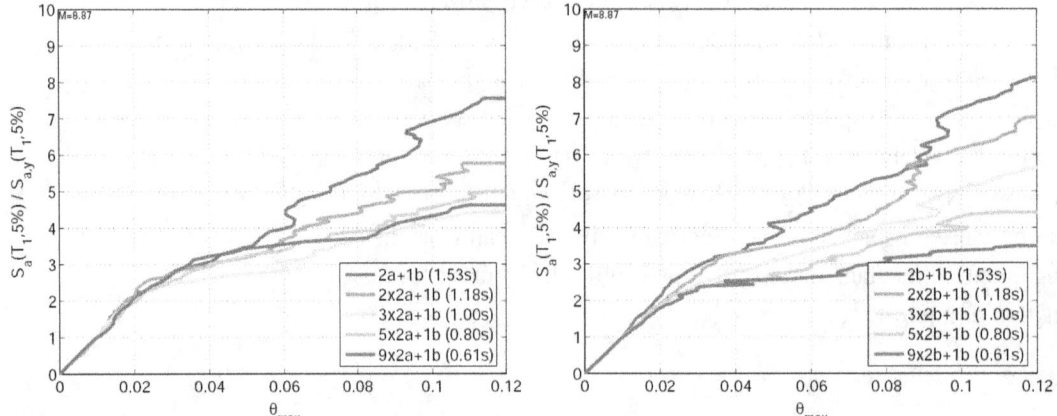

Figure C-3   Median IDA curves plotted versus the normalized intensity measure $R = S_a(T_1,5\%)/S_{ay}(T_1,5\%)$ for systems Nx2a+1b and Nx2b+1b with mass M=8.87ton.

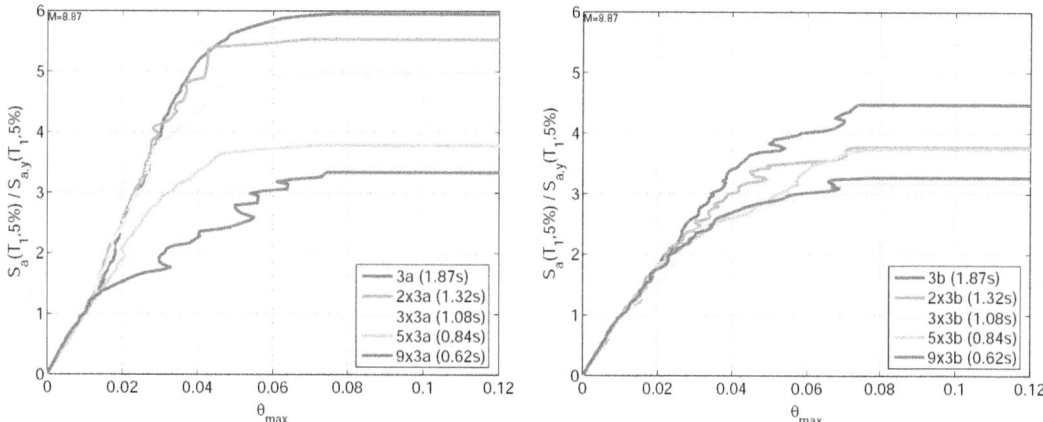

Figure C-4   Median IDA curves plotted versus the normalized intensity measure $R = S_a(T_1, 5\%)/S_{ay}(T_1, 5\%)$ for systems Nx3a and Nx3b with mass M=8.87ton.

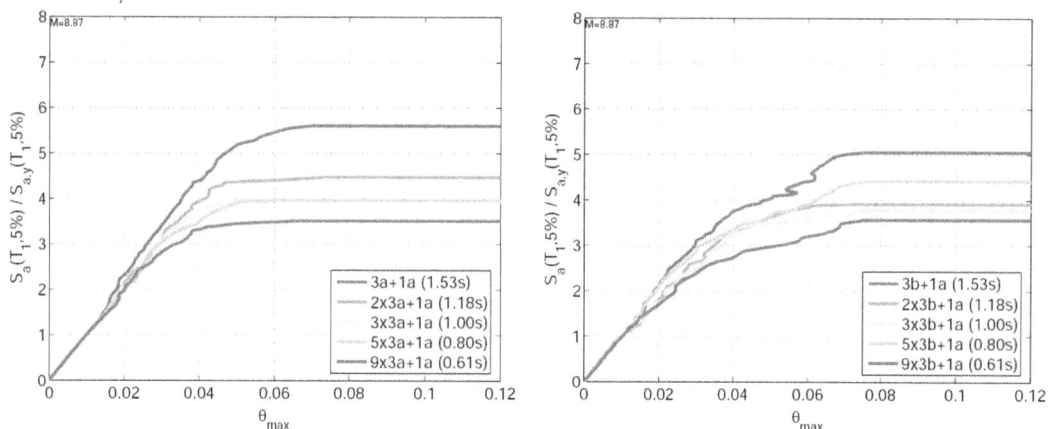

Figure C-5   Median IDA curves plotted versus the normalized intensity measure $R = S_a(T_1, 5\%)/S_{ay}(T_1, 5\%)$ for systems Nx3a+1a and Nx3b+1a with mass M=8.87ton.

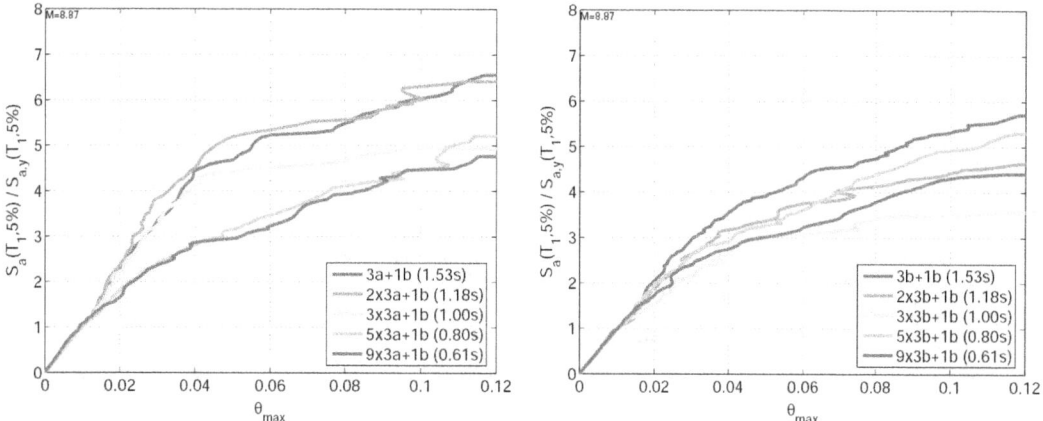

Figure C-6   Median IDA curves plotted versus the normalized intensity measure $R = S_a(T_1, 5\%)/S_{ay}(T_1, 5\%)$ for systems Nx3a+1b and Nx3b+1b with mass M=8.87ton.

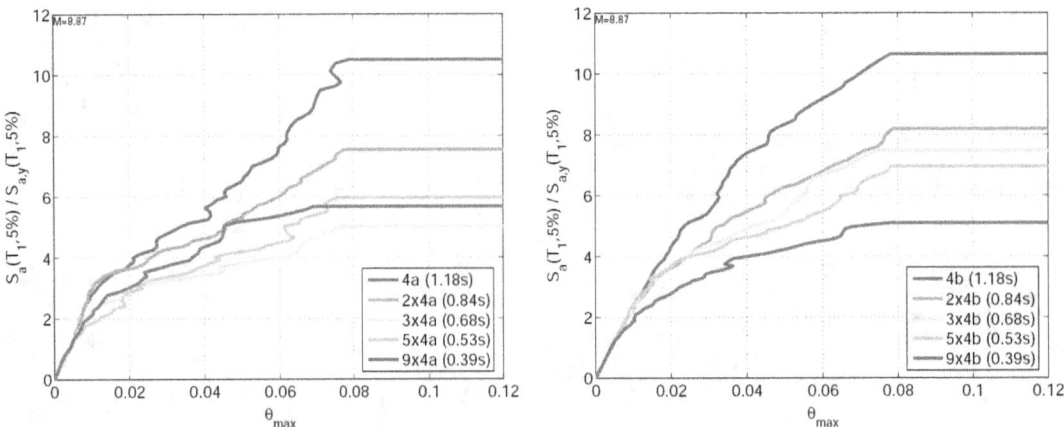

Figure C-7   Median IDA curves plotted versus the normalized intensity measure $R = S_a(T_1,5\%)/S_{ay}(T_1,5\%)$ for systems Nx4a and Nx4b with mass M=8.87ton.

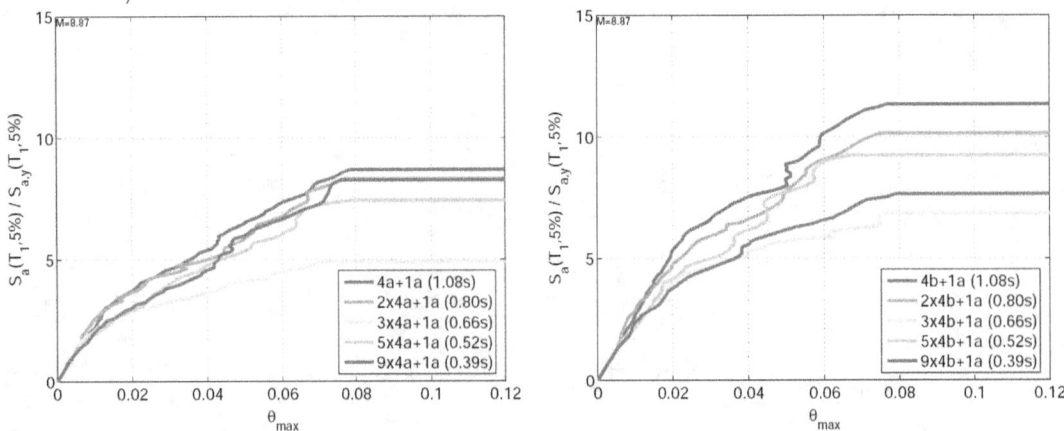

Figure C-8   Median IDA curves plotted versus the normalized intensity measure $R = S_a(T_1,5\%)/S_{ay}(T_1,5\%)$ for systems Nx4a+1a and Nx4b+1a with mass M=8.87ton.

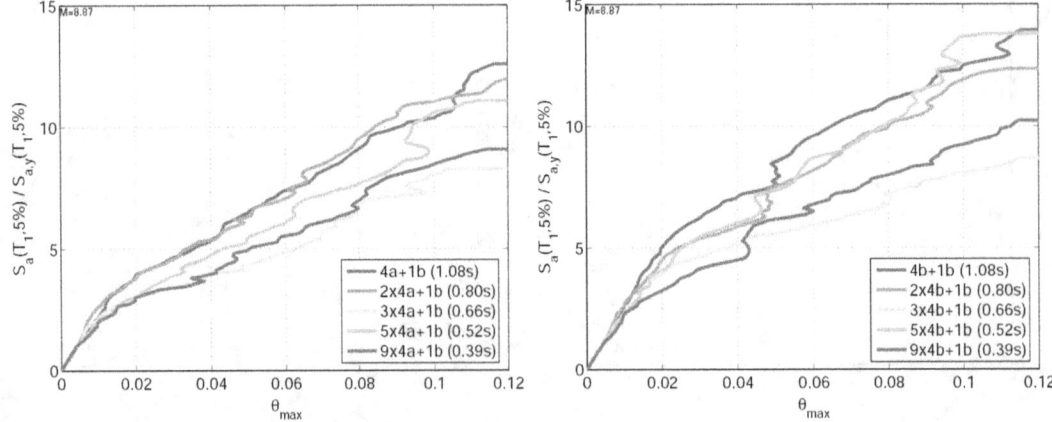

Figure C-9   Median IDA curves plotted versus the normalized intensity measure $R = S_a(T_1,5\%)/S_{ay}(T_1,5\%)$ for systems Nx4a+1b and Nx4b+1b with mass M=8.87ton.

C: Median IDA Curves for Multi-Spring Systems
versus Normalized Intensity Measures

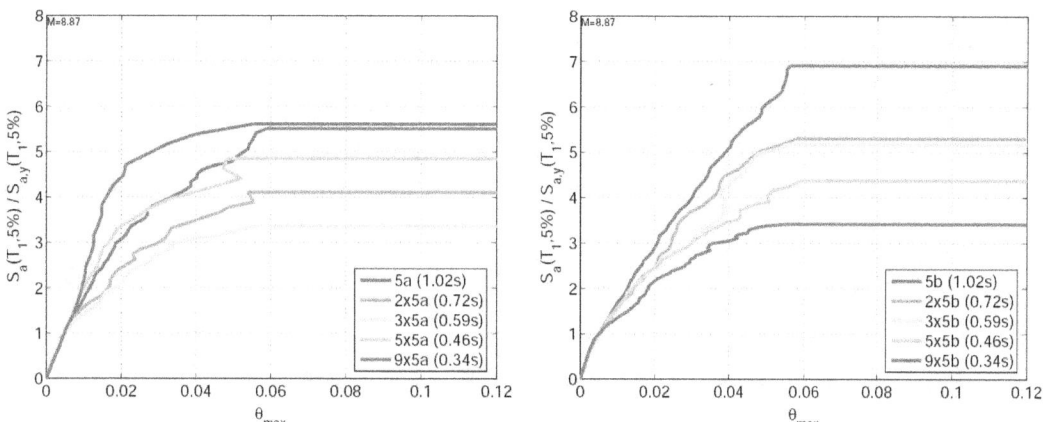

Figure C-10   Median IDA curves plotted versus the normalized intensity measure $R = S_a(T_1, 5\%)/S_{ay}(T_1, 5\%)$ for systems Nx5a and Nx5b with mass M=8.87ton.

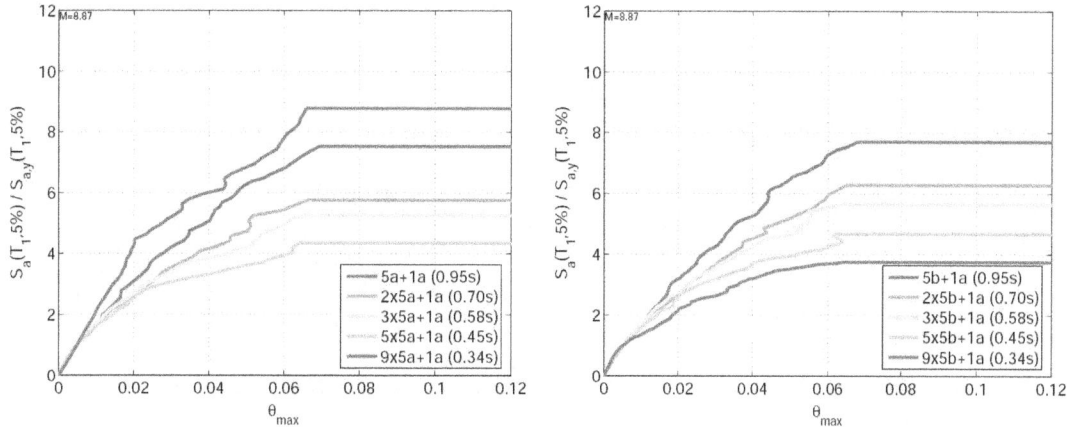

Figure C-11   Median IDA curves plotted versus the normalized intensity measure $R = S_a(T_1, 5\%)/S_{ay}(T_1, 5\%)$ for systems Nx5a+1a and Nx5b+1a with mass M=8.87ton.

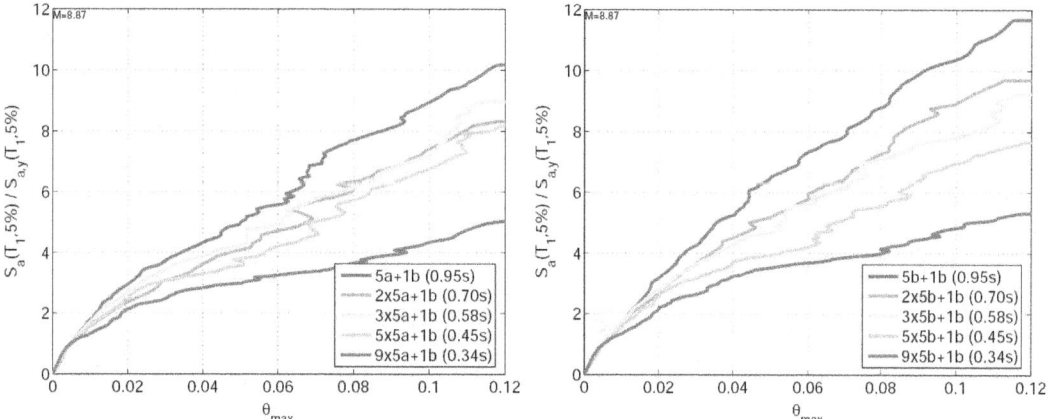

Figure C-12   Median IDA curves plotted versus the normalized intensity measure $R = S_a(T_1, 5\%)/S_{ay}(T_1, 5\%)$ for systems Nx5a+1b and Nx5b+1b with mass M=8.87ton.

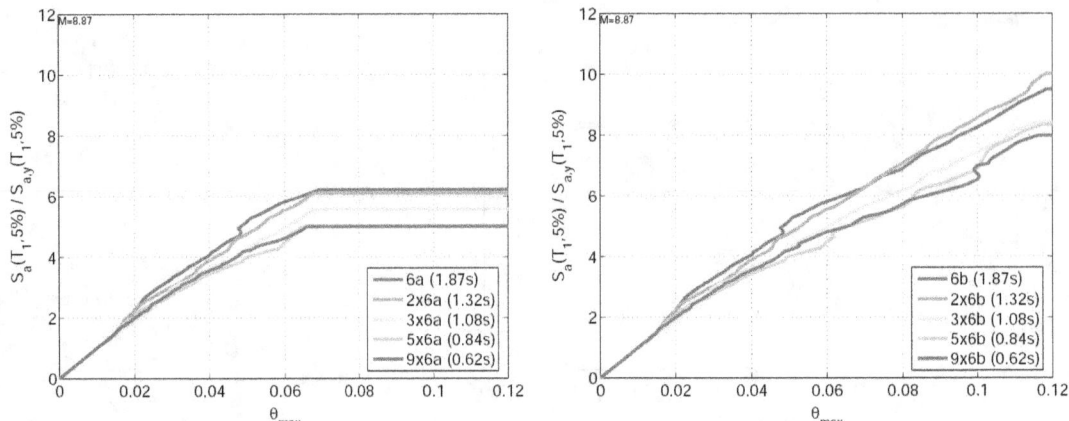

Figure C-13   Median IDA curves plotted versus the normalized intensity measure $R = S_a(T_1,5\%)/S_{a,y}(T_1,5\%)$ for systems Nx6a and Nx6b with mass M=8.87ton.

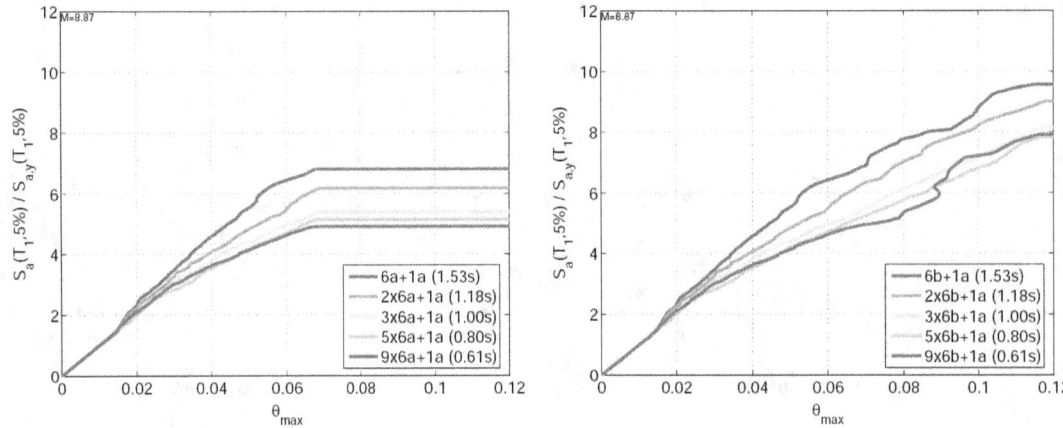

Figure C-14   Median IDA curves plotted versus the normalized intensity measure $R = S_a(T_1,5\%)/S_{a,y}(T_1,5\%)$ for systems Nx6a+1a and Nx6b+1a with mass M=8.87ton.

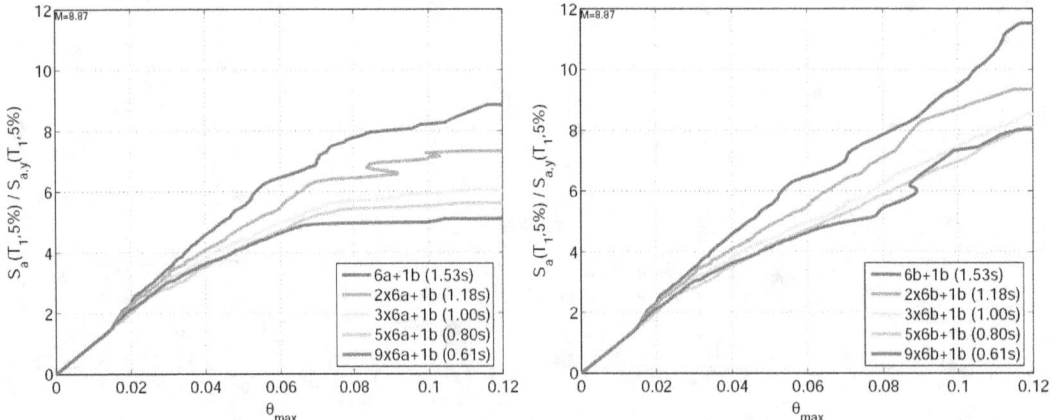

Figure C-15   Median IDA curves plotted versus the normalized intensity measure $R = S_a(T_1,5\%)/S_{a,y}(T_1,5\%)$ for systems Nx6a+1b and Nx6b+1b with mass M=8.87ton.

C: Median IDA Curves for Multi-Spring Systems   **FEMA P440A**
                                        versus Normalized Intensity Measures

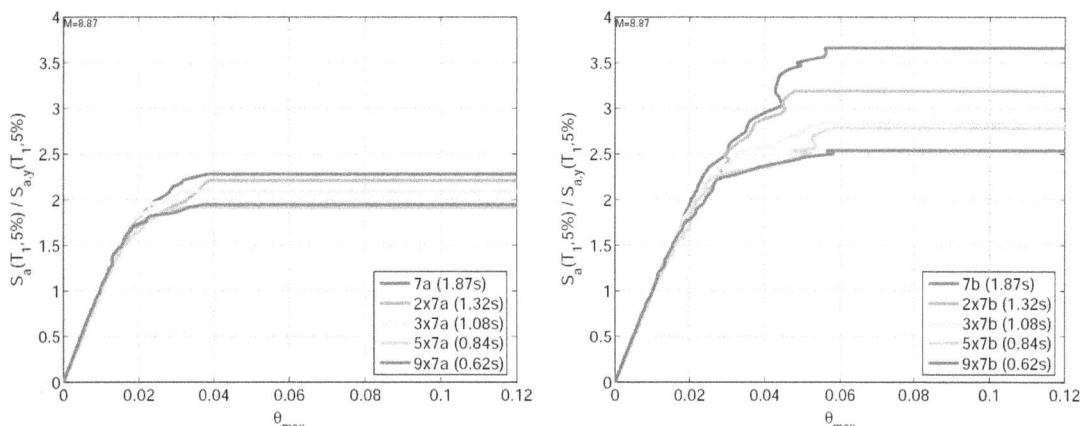

Figure C-16   Median IDA curves plotted versus the normalized intensity measure $R = S_a(T_1,5\%)/S_{a,y}(T_1,5\%)$ for systems Nx7a and Nx7b with mass M=8.87ton.

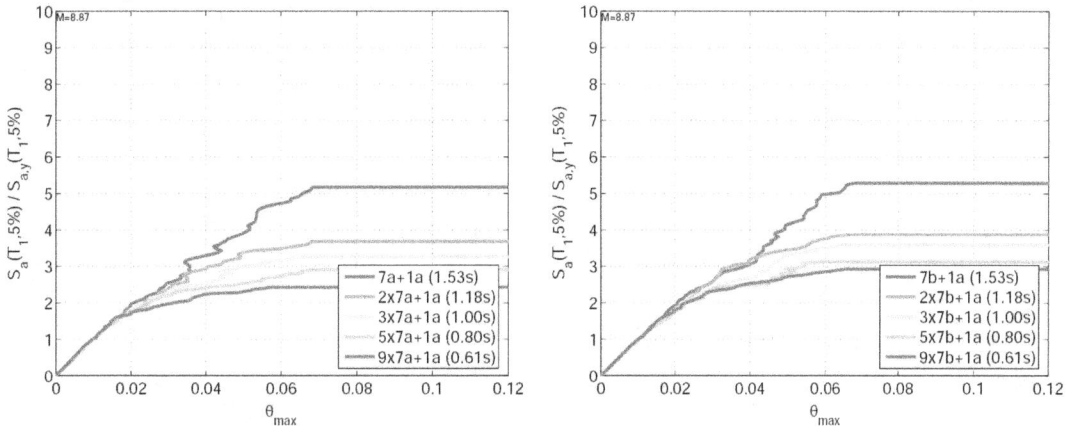

Figure C-17   Median IDA curves plotted versus the normalized intensity measure $R = S_a(T_1,5\%)/S_{a,y}(T_1,5\%)$ for systems Nx7a+1a and Nx7b+1a with mass M=8.87ton.

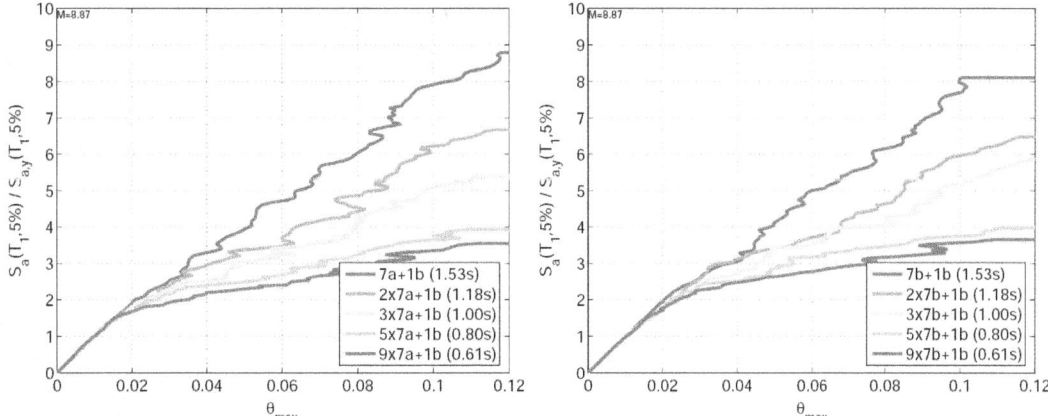

Figure C-18   Median IDA curves plotted versus the normalized intensity measure $R = S_a(T_1,5\%)/S_{a,y}(T_1,5\%)$ for systems Nx7a+1b and Nx7b+1b with mass M=8.87ton.

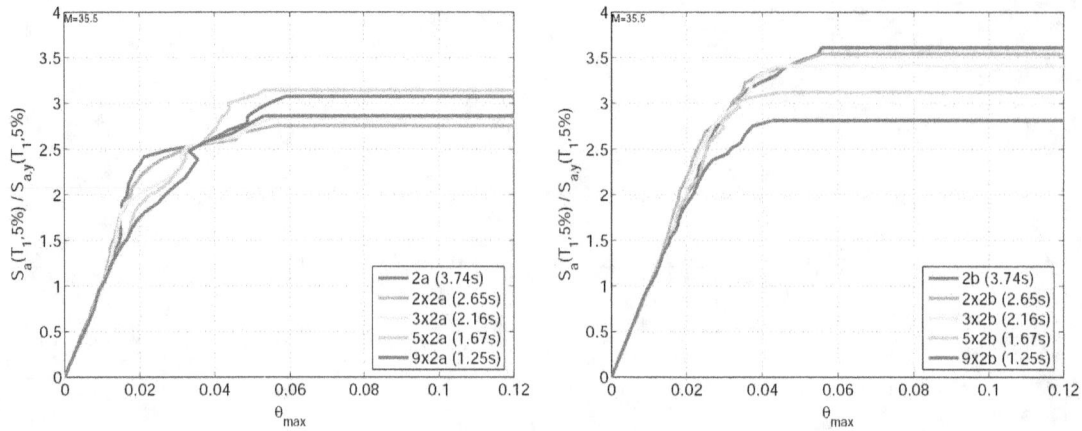

Figure C-19. Median IDA curves plotted versus the normalized intensity measure $R = S_a(T_1,5\%)/S_{ay}(T_1,5\%)$ for systems Nx2a and Nx2b with mass M=35.46ton.

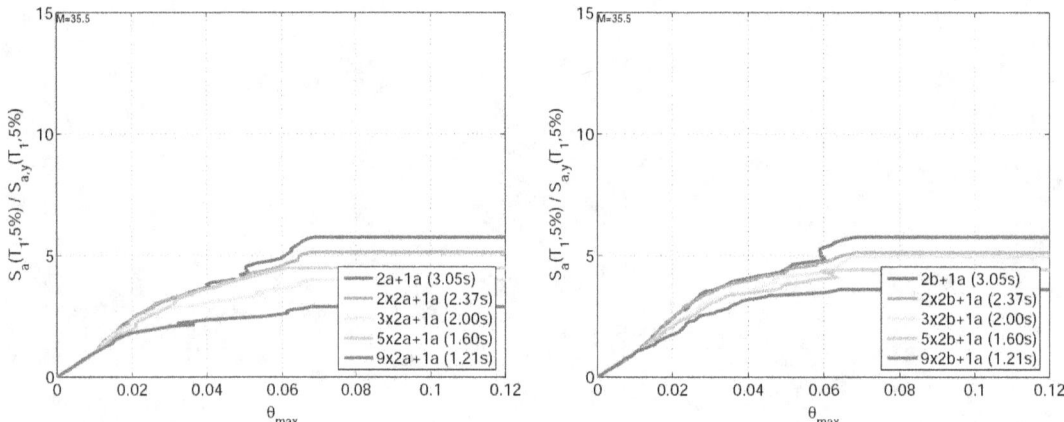

Figure C-20   Median IDA curves plotted versus the normalized intensity measure $R = S_a(T_1,5\%)/S_{ay}(T_1,5\%)$ for systems Nx2a+1a and Nx2b+1a with mass M=35.46ton.

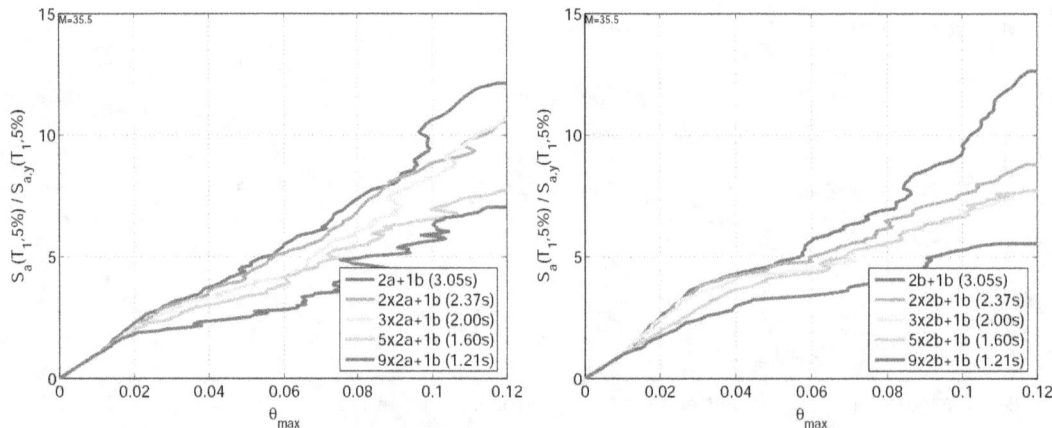

Figure C-21   Median IDA curves plotted versus the normalized intensity measure $R = S_a(T_1,5\%)/S_{ay}(T_1,5\%)$ for systems Nx2a+1b and Nx2b+1b with mass M=35.46ton.

C: Median IDA Curves for Multi-Spring Systems
versus Normalized Intensity Measures

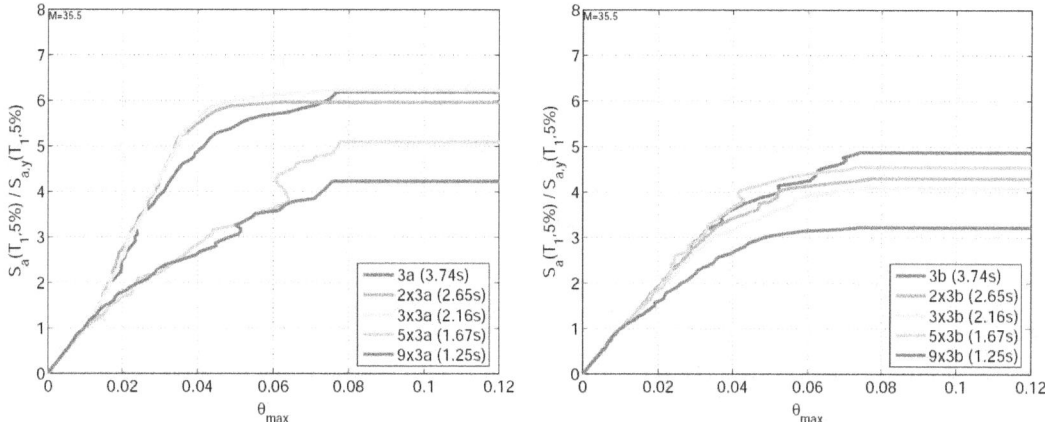

Figure C-22   Median IDA curves plotted versus the normalized intensity measure $R = S_a(T_1,5\%)/S_{ay}(T_1,5\%)$ for systems Nx3a and Nx3b with mass M=35.46ton.

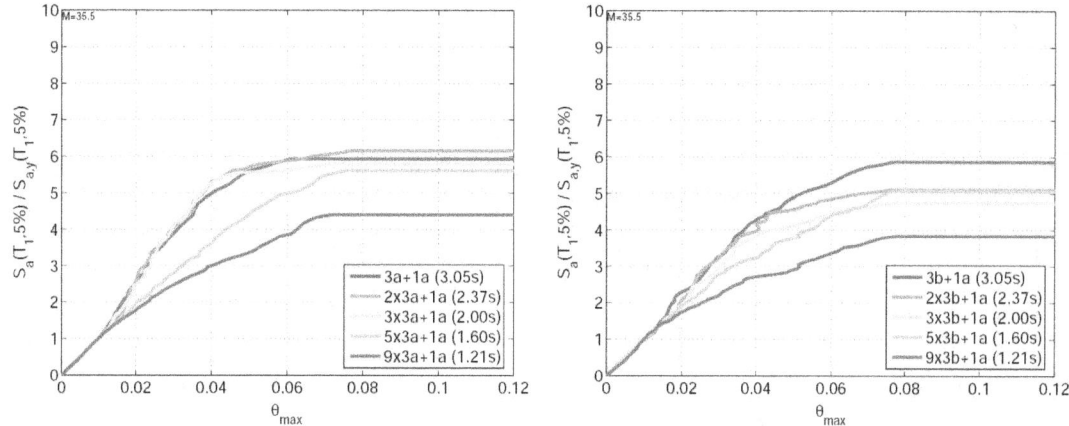

Figure C-23   Median IDA curves plotted versus the normalized intensity measure $R = S_a(T_1,5\%)/S_{ay}(T_1,5\%)$ for systems Nx3a+1a and Nx3b+1a with mass M=35.46ton.

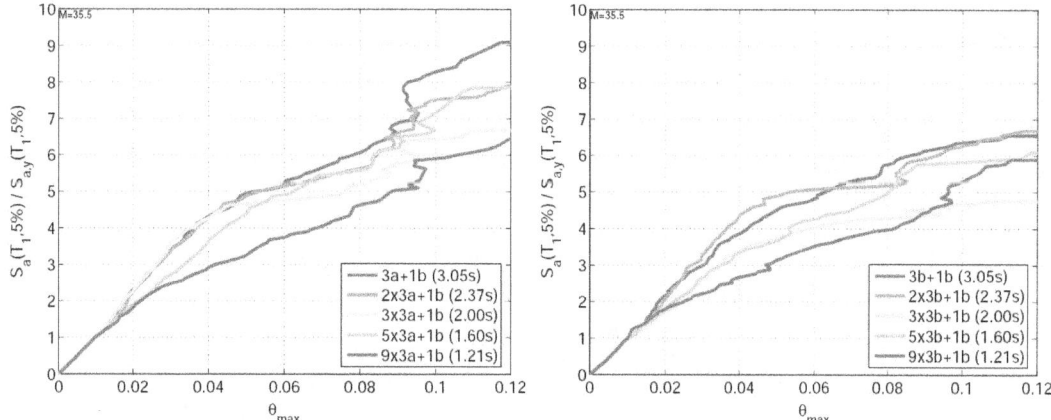

Figure C-24   Median IDA curves plotted versus the normalized intensity measure $R = S_a(T_1,5\%)/S_{ay}(T_1,5\%)$ for systems Nx3a+1b and Nx3b+1b with mass M=35.46ton.

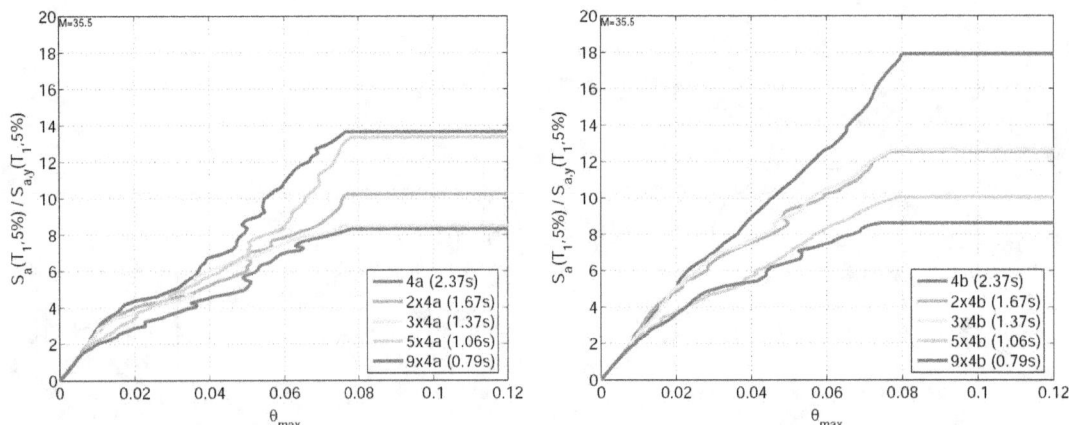

Figure C-25   Median IDA curves plotted versus the normalized intensity measure $R = S_a(T_1,5\%)/S_{ay}(T_1,5\%)$ for systems Nx4a and Nx4b with mass M=35.46ton.

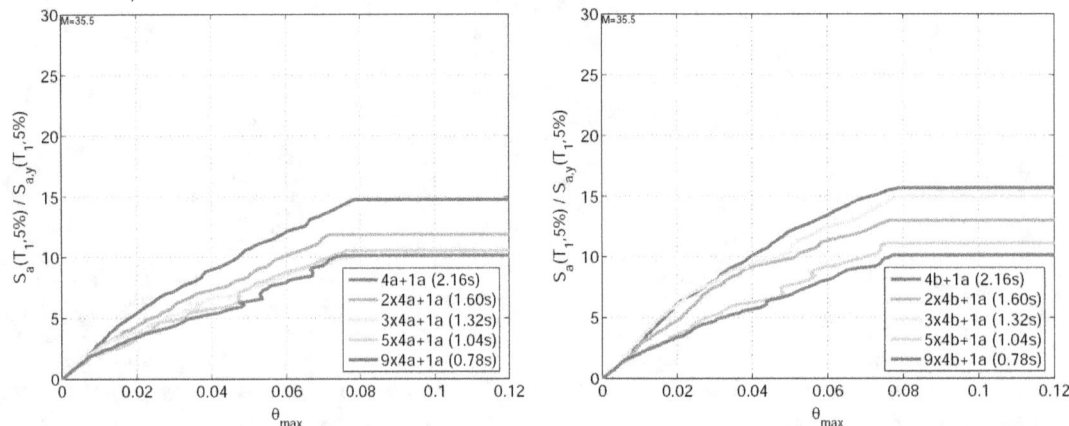

Figure C-26   Median IDA curves plotted versus the normalized intensity measure $R = S_a(T_1,5\%)/S_{ay}(T_1,5\%)$ for systems Nx4a+1a and Nx4b+1a with mass M=35.46ton.

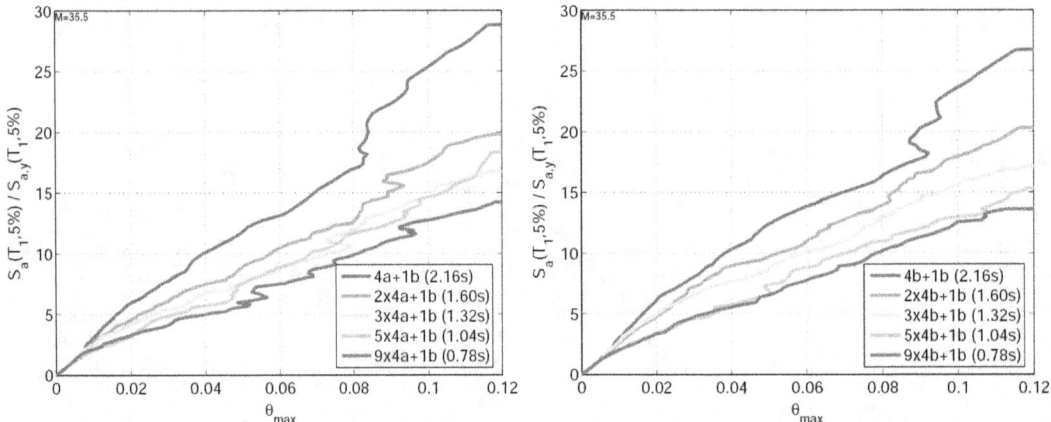

Figure C-27   Median IDA curves plotted versus the normalized intensity measure $R = S_a(T_1,5\%)/S_{ay}(T_1,5\%)$ for systems Nx4a+1b and Nx4b+1b with mass M=35.46ton.

C: Median IDA Curves for Multi-Spring Systems versus Normalized Intensity Measures

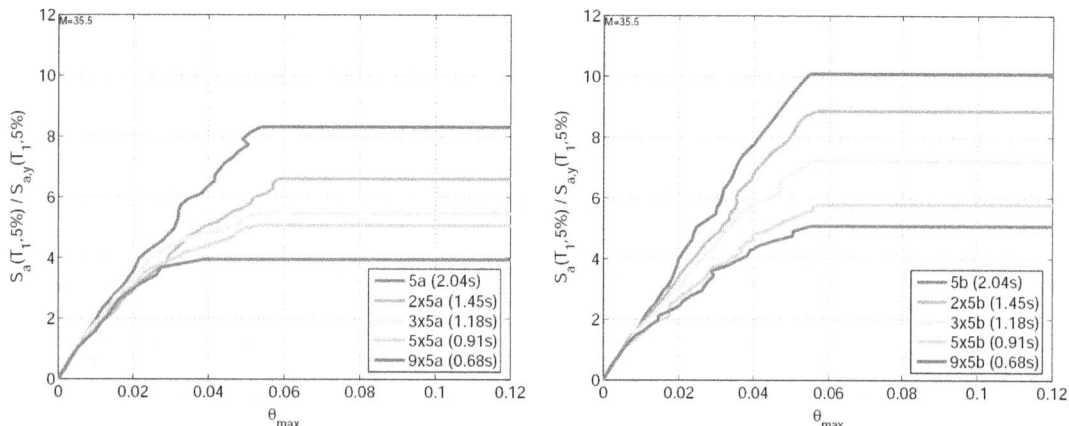

Figure C-28   Median IDA curves plotted versus the normalized intensity measure $R = S_a(T_1,5\%)/S_{ay}(T_1,5\%)$ for systems Nx5a and Nx5b with mass M=35.46ton.

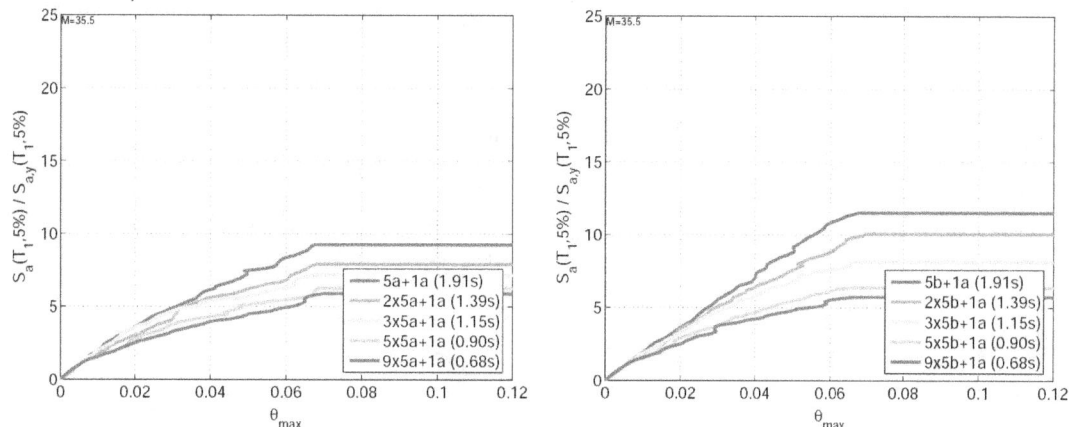

Figure C-29   Median IDA curves plotted versus the normalized intensity measure $R = S_a(T_1,5\%)/S_{ay}(T_1,5\%)$ for systems Nx5a+1a and Nx5b+1a with mass M=35.46ton.

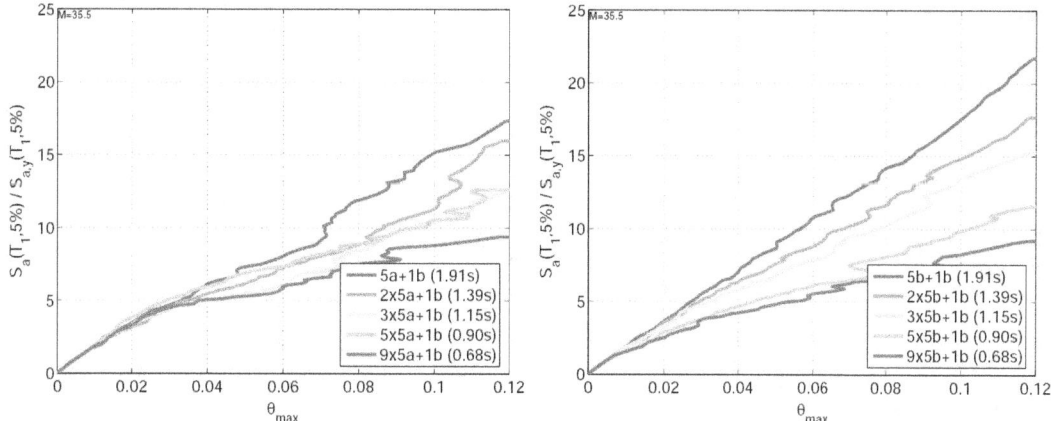

Figure C-30   Median IDA curves plotted versus the normalized intensity measure $R = S_a(T_1,5\%)/S_{ay}(T_1,5\%)$ for systems Nx5a+1b and Nx5b+1b with mass M=35.46ton.

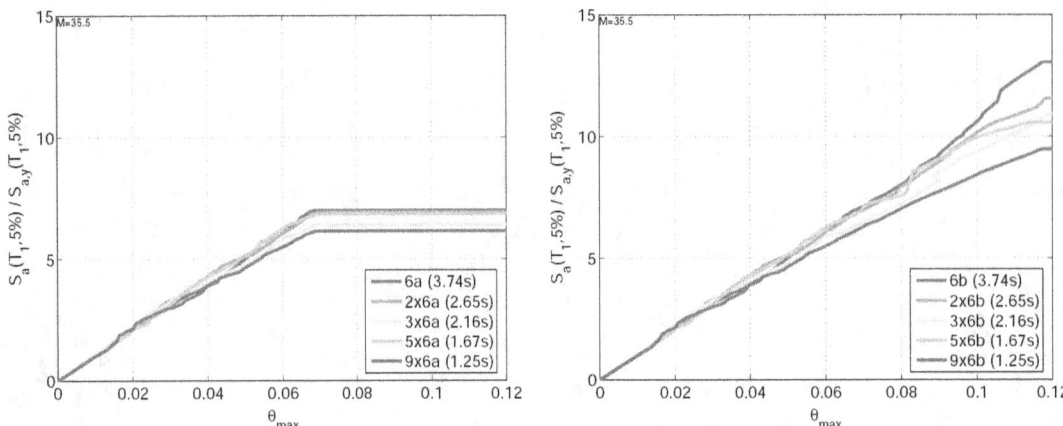

Figure C-31   Median IDA curves plotted versus the normalized intensity measure $R = S_a(T_1,5\%)/S_{a,y}(T_1,5\%)$ for systems Nx6a and Nx6b with mass M=35.46ton.

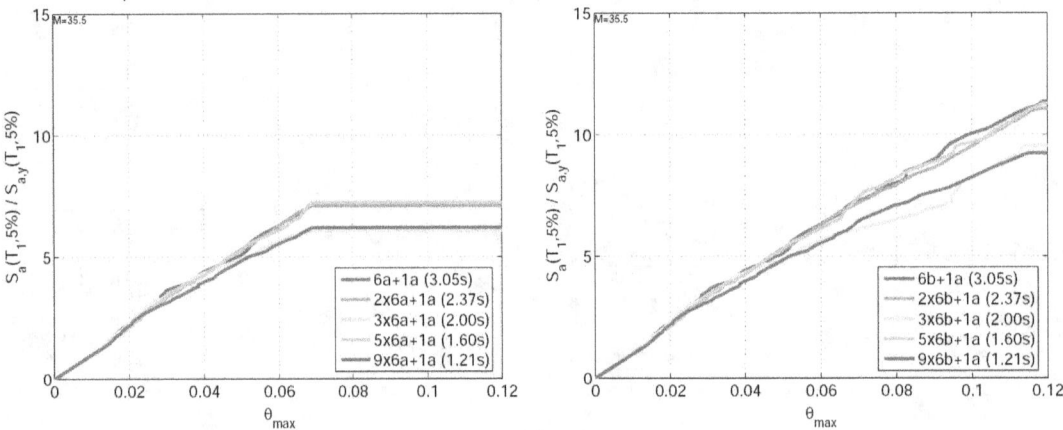

Figure C-32   Median IDA curves plotted versus the normalized intensity measure $R = S_a(T_1,5\%)/S_{a,y}(T_1,5\%)$ for systems Nx6a+1a and Nx6b+1a with mass M=35.46ton.

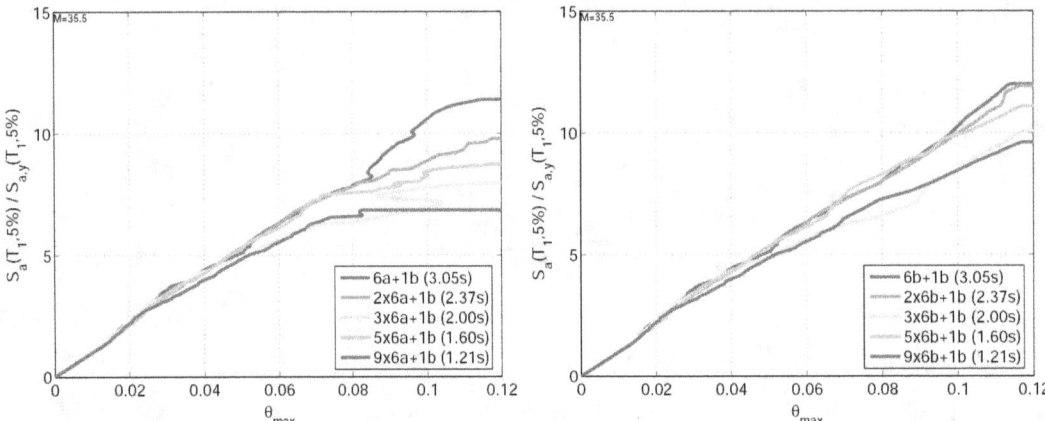

Figure C-33   Median IDA curves plotted versus the normalized intensity measure $R = S_a(T_1,5\%)/S_{ay}(T_1,5\%)$ for systems Nx6a+1b and Nx6b+1b with mass M=35.46ton.

**C: Median IDA Curves for Multi-Spring Systems
versus Normalized Intensity Measures**

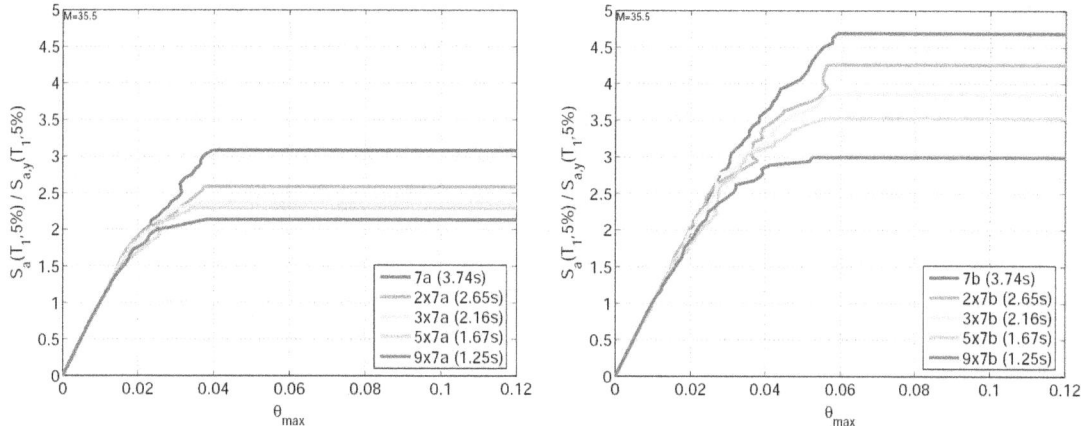

Figure C-34   Median IDA curves plotted versus the normalized intensity measure $R = S_a(T_1,5\%)/S_{ay}(T_1,5\%)$ for systems Nx7a and Nx7b with mass M=35.46ton.

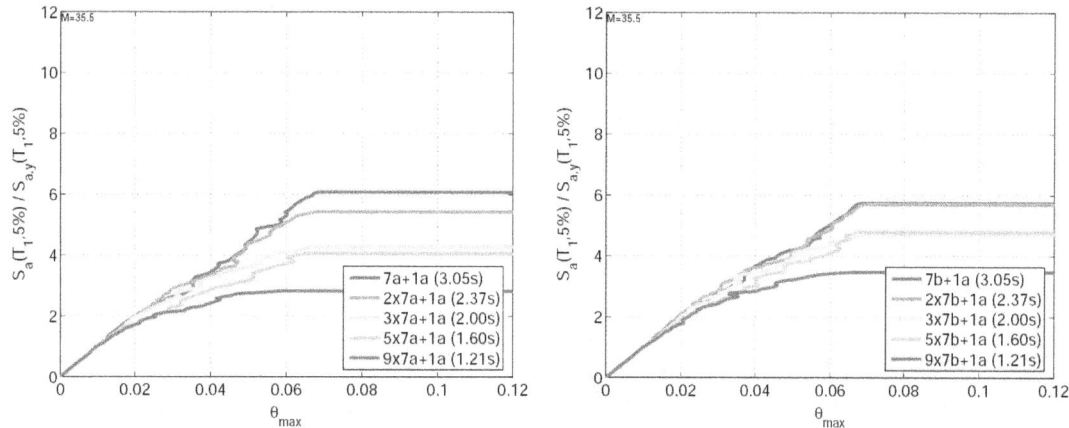

Figure C-35   Median IDA curves plotted versus the normalized intensity measure $R = S_a(T_1,5\%)/S_{ay}(T_1,5\%)$ for systems Nx7a+1a and Nx7b+1a with mass M=35.46ton.

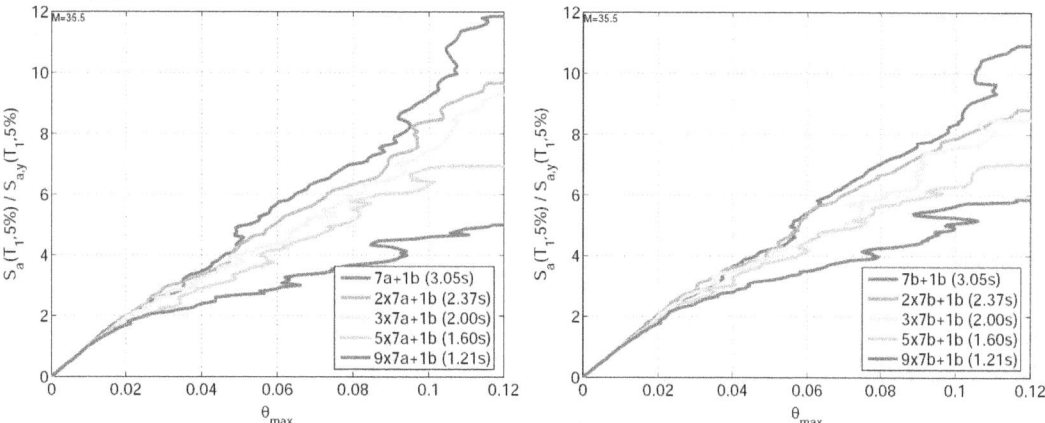

Figure C-36   Median IDA curves plotted versus the normalized intensity measure $R = S_a(T_1,5\%)/S_{ay}(T_1,5\%)$ for systems Nx7a+1b and Nx7b+1b with mass M=35.46ton.

Appendix D

Median IDA Curves for Multi-Spring Systems versus Non-Normalized Intensity Measures

This appendix contains non-normalized plots of median incremental dynamic analysis (IDA) curves from focused analytical studies on multi-spring single-degree-of-freedom (SDOF) systems. All systems are composed of two springs representing a primary lateral-force-resisting system and a secondary gravity system with the characteristics described in Chapter 3. Multi-spring systems carry a designation of "NxJa+1a" or "NxJa+1b" where "N" is the peak strength multiplier (N = 1, 2, 3, 5, or 9), "J" is the lateral-force-resisting spring number (J = 2, 3, 4, 5, 6, or 7), and 1a or 1b is the gravity system identifier. In the figures, the vertical axis is one of two ground motion intensities IM = $S_a(1s,5\%)$ or $S_a(2s,5\%)$, which are not normalized, and the horizontal axis is the maximum story drift ratio, θ_{max}, in radians. The period of vibration for each system is indicated in parentheses.

D.1 Visualization Tool

Given the large volume of analytical data, customized algorithms were developed for post-processing, statistical analysis, and visualization of results. The accompanying CD includes an electronic visualization tool that was developed to view results of multi-spring studies. The tool is a Microsoft Excel based application with a user-interface that accesses a database of all available multi-spring data. By selecting a desired spring combination ("NxJa+1a" or "NxJa+1b"), stiffness level (stiff or flexible), and intensity measure (normalized or non-normalized), users can view the resulting quantile (median, 16th, and 84th percentile) IDA curves for the combination of interest.

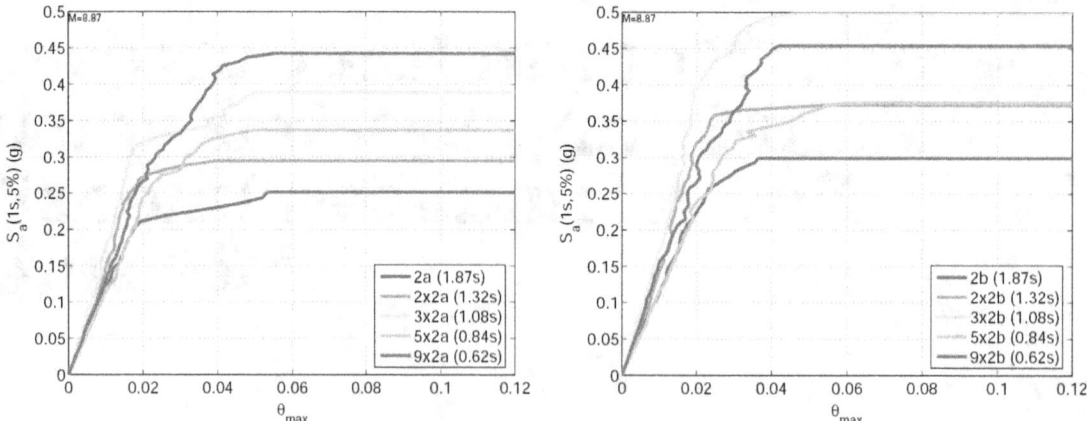

Figure D-1   Median IDA curves plotted versus the intensity measure $S_a(1s,5\%)$ for systems Nx2a and Nx2b with mass M=8.87ton.

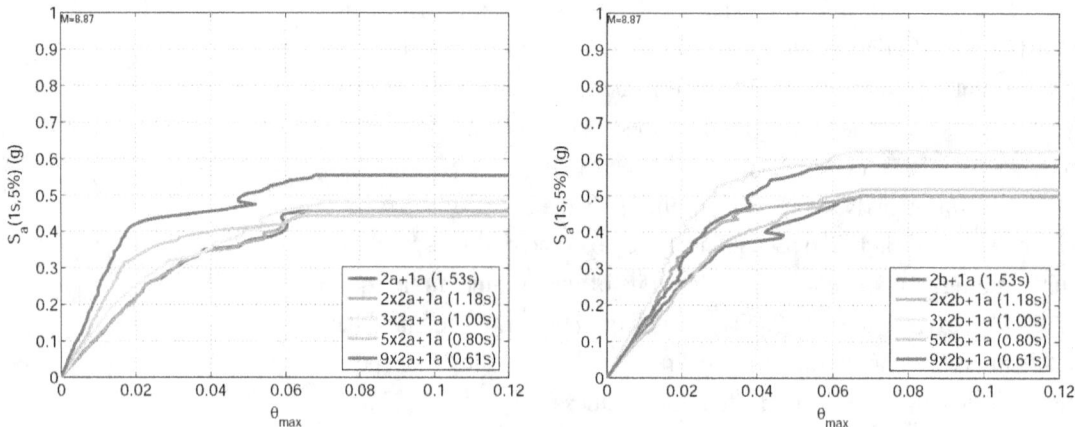

Figure D-2   Median IDA curves plotted versus the intensity measure $S_a(1s,5\%)$ for systems Nx2a+1a and Nx2b+1a with mass M=8.87ton.

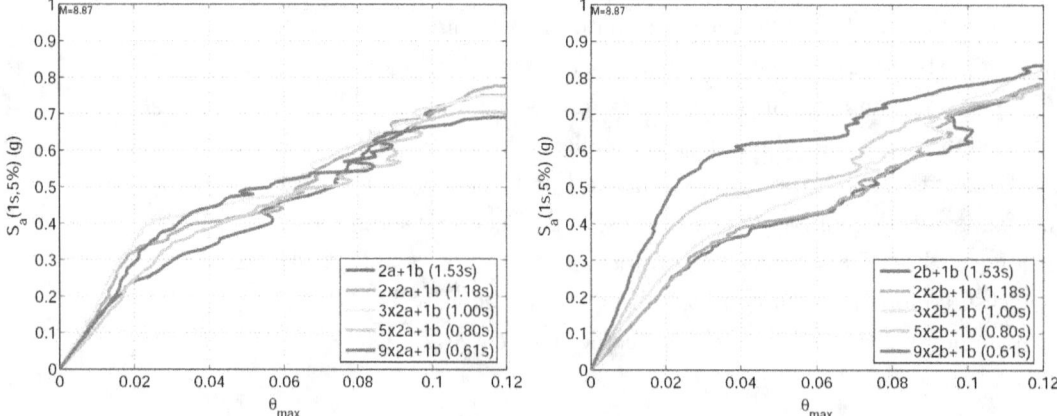

Figure D-3   Median IDA curves plotted versus the intensity measure $S_a(1s,5\%)$ for systems Nx2a+1b and Nx2b+1b with mass M=8.87ton.

D: Median IDA Curves for Multi-Spring Systems versus Non-Normalized Intensity Measures

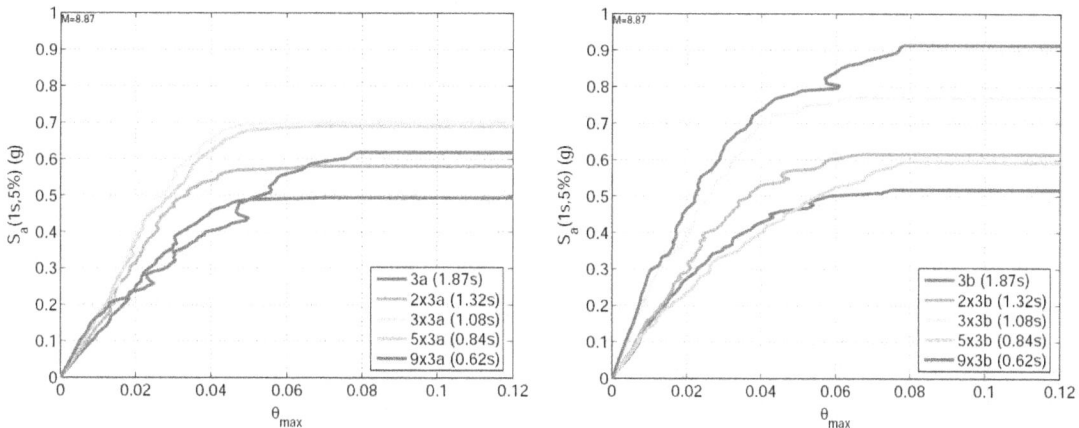

Figure D-4   Median IDA curves plotted versus the intensity measure $S_a(1s,5\%)$ for systems Nx3a and Nx3b with mass M=8.87ton.

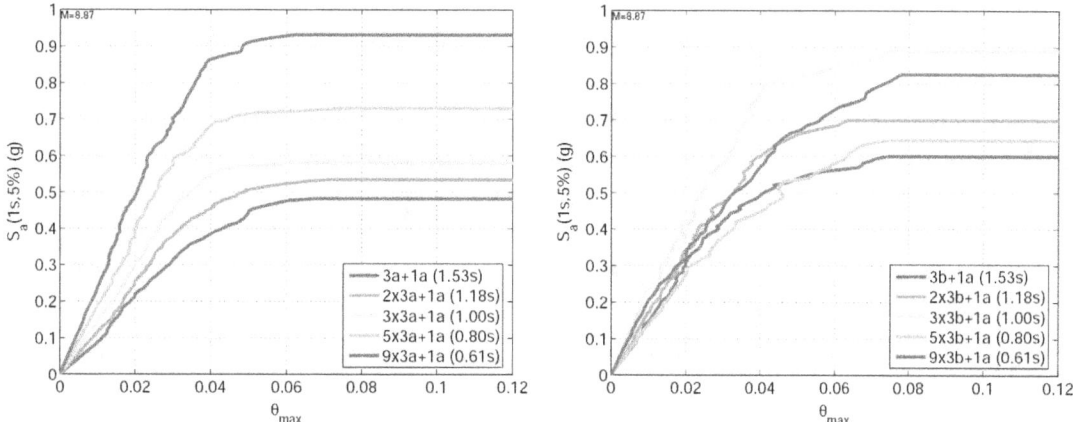

Figure D-5   Median IDA curves plotted versus the intensity measure $S_a(1s,5\%)$ for systems Nx3a+1a and Nx3b+1a with mass M=8.87ton.

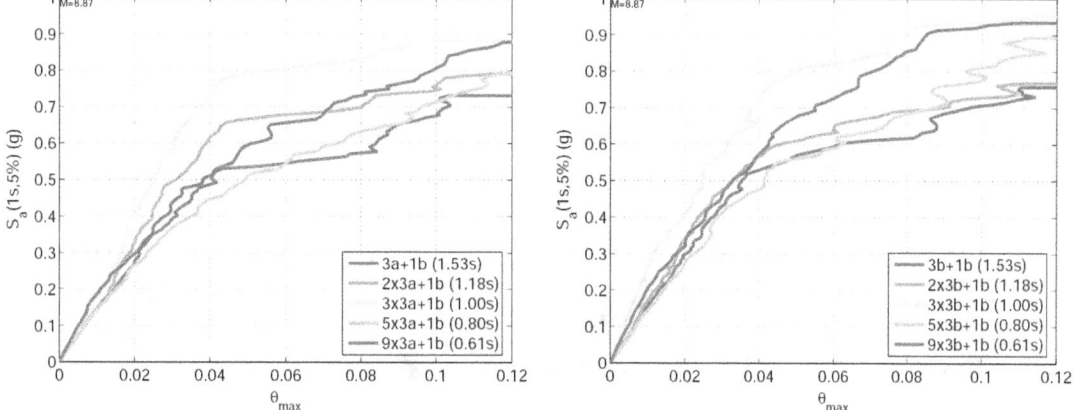

Figure D-6   Median IDA curves plotted versus the intensity measure $S_a(1s,5\%)$ for systems Nx3a+1b and Nx3b+1b with mass M=8.87ton.

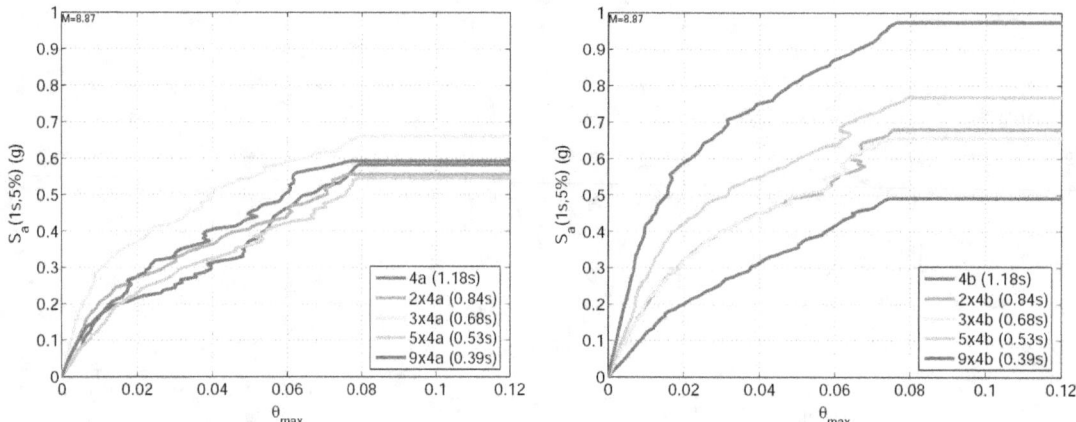

Figure D-7   Median IDA curves plotted versus the intensity measure $S_a(1s,5\%)$ for systems Nx4a and Nx4b with mass M=8.87ton.

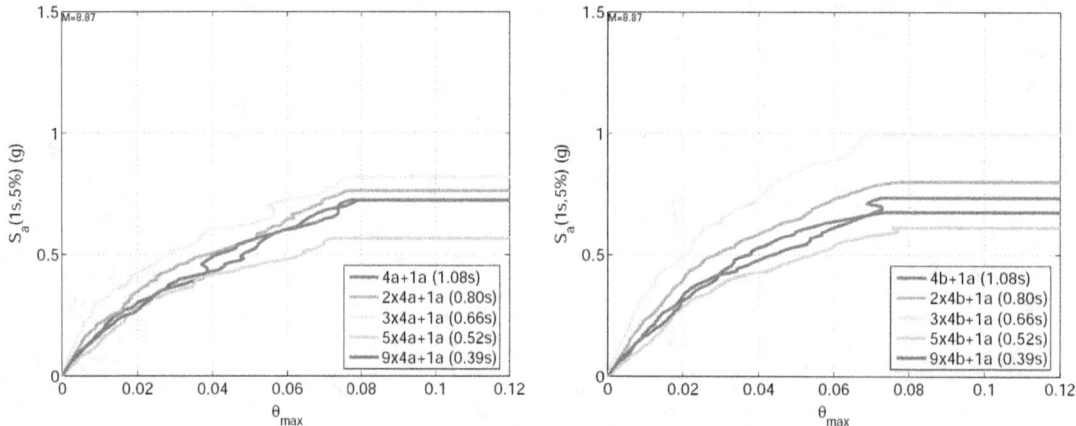

Figure D-8   Median IDA curves plotted versus the intensity measure $S_a(1s,5\%)$ for systems Nx4a+1a and Nx4b+1a with mass M=8.87ton.

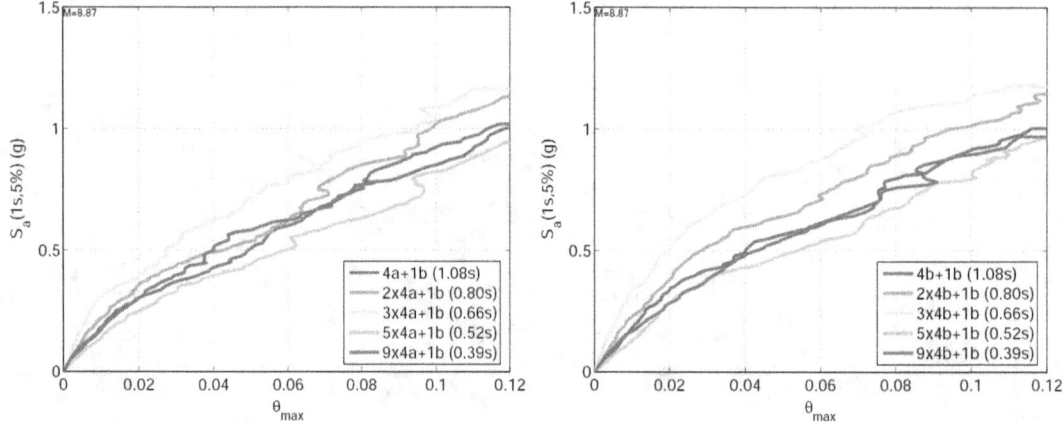

Figure D-9   Median IDA curves plotted versus the intensity measure $S_a(1s,5\%)$ for systems Nx4a+1b and Nx4b+1b with mass M=8.87ton.

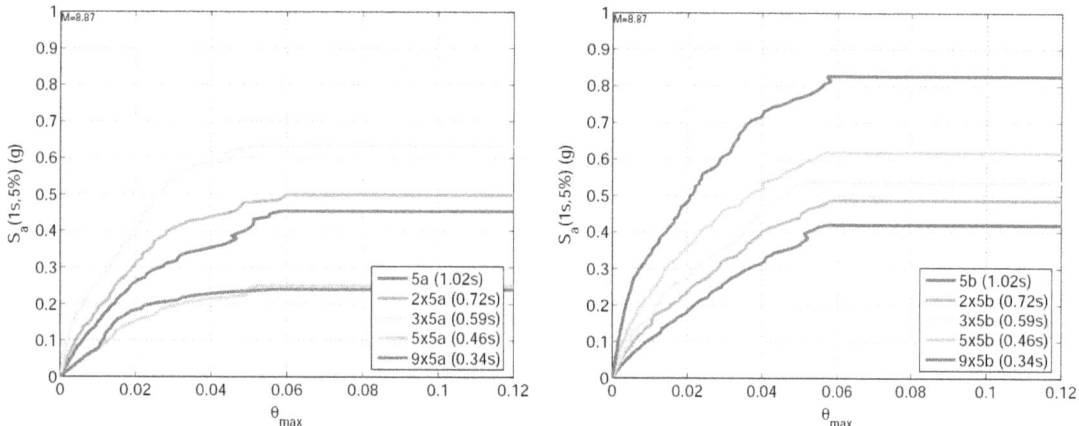

Figure D-10   Median IDA curves plotted versus the intensity measure $S_a(1s,5\%)$ for systems Nx5a and Nx5b with mass M=8.87ton.

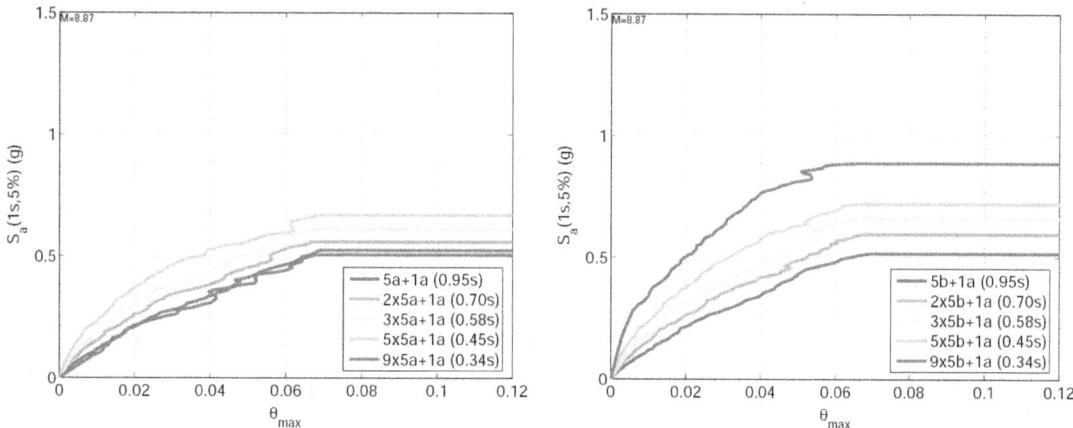

Figure D-11   Median IDA curves plotted versus the intensity measure $S_a(1s,5\%)$ for systems Nx5a+1a and Nx5b+1a with mass M=8.87ton.

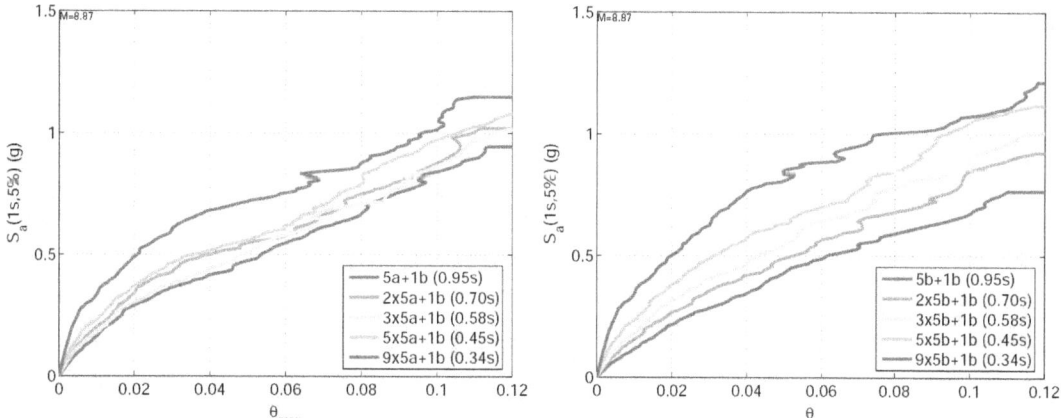

Figure D-12   Median IDA curves plotted versus the intensity measure $S_a(1s,5\%)$ for systems Nx5a+1b and Nx5b+1b with mass M=8.87ton.

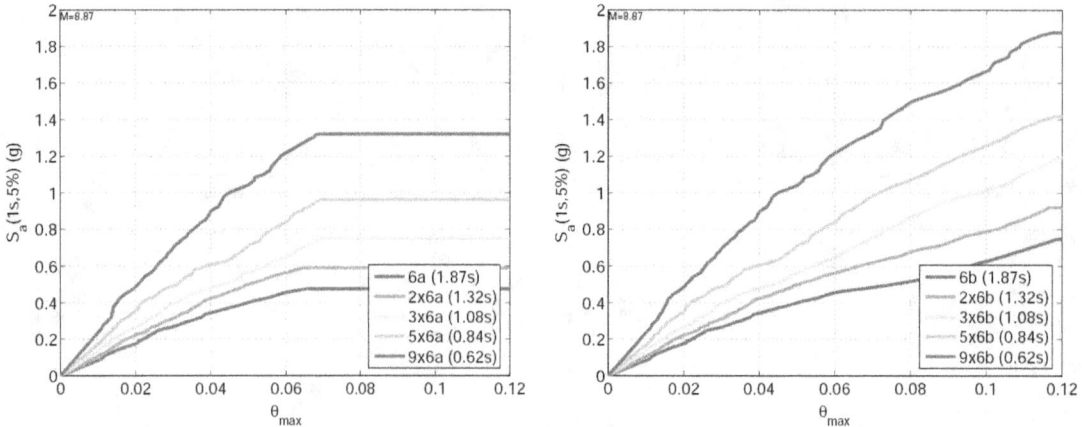

Figure D-13   Median IDA curves plotted versus the intensity measure S_a(1s,5%) for systems Nx6a and Nx6b with mass M=8.87ton.

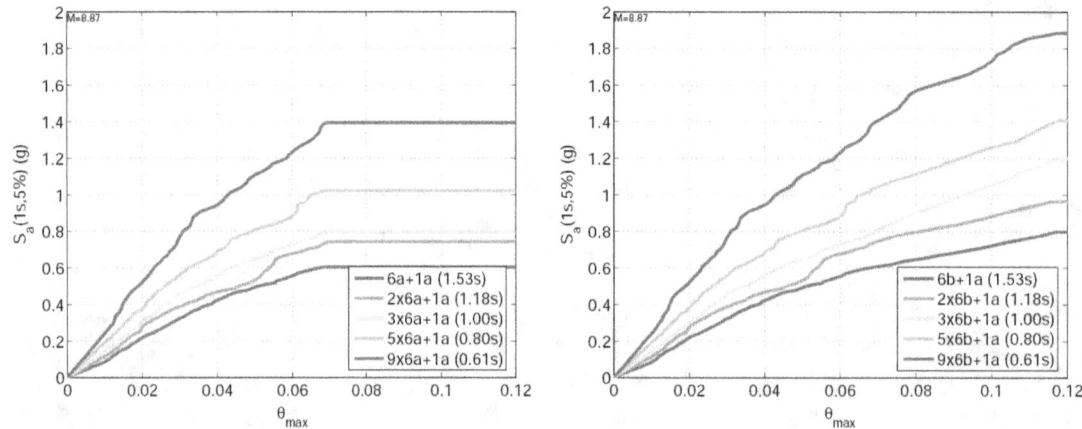

Figure D-14   Median IDA curves plotted versus the intensity measure S_a(1s,5%) for systems Nx6a+1a and Nx6b+1a with mass M=8.87ton.

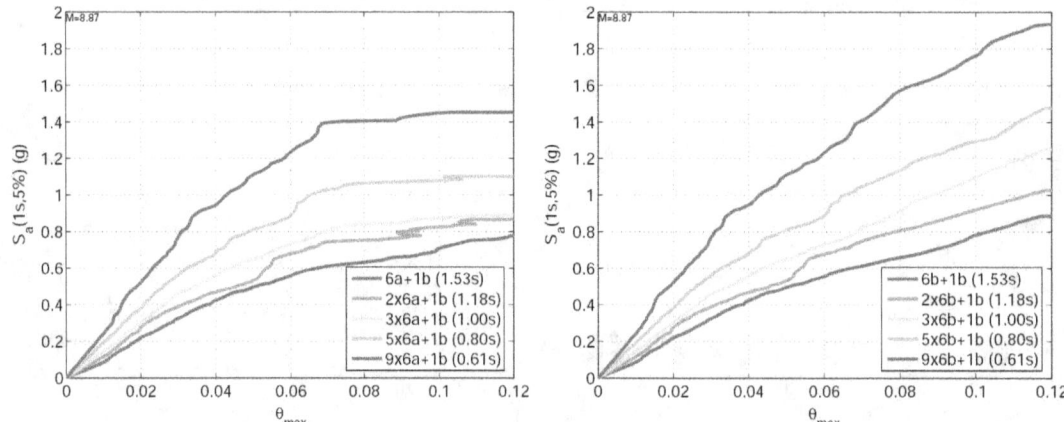

Figure D-15   Median IDA curves plotted versus the intensity measure S_a(1s,5%) for systems Nx6a+1b and Nx6b+1b with mass M=8.87ton.

D: Median IDA Curves for Multi-Spring Systems
versus Non-Normalized Intensity Measures

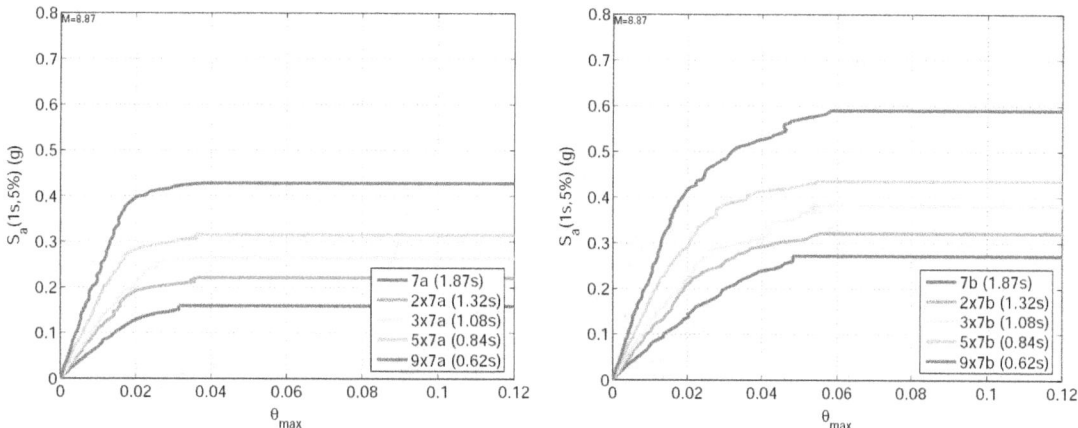

Figure D-16   Median IDA curves plotted versus the intensity measure $S_a(1s,5\%)$ for systems Nx7a and Nx7b with mass M=8.87ton.

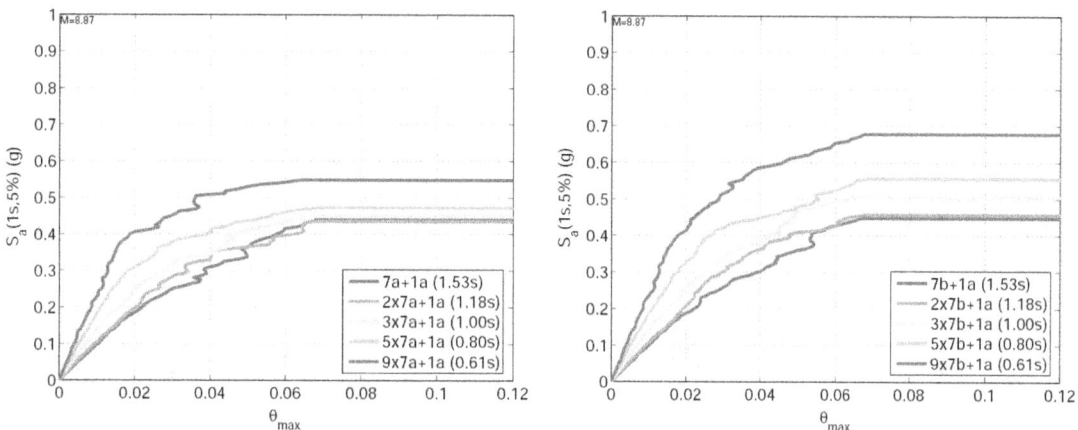

Figure D-17   Median IDA curves plotted versus the intensity measure $S_a(1s,5\%)$ for systems Nx7a+1a and Nx7b+1a with mass M=8.87ton.

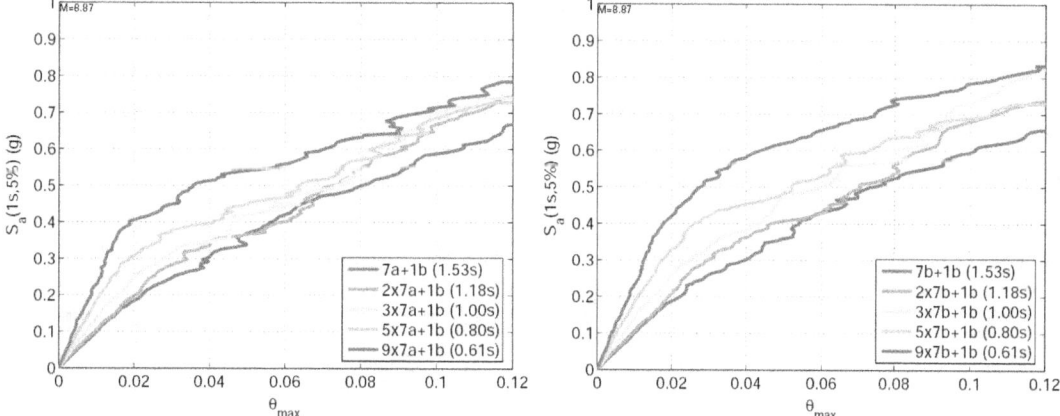

Figure D-18   Median IDA curves plotted versus the intensity measure $S_a(1s,5\%)$ for systems Nx7a+1b and Nx7b+1b with mass M=8.87ton.

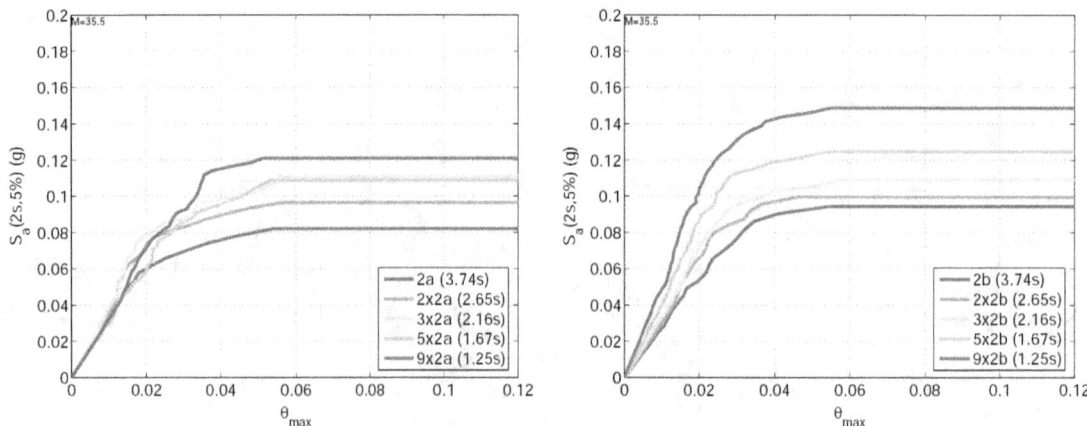

Figure D-19   Median IDA curves plotted versus the intensity measure $S_a(2s,5\%)$ for systems Nx2a and Nx2b with mass $M=35.46$ton.

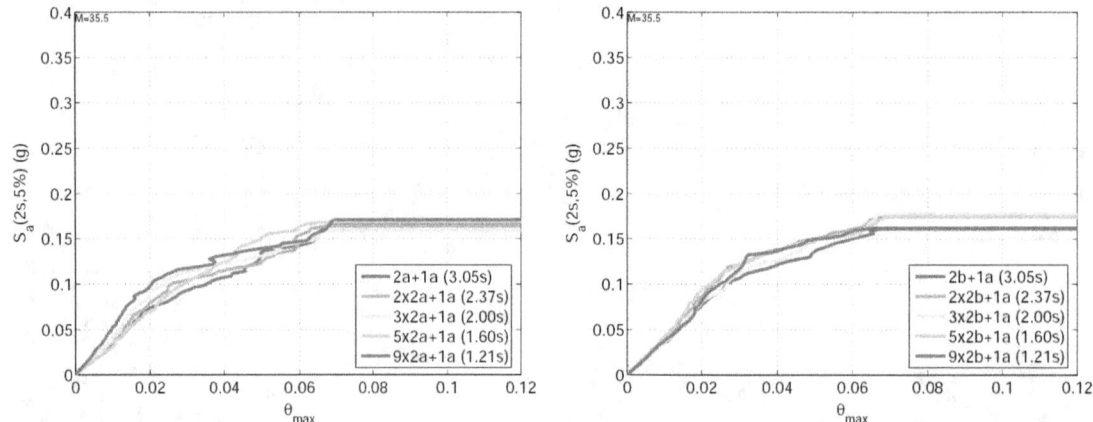

Figure D-20   Median IDA curves plotted versus the intensity measure $S_a(2s,5\%)$ for systems Nx2a+1a and Nx2b+1a with mass $M=35.46$ton.

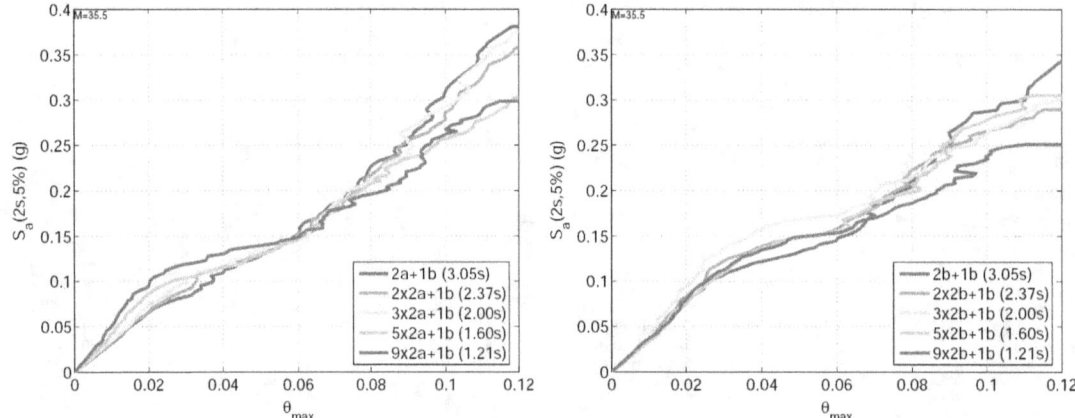

Figure D-21   Median IDA curves plotted versus the intensity measure $S_a(2s,5\%)$ for systems Nx2a+1b and Nx2b+1b with mass $M=35.46$ton.

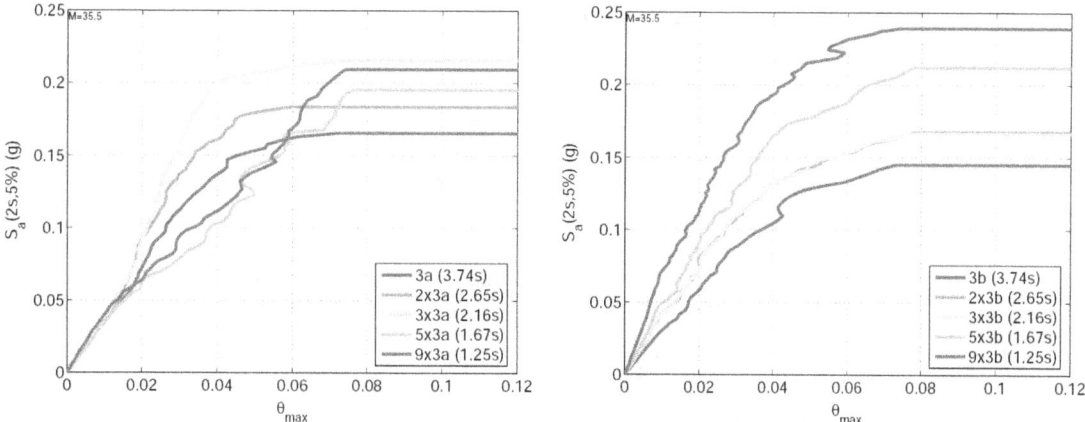

Figure D-22   Median IDA curves plotted versus the intensity measure $S_a(2s,5\%)$ for systems Nx3a and Nx3b with mass M=35.46ton.

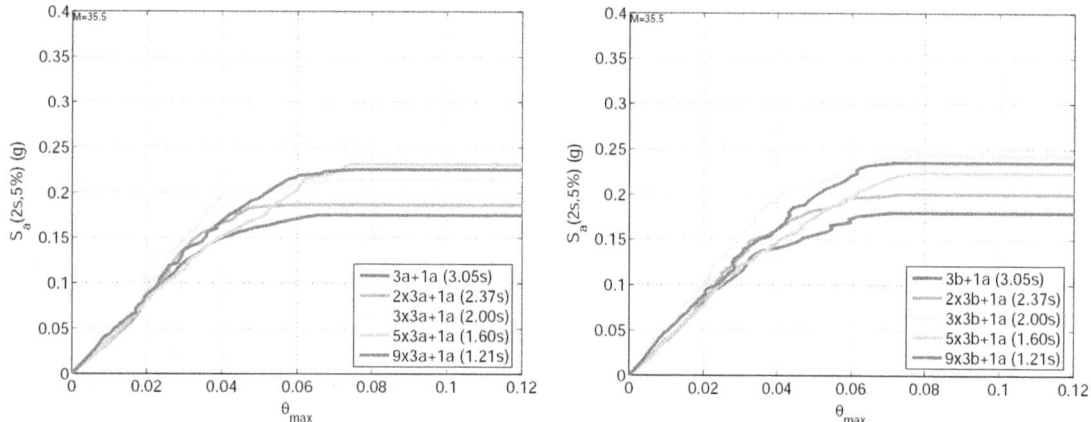

Figure D-23   Median IDA curves plotted versus the intensity measure $S_a(2s,5\%)$ for systems Nx3a+1a and Nx3b+1a with mass M=35.46ton.

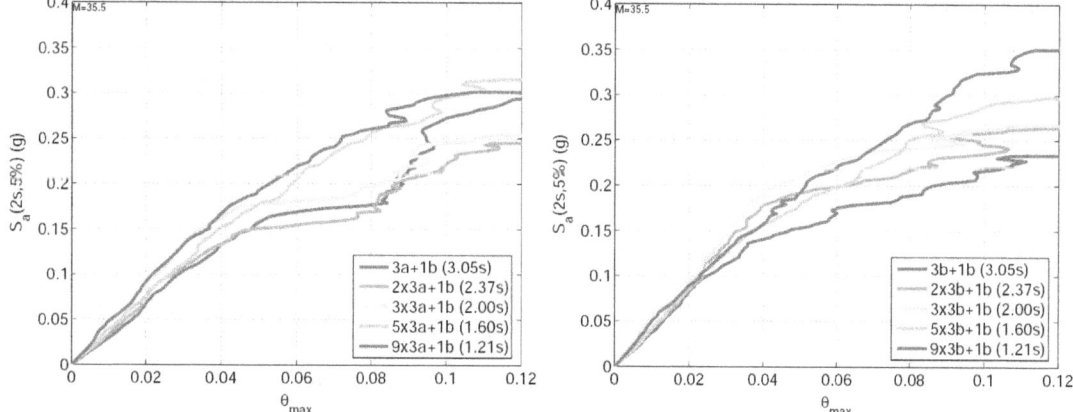

Figure D-24   Median IDA curves plotted versus the intensity measure $S_a(2s,5\%)$ for systems Nx3a+1b and Nx3b+1b with mass M=35.46ton.

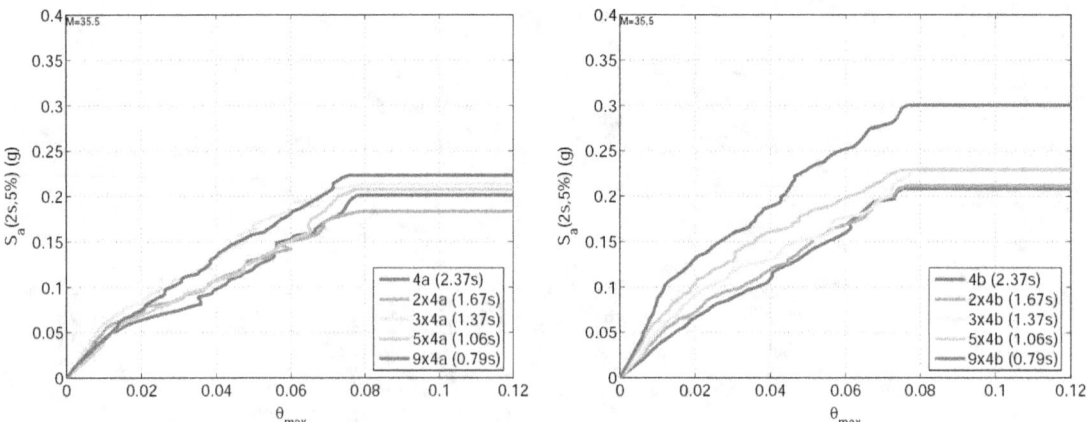

Figure D-25   Median IDA curves plotted versus the intensity measure $S_a(2s,5\%)$ for systems Nx4a and Nx4b with mass M=35.46ton.

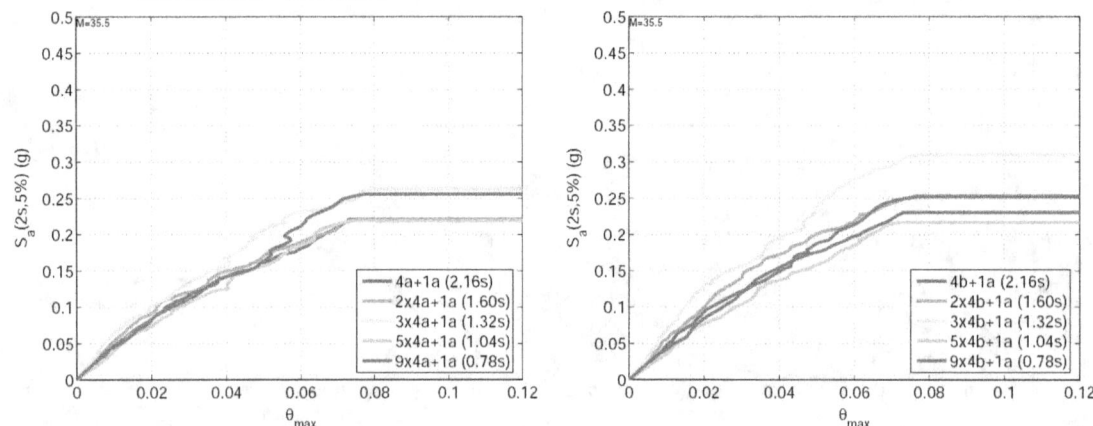

Figure D-26   Median IDA curves plotted versus the intensity measure $S_a(2s,5\%)$ for (systems Nx4a+1a and Nx4b+1a with mass M=35.46ton.

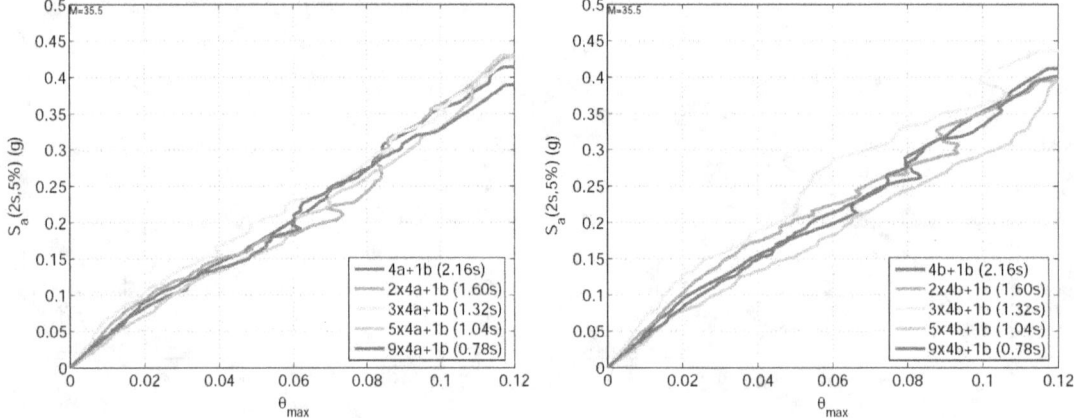

Figure D-27   Median IDA curves plotted versus the intensity measure $S_a(2s,5\%)$ for systems Nx4a+1b and Nx4b+1b with mass M=35.46ton.

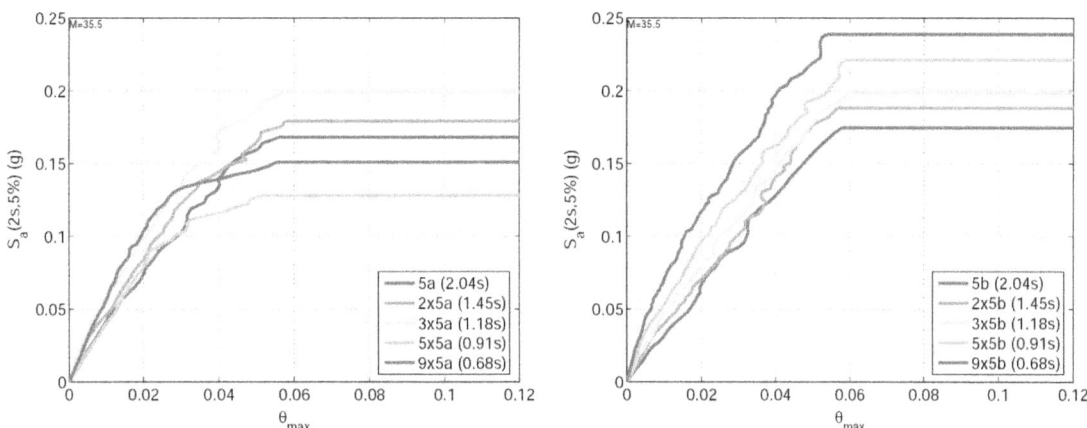

Figure D-28   Median IDA curves plotted versus the intensity measure $S_a(2s,5\%)$ for systems Nx5a and Nx5b with mass M=35.46ton.

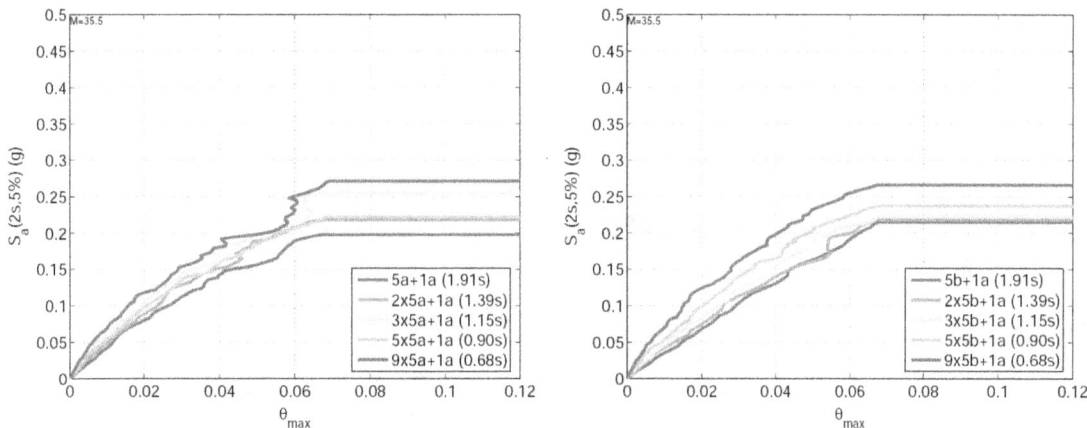

Figure D-29   Median IDA curves plotted versus the intensity measure $S_a(2s,5\%)$ for systems Nx5a+1a and Nx5b+1a with mass M=35.46ton.

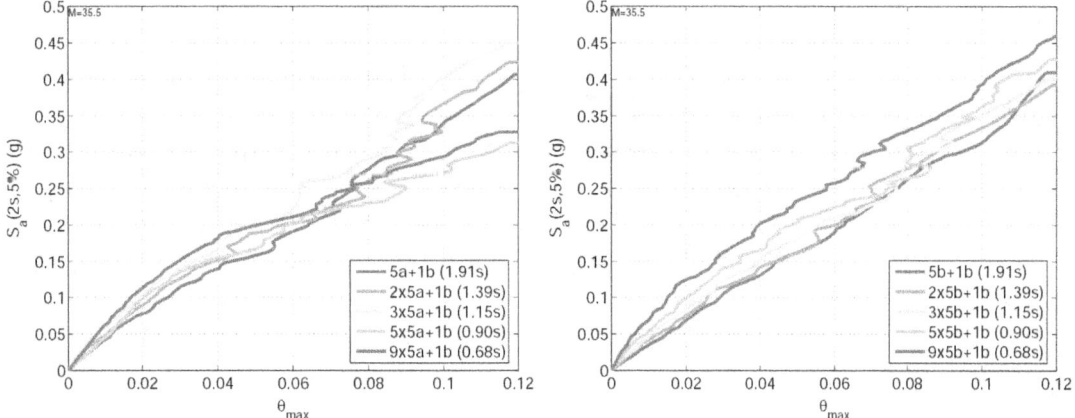

Figure D-30   Median IDA curves plotted versus the intensity measure $S_a(2s,5\%)$ for systems Nx5a+1b and Nx5b+1b with mass M=35.46ton.

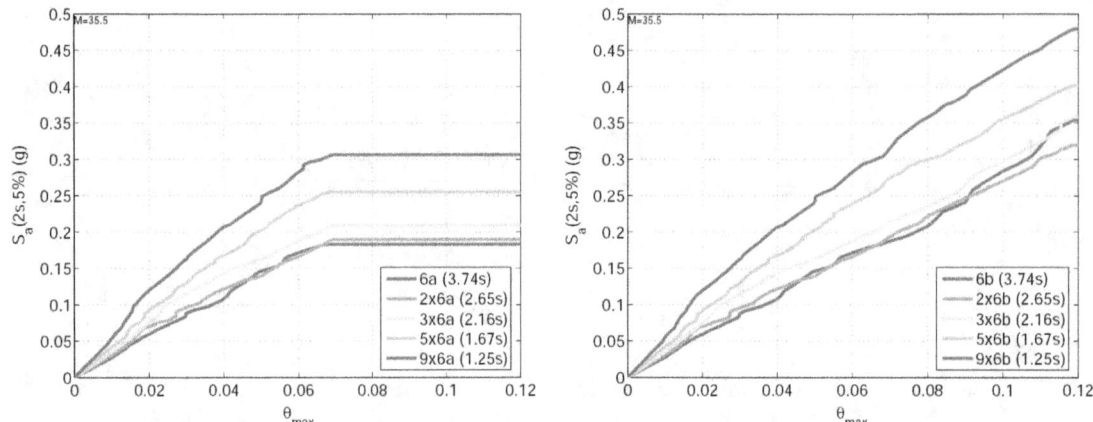

Figure D-31   Median IDA curves plotted versus the intensity measure $S_a(2s,5\%)$ for systems Nx6a and Nx6b with mass M=35.46ton.

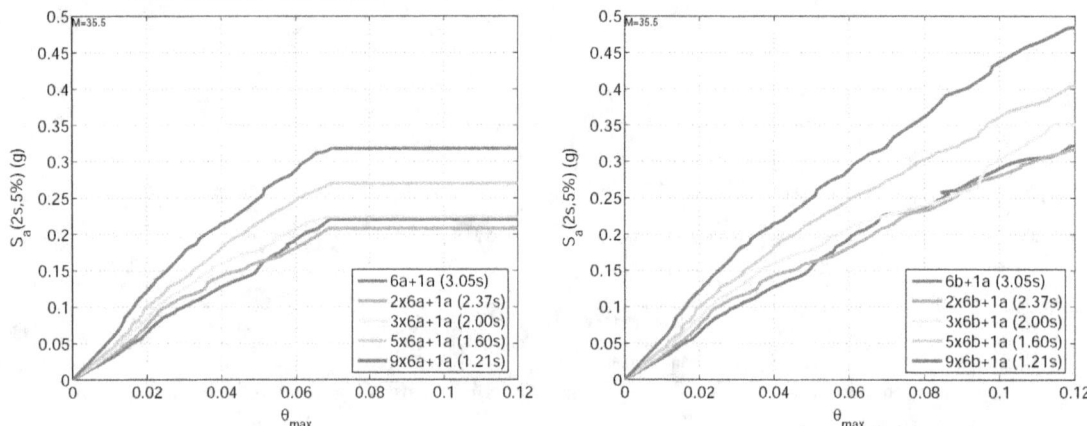

Figure D-32   Median IDA curves plotted versus the intensity measure $S_a(2s,5\%)$ for systems Nx6a+1a and Nx6b+1a with mass M=35.46ton.

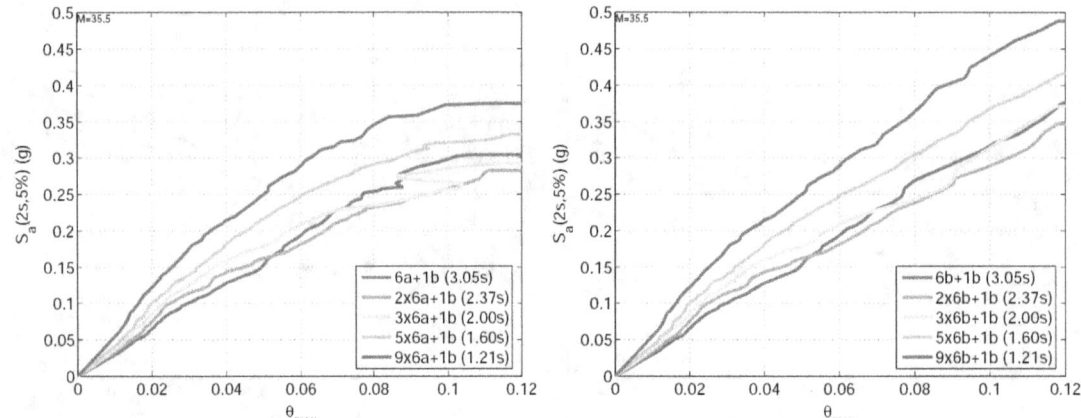

Figure D-33   Median IDA curves plotted versus the intensity measure $S_a(2s,5\%)$ for systems Nx6a+1b and Nx6b+1b with mass M=35.46ton.

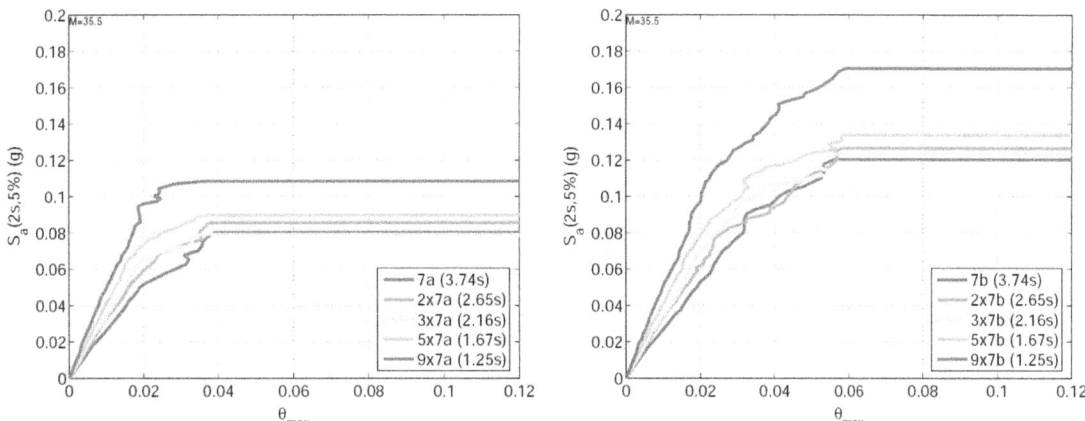

Figure D-34   Median IDA curves plotted versus the intensity measure $S_a(2s,5\%)$ for systems Nx7a and Nx7b with mass M=35.46ton.

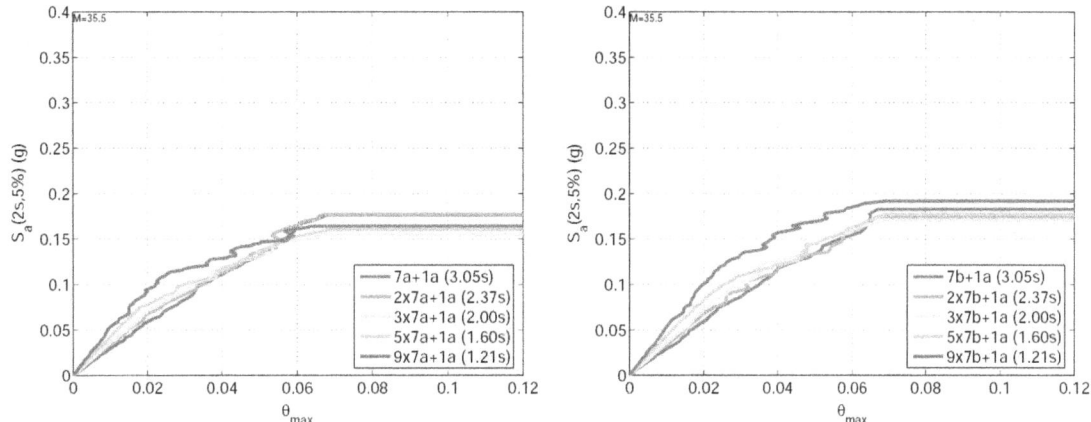

Figure D-35   Median IDA curves plotted versus the intensity measure $S_a(2s,5\%)$ for systems Nx7a+1a and Nx7b+1a with mass M=35.46ton.

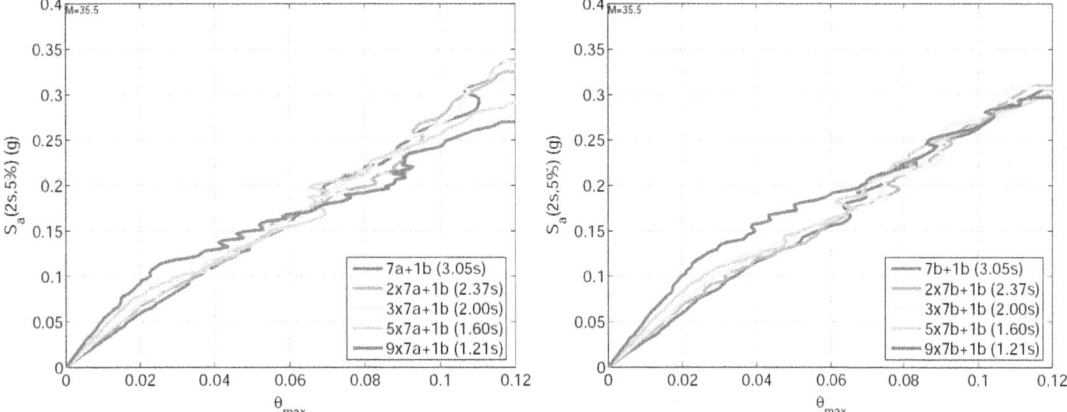

Figure D-36   Median IDA curves plotted versus the intensity measure $S_a(2s,5\%)$ for systems Nx7a+1b and Nx7b+1b with mass M=35.46ton.

Appendix E

Uncertainty, Fragility, and Probability

The concepts presented in this report are compatible with current probabilistic trends in performance-based seismic design. A probabilistic context allows explicit consideration of the variability and uncertainty associated with each of the contributing parameters. An important concept in probabilistic procedures is the development and use of fragility curves.

Use of fragility curves, and explicit consideration of uncertainty in performance assessment, is described in ATC-58 *Guidelines for Seismic Performance Assessment of Buildings* (ATC, 2007). Fragility curves are also used to determine the margin of safety against collapse in FEMA P695 *Quantification of Building Seismic Performance Factors* (FEMA, 2009).

This appendix explains the conversion of incremental dynamic analysis (IDA) results into fragilities, and presents equations that could be used to calculate annual probabilities for collapse, or any other limit state of interest.

E.1 Conversion of IDA Results to Fragilities

Incremental dynamic analysis results can be readily converted to fragilities. Figure E-1 shows an example of IDA results for a single structure subjected to a suite of ground motions of varying intensities.

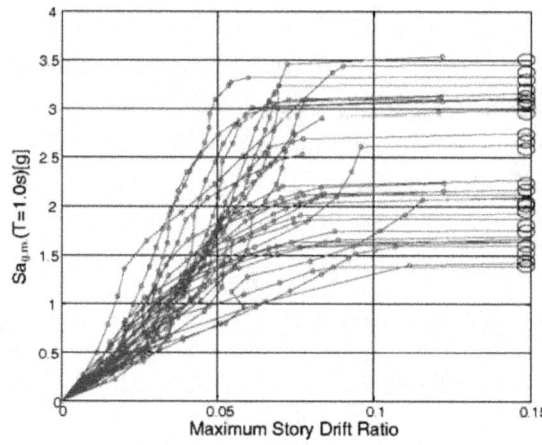

Figure E-1 IDA results for a single structure subjected to a suite of ground motions of varying intensities.

In this illustration, sidesway collapse is the governing mechanism, and collapse prediction is based on dynamic instability or excessive lateral displacements. Using collapse data obtained from IDA results, a collapse fragility can be defined through a cumulative distribution function (CDF), which relates the ground motion intensity to the probability of collapse (Ibarra et al., 2002). Studies have shown that this cumulative distribution function can be assumed to be lognormally distributed. Figure E-2 shows an example of a cumulative distribution plot obtained by fitting a lognormal distribution to the collapse data from Figure E-1.

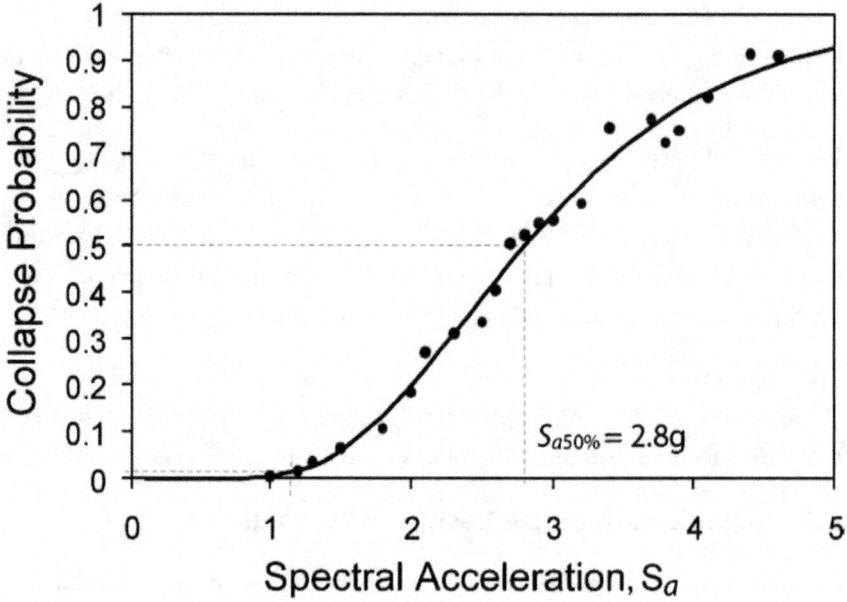

Figure E-2   Cumulative distribution plot obtained by fitting a lognormal distribution to collapse data from IDA results.

Lognormal distributions are defined by a median value and a dispersion parameter. The median collapse capacity, $S_{a50\%}$, indicates a ground motion intensity that has a 50% chance of producing collapse in the system. It also indicates the point at which half of the ground motions will produce collapse at higher intensities, and half will produce collapse at lower intensities. For each mode of collapse, the record-to-record dispersion can be estimated as:

$$\beta_{RTR} = \frac{ln(S_{a84\%}) - ln(S_{a16\%})}{2} \qquad \text{(E-1)}$$

Figure E-3 provides conceptual collapse fragility curves showing the probability of collapse due to loss of vertical-load-carrying capacity (LVCC) or lateral dynamic instability (LDI). These are events are mutually exclusive, meaning that either one or the other can occur, but both events cannot occur at the same time.

E: Uncertainty, Fragility, and Probability

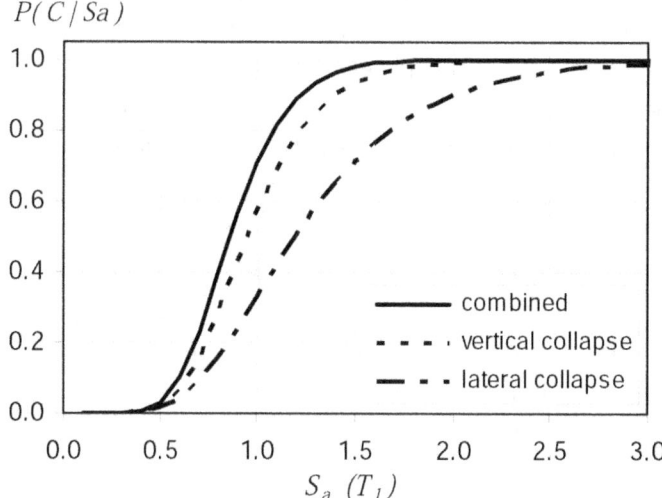

Figure E-3   Conceptual collapse fragility curves for sidesway (lateral) collapse, vertical collapse, and a combination of both.

The total probability of collapse due to either mode can then be represented (Cornell et al., 2005) as:

$$P(C \mid s_a) = P(C_{LDI} \mid s_a) + P(C_{LVCC} \mid NC_{LDI}, s_a) \cdot P(NC_{LDI} \mid s_a \qquad \text{(E-2)}$$

where:

$P(C_{LDI} \mid s_a)$ is the probability of collapse due to lateral dynamic instability at a ground motion intensity level s_a ,

$P(C_{LVCC} \mid NC_{LDI}, s_a)$ is the probability of collapse due to loss of vertical-load-carrying capacity, given that collapse due to lateral dynamic instability has not occurred at intensity s_a , and

$P(NC_{LDI} \mid s_a)$ is the probability of no collapse due to lateral dynamic instability at a ground motion intensity level s_a.

Since $P(NC_{LDI} \mid s_a)$ is equal to $1 - P(C_{LDI} \mid s_a)$, then Equation E-2 can also be written as:

$$P(C \mid s_a) = P(C_{LVCC} \mid NC_{LDI}, s_a) + P(C_{LDI} \mid s_a) - P(C_{LVCC} \mid NC_{LDI} s_a) \cdot P(C_{LDI} \mid s_a) \qquad \text{(E-3)}$$

E.2   Calculation of Annualized Probability

The results of an incremental dynamic analysis expressed as a cumulative distribution function can be used in combination with a seismic hazard curve to generate mean annual frequencies (MAF) for collapse (or for other limit states of interest). This process is the integration of the limit state CDF (e.g., fragility representing the probability of collapse as a function of spectral acceleration) with respect to the probability of occurrence of the intensity measure (e.g., hazard curve representing the annual probability of exceeding a full range of spectral accelerations). The mean annual frequency of collapse, λ_{col}, or other limit state of interest, can be approximated (Cornell, 2002) as:

$$\lambda_{col} = \lambda_{S_a}\left(\eta_C\right)\exp\left(\frac{1}{2}k^2\beta_{RTR}^2\right) \qquad \text{(E-4)}$$

where $\lambda_{S_a}\left(\eta_C\right)$ is the mean annual probability of the median spectral acceleration associated with collapse. The parameter k is the slope of the hazard curve, and can be calculated as:

$$k = \frac{\ln\left(\dfrac{H_{S_{aT}(10/50)}}{H_{S_{aT}(2/50)}}\right)}{\ln\left(\dfrac{S_{aT(2/50)}}{S_{aT(10/50)}}\right)} = \frac{1.65}{\ln\left(\dfrac{S_{aT(2/50)}}{S_{aT(10/50)}}\right)} \qquad \text{(E-5)}$$

Appendix F

Example Application

This appendix presents an example application of a simplified nonlinear dynamic analysis procedure. The concept originated during the conduct of focused analytical studies comparing force-displacement capacity boundaries to incremental dynamic analysis results. In this procedure, a nonlinear static analysis is used to generate an idealized force-deformation curve (i.e., static pushover curve). The resulting curve is then used as a force-displacement capacity boundary to constrain the hysteretic behavior of an equivalent single-degree-of-freedom (SDOF) oscillator. This SDOF oscillator is then subjected to incremental dynamic analysis.

The steps for conducting a simplified nonlinear dynamic analysis are outlined in the following section, and illustrated using an example building. Alternative retrofit strategies are evaluated using the same procedure. Use of the procedure to develop probabilistic estimates of performance for use in making design decisions is also illustrated.

F.1 Simplified Nonlinear Dynamic Analysis Procedure

The concept of a simplified nonlinear dynamic analysis procedure includes the following steps:

- **Develop an analytical model of the system.**

  Models can be developed in accordance with prevailing practice for seismic evaluation, design, and rehabilitation of buildings described in ASCE/SEI Standard 41-06 *Seismic Rehabilitation of Existing Buildings* (ASCE, 2006b). Component properties should be based on force-displacement capacity boundaries, rather than cyclic envelopes.

- **Perform a nonlinear static pushover analysis.**

  Subject the model to a conventional pushover analysis in accordance with prevailing practice. Lateral load increments and resulting displacements are recorded to generate an idealized force-deformation curve.

- **Conduct an incremental dynamic analysis of the system based on an equivalent SDOF model.**

  The idealized force-deformation curve is, in effect, a system force-displacement capacity boundary that can be used to constrain a hysteretic model of an equivalent SDOF oscillator. This SDOF oscillator is then subjected to incremental dynamic analysis to check for lateral dynamic instability and other limit states of interest. Alternatively, approximate incremental dynamic analysis can be accomplished using the idealized force-deformation curve and the *Static Pushover 2 Incremental Dynamic Analysis* open source software tool, SPO2IDA (Vamvatsikos and Cornell, 2006).

- **Determine probabilities associated with limit states of interest.**

  Results from incremental dynamic analysis can be used to obtain response statistics associated with limit states of interest in addition to lateral dynamic instability. SPO2IDA can also be used to obtain median, 16th, and 84th percentile IDA curves relating displacements to intensity. Using the fragility relationships described in Appendix E in conjunction with a site hazard curve, this information can be converted into annual probabilities of exceedance for each limit state. Probabilistic information in this form can be used to make enhanced decisions based on risk and uncertainty, rather than on discrete threshold values of acceptance.

F.2   Example Building

The example building is a five-story reinforced concrete frame residential structure with interior unreinforced masonry infill partitions in the upper stories, and a soft/weak first-story. An exterior elevation of the building is shown in Figure F-1 and first floor plan is shown in Figure F-2. Reinforced concrete columns in each orthogonal direction provide lateral resistance to seismic forces. As indicated in Figure F-2, the first story includes a mixture of components with column-like proportions and components that are more like slender shear walls.

This building is a prototypical example of a soft/weak story structure. Concentration of inelastic deformations in the first story presents an obvious potential story collapse mechanism.

F.3   Structural Analysis Model

To investigate the potential for collapse in this structure, it is reasonable to assume that the response can be represented by a SDOF model. The first story column components are classified for modeling purposes in accordance with ASCE/SEI 41-06. Most of the columns in this example are classified as shear-controlled or flexure-shear controlled. Wall-like column components are shear-controlled along the strong axis of the member.

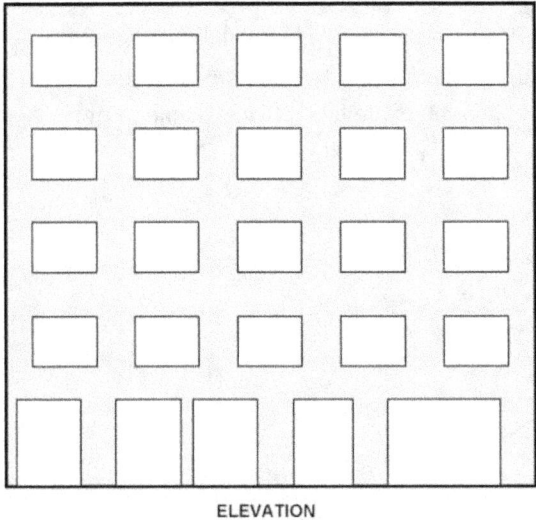

ELEVATION

Figure F-1   Example building exterior elevation.

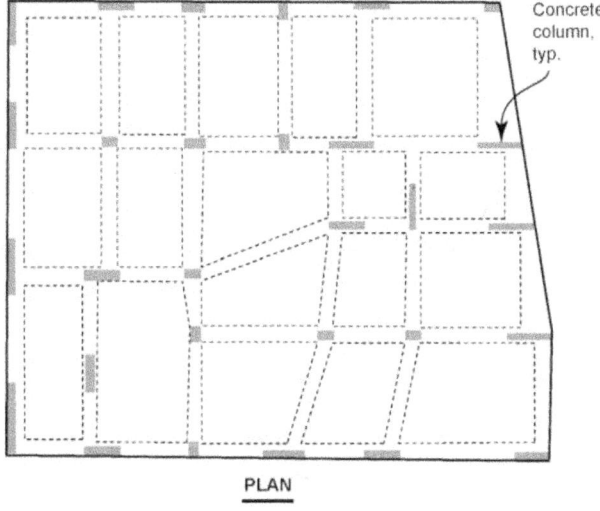

PLAN

Figure F-2   Example building first floor plan.

Modeling parameters for the column components can be characterized by the conceptual force-displacement relationship ("backbone") specified in ASCE/SEI 41-06. The modeling parameters selected for the components in this example are taken from Chapter 6 of ASCE/SEI 41-06, and depicted in Figure F-3. In both cases, the residual strength, c, is taken as zero.

The column components are assembled into a model of the structural system as shown in Figure F-4. Inelastic response is assumed to occur predominantly in the first story. First-story columns are taken as fixed at the base on a rigid foundation. The stiffness of the column components are based on elastic properties in flexure and shear. The model includes soil flexibility, allowing for rigid body rotation due to the response of the structure above. Soil stiffness parameters are taken from Chapter 4 of ASCE/SEI 41-06, assuming a relatively soft soil site (site Class E).

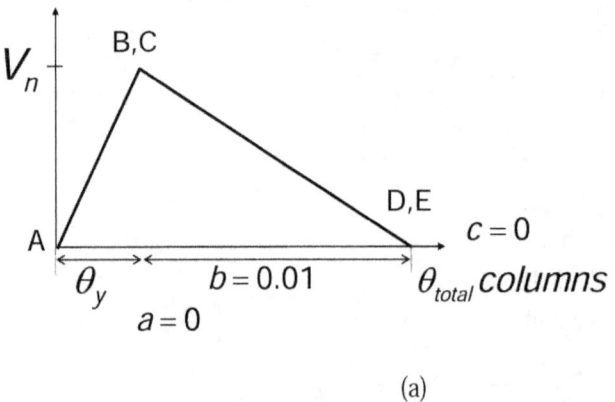

(a)

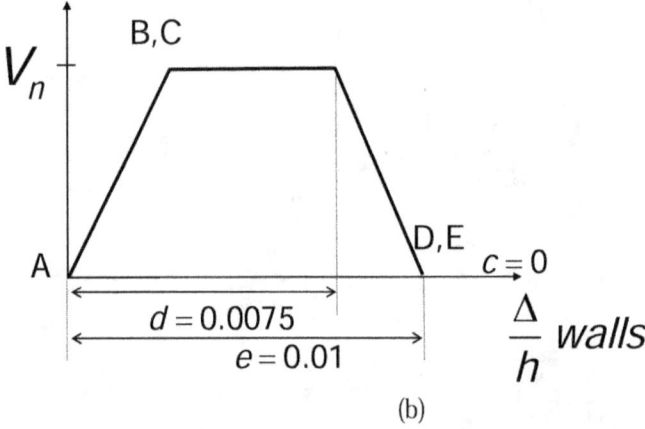

(b)

Figure F-3   Force-displacement modeling parameters for: (a) column components; and (b) wall-like column components.

When developing a SDOF representation of a system, it is important to account for foundation rotation in assessing column distortions. The resulting SDOF model represents the relationship between the total first floor drift, including contributions from the foundation, $\theta_{sys} = \theta_{fdn} + \theta_{cols}$, and the applied inertial loads, V.

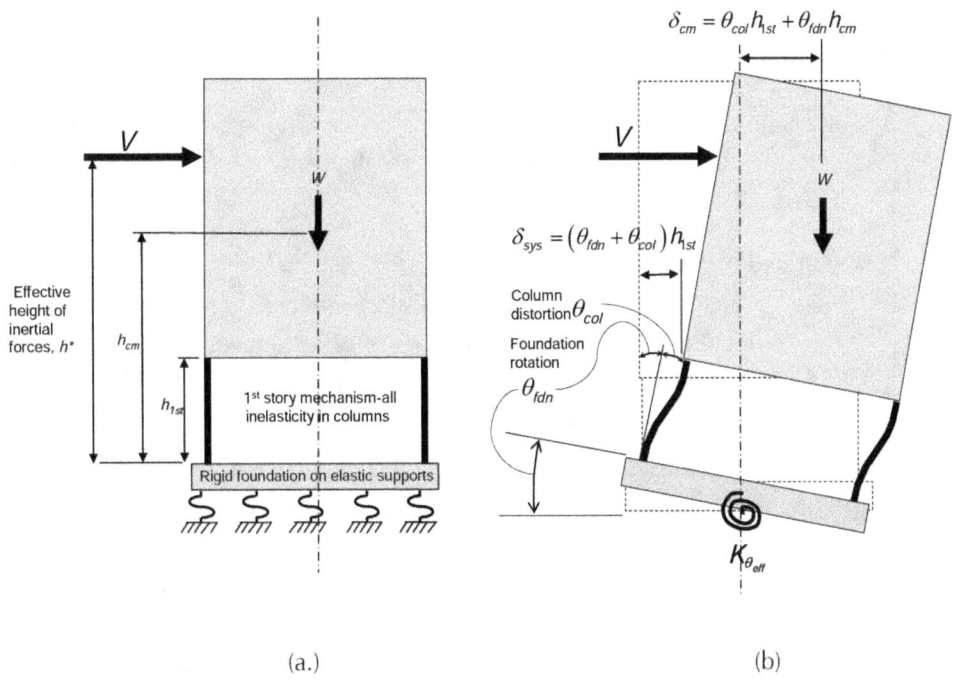

$$\delta_{cm} = \theta_{col}h_{1st} + \theta_{fdn}h_{cm}$$

$$\delta_{sys} = \left(\theta_{fdn} + \theta_{col}\right)h_{1st}$$

(a.)   (b)

Figure F-4   Structural analysis model showing: (a) assumptions; and (b) distortions.

F.4   Nonlinear Static Pushover Analysis

The analytical model is subjected to a conventional pushover analysis. Results are shown in Figure F-5.

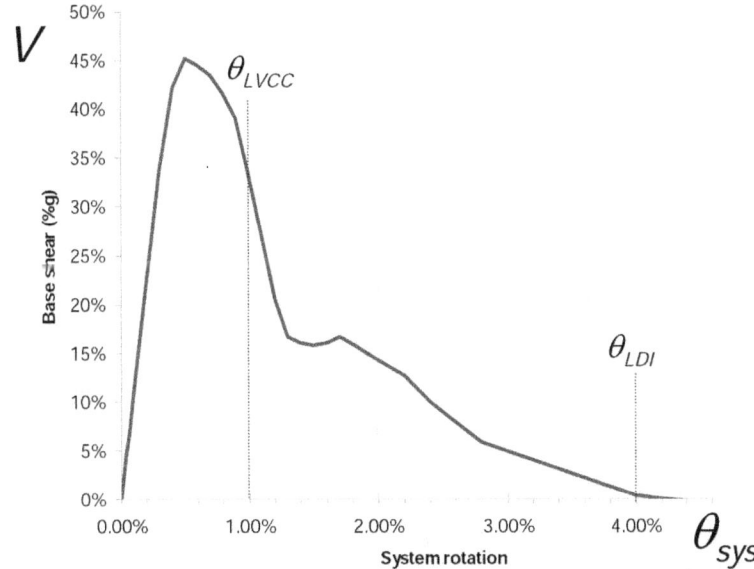

Figure F-5   Pushover curve from nonlinear static analysis.

F.5 Evaluation of Limit States of Interest

Collapse in real structures can be caused by sidesway collapse (lateral dynamic instability) or by loss of vertical-load-carrying capacity. In this example, the following two limit states are defined (both are shown in Figure F-5):

θ_{LVCC} the total system rotation at which loss of vertical-load-carrying capacity occurs (i.e., when first story columns fail due to shear distortion). In this example, the critical column distortion for loss of vertical-load-carrying capacity is taken as 1% inelastic rotation, which occurs when the total system rotation $\theta_{sys} = 1.2\%$.

θ_{LDI} the total system rotation at which lateral dynamic instability occurs (i.e., when first story columns lose all lateral-force-resisting capacity). In this example, this is taken to occur when the total system rotation $\theta_{sys} = 4.0\%$.

The target displacement for a given intensity is estimated using the Coefficient Method:

$$\delta_t = C_0 C_1 C_2 S_a \frac{T_e^2}{4\pi^2} g$$

Uniform hazard spectra for the example site are shown in Figure F-6 and Figure F-7.

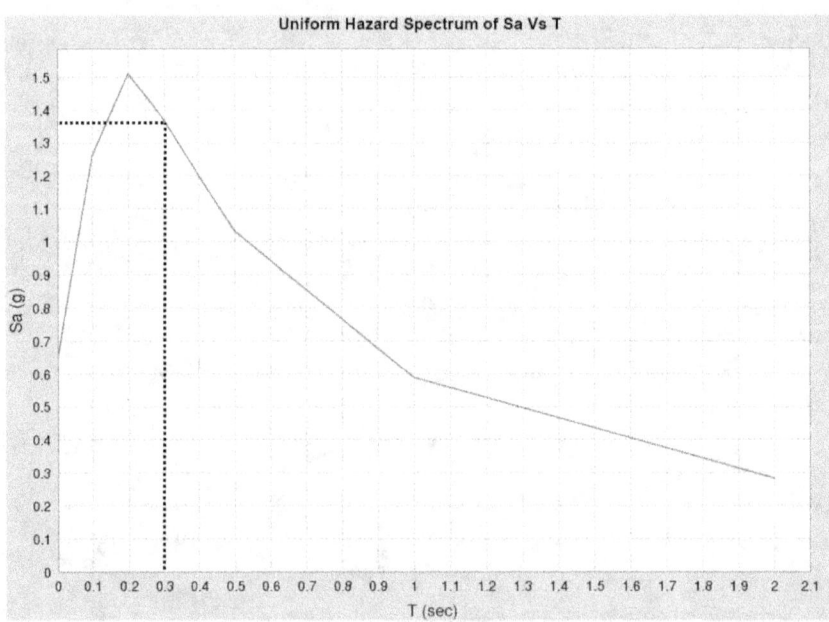

Figure F-6   Uniform hazard spectrum for intensity corresponding to 10% chance of exceedance in 50 years (from USGS).

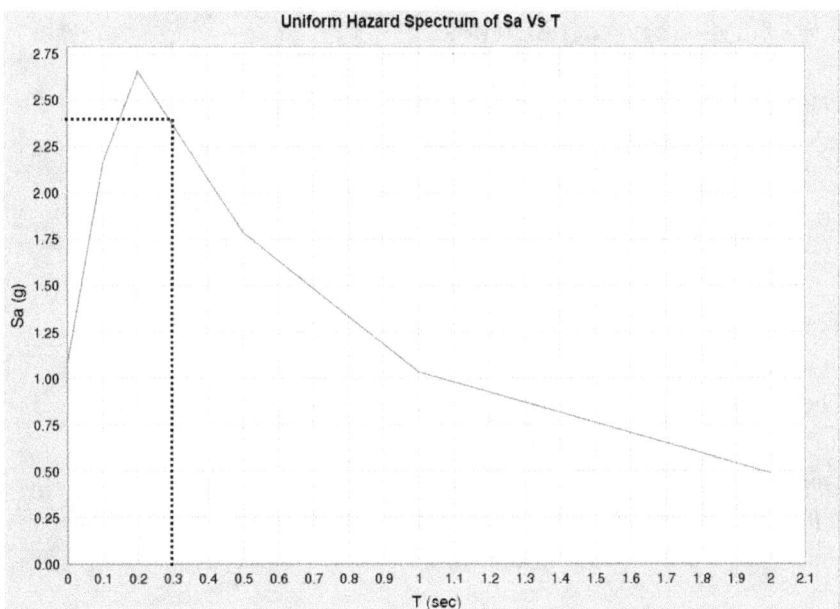

Figure F-7   Uniform hazard spectrum for intensities corresponding to 2% chance of exceedance in 50 years (from USGS).

For an intensity corresponding to a 10% chance of exceedance in 50 years (i.e., 475-year return period), and a period T = 0.3s:

$$S_a = 1.36g$$

The strength of the model is:

$$F_y = 0.45g$$

which results in:

$$R = \frac{S_{aT}g}{F_y} = 3.0 \text{ for the 10\%/50 year hazard level.}$$

The coefficients are:

C_0 = first mode participation factor = 1.0,

$$C_1 = 1 + \frac{R-1}{aT_e^2} = 1.3,$$

where $a = 50$, and

$$C_2 = 1 + \frac{1}{800}\left(\frac{R-1}{T_e}\right)^2 = 1.04.$$

This results in a target displacement of:

$\delta_t = 1.6$ inches, or $\theta_{sys} = 1.6/100 = 1.6\%$.

This is greater than the acceptable limit for loss of vertical-load-carrying capacity (θ_{LVCC}) taken as $\theta_{sys} = 1.2\%$.

To check for lateral dynamic instability, the proposed equation for R_{di} is:

$$R_{di} = \left(\frac{\Delta_c}{\Delta_y} \right)^a + b \frac{T_e}{3|\gamma|} + \frac{F_r}{F_c} \left(\frac{\Delta_u - \Delta_r}{\Delta_y} \right) \sqrt[3]{T_e}$$

where T_e is the effective fundamental period of vibration of the structure, Δ_y, Δ_c, Δ_r, and Δ_u are displacements corresponding to the yield strength, F_y, capping strength, F_c, residual strength, F_r, and ultimate deformation capacity at the end of the residual strength plateau. Determination of these parameters requires a multi-linear idealization of the pushover curve, as shown in Figure F-8.

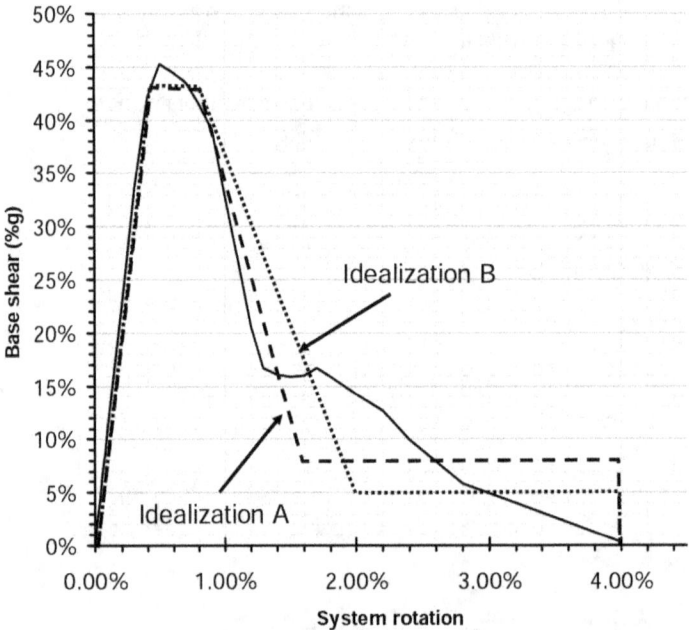

Figure F-8   Pushover curve from nonlinear static analysis and two idealized system force-displacement capacity boundaries.

Parameters a and b are functions given by:

$$a = 1 - \exp(-dT_e)$$

$$b = 1 - \left(\frac{F_r}{F_c} \right)^2$$

and parameter d is a constant equal to 4 for the example building (assuming the presence of stiffness degradation).

Using the above expressions along with parameters from Idealization 'A' in Figure F-8, results in:

$$R_{di} = 2.6$$

which is less that the calculated value of $R = 3.0$ for intensities corresponding to the 10%/50 year hazard level. The parameter R is a ratio equal to the strength necessary to keep a system elastic for a given intensity, divided by the yield strength of the system. Higher values of R imply lower values of system yield strength. Values of R that exceed R_{di} mean that the structure does not meet the minimum strength necessary to avoid lateral dynamic instability at this hazard level.

At higher intensities (e.g., 2%/50 year hazard level) the calculated value of R would be even higher ($R = 5.3 >> R_{di} = 2.6$), illustrating how the comparison between system strength and the limit on lateral dynamic instability would change for a different hazard level.

In summary, the example structure does not meet acceptability criteria for loss of vertical-load-carrying capacity and lateral dynamic instability at the 10%/50 year hazard level. Thus, a nonlinear response-history analysis must be performed.

F.6   Incremental Dynamic Analysis

The resulting force-displacement relationship from the pushover curve can be used to generate a force-displacement capacity boundary for the system. An incremental dynamic analysis (IDA) can then be applied to a SDOF oscillator constrained by the resulting force-displacement capacity boundary. Performing an incremental dynamic analysis will allow determination of the ground motion intensity at which various limit state deformations occur.

For the example building, an approximate incremental dynamic analysis is performed using the open source software tool, SPO2IDA. Use of SPO2IDA along with Idealization 'A' in Figure F-8 results in the median, 16th, and 84th percentile IDA curves shown in Figure F-9. The figure also includes the estimate of R_{di} for lateral dynamic instability.

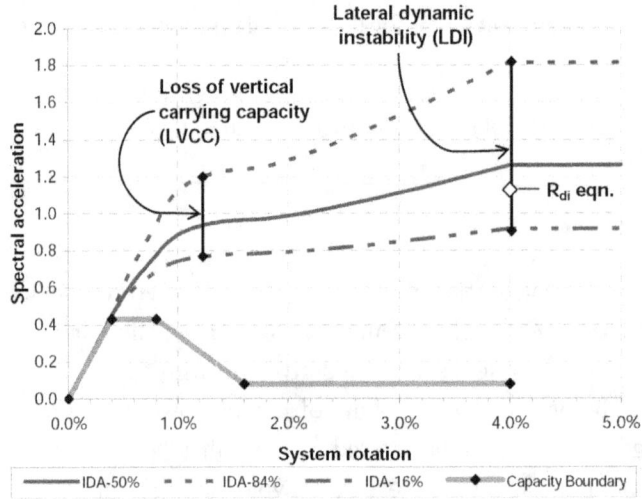

Figure F-9   Results of approximate incremental dynamic analysis using
                    SPO2IDA.

F.7   Determination of Probabilities Associated with Limit States of Interest

From Figure F-9, median values of intensity causing loss of vertical-load-carrying capacity (LVCC) or lateral dynamic instability (LDI) in the example building can be obtained. Using the expressions in Appendix E, the dispersion and mean annual frequencies (MAF) associated with these limit states can be determined. The resulting data is presented Table F-1.

Table F-1 Mean Annual Frequencies for Collapse Limit States				
Limit state/collapse mode	S_{a50}	MAF S_{a50}	β	MAF collapse
Loss of vertical load carrying capability	0.92	0.0050	0.20	0.0060
Lateral dynamic instability	1.26	0.0025	0.32	0.0040
LVCC or LDI	0.88	0.0060	0.17	0.0069

For the example building, fragilities associated with loss of vertical-load-carrying capacity (LVCC) or lateral dynamic instability (LDI) are derived from the median values of spectral acceleration and dispersions in Table F-1, as illustrated in Figure F-10.

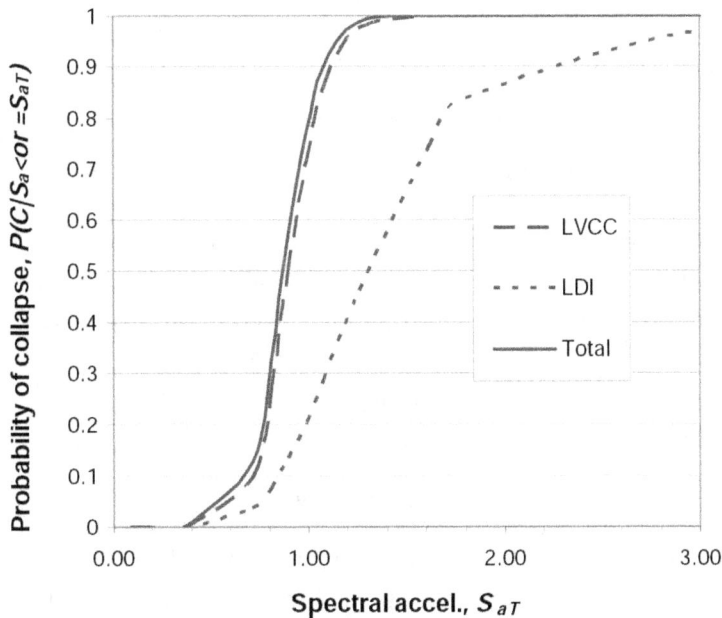

Figure F-10   Example building collapse fragilities for loss of vertical-load-carrying capacity (LVCC), lateral dynamic instability (LDI), and a combination of both.

The probabilities of experiencing the limit states of interest for the example building are derived in combination with a site-specific seismic hazard curve. For a representative soft site, the USGS hazard curve for a period of 0.3 seconds is shown in Figure F-11.

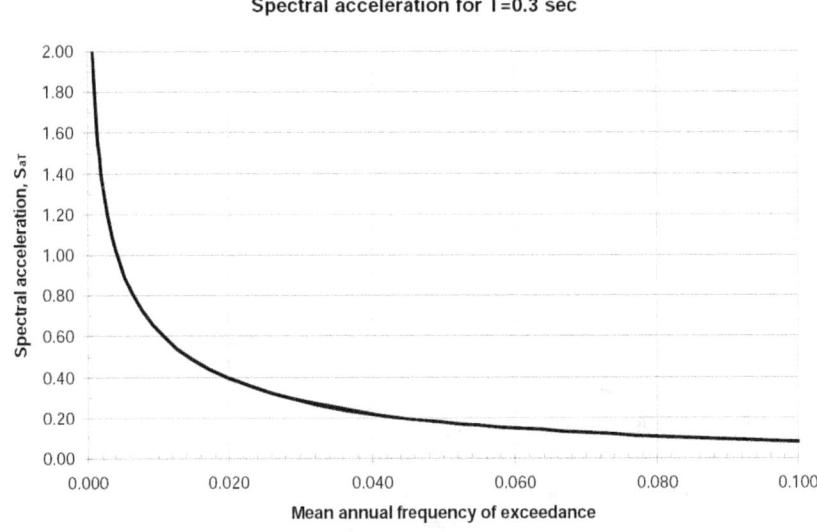

Figure F-11   Hazard curve for representative soft site (from USGS).

Using the expressions in Appendix E, the mean annual frequency (MAF) of collapse of the example building can be calculated. The MAF of collapse, shown in Table F-1, is approaching 1% annually, which can be considered high.

F.8   Retrofit strategies

As indicated by results of focused analytical studies presented in Chapter 4, the probability of collapse, as measured by the potential for lateral dynamic instability, can be reduced by adjusting the force-deformation characteristics of the system. Retrofit strategies to reduce the annual probability of collapse could include: (1) the addition of secondary lateral system; or (2) improvement of primary system strength and ductility.

F.8.1   Addition of a Secondary Lateral System

Lateral dynamic instability can be improved through the addition of a flexible and ductile secondary lateral system. In order to be effective, the secondary system does not need to provide much additional lateral strength to the overall system.

Figure F-12 shows a revised pushover curve for the example building after retrofit using a ductile moment frame system with a lateral strength approximately equal to 10% of the building weight. Calculations similar to the above show that this strategy would reduce the mean annual frequency of collapse due to lateral dynamic instability to about 0.2%.

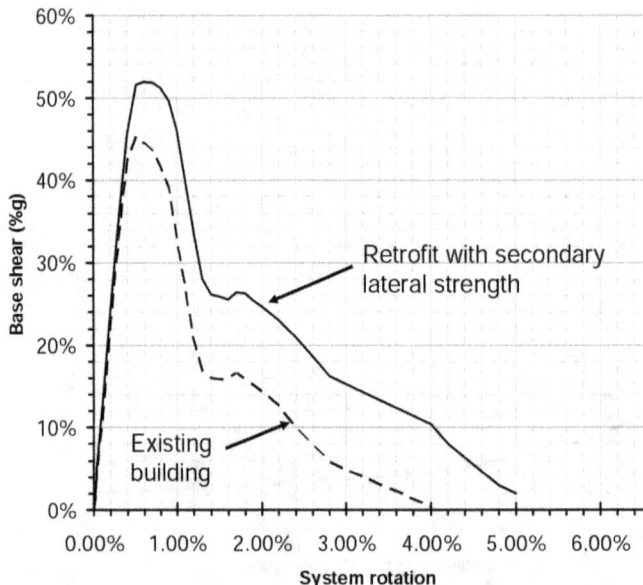

Figure F-12   Revised pushover curve for the example building after retrofit with a secondary lateral system.

F.8.2 Improvement of Primary System Strength and Ductility

Lateral dynamic instability can also be improved by providing additional deformation capacity in the system. This can be accomplished by wrapping the concrete columns with fiber reinforced polymers (FRP) to increase strength and ductility of the primary seismic-force-resisting system.

Figure F-13 shows a revised pushover curve for the example building after retrofitting the existing columns with FRP. Calculations similar to the above show that this strategy would reduce the mean annual frequency of collapse due to lateral dynamic instability to about 0.03%, which is an order of magnitude improvement over the existing structure.

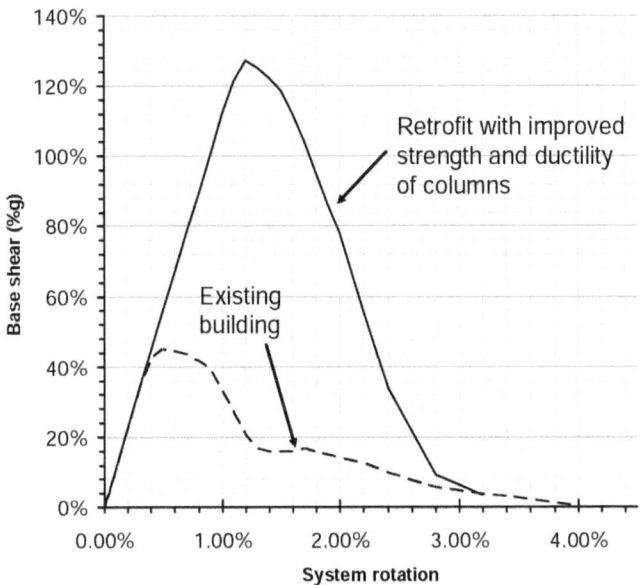

Figure F-13   Revised pushover curve for the example building after retrofit for improved strength and ductility of columns.

Appendix G
Preliminary Multiple-Degree-of-Freedom System Studies

In focused analytical studies on single-degree-of-freedom (SDOF) systems, it was observed that nonlinear response of a system depends on the characteristics of the force-displacement capacity boundary. It was demonstrated that lateral dynamic instability of SDOF systems could be evaluated through the use of approximate equations or simplified nonlinear dynamic analyses based on the characteristics of the system force-displacement capacity boundary.

Multiple-degree-of-freedom (MDOF) systems are more complex, and their dynamic response is more difficult to estimate than that of SDOF systems. Recent studies have suggested that it may be possible to estimate the collapse capacity of MDOF systems by using static pushover analyses and performing dynamic analysis on equivalent SDOF systems (Bernal, 1998; Vamvatsikos, 2002; Vamvatsikos and Cornell, 2005a, 2005b). In particular, Vamvatsikos and Cornell (2005b) suggested that the seismic response of MDOF systems could be estimated through the use of incremental dynamic analyses on a reference SDOF system whose properties are determined through a nonlinear static (pushover) analysis.

This appendix presents the results of preliminary studies of multiple-degree-of-freedom (MDOF) systems. It explores the application of nonlinear static analyses combined with dynamic analyses of SDOF systems to evaluate the lateral dynamic instability of MDOF systems. On a preliminary basis, it tests how approximate measures of lateral dynamic instability developed for SDOF systems might work on more complex MDOF systems. These approximate measures include the proposed equation for R_{di} (Equation 5-8) and the open source software tool *Static Pushover 2 Incremental Dynamic Analysis*, SPO2IDA (Vamvatsikos and Cornell, 2006).

A total of six buildings ranging in height from 4 to 20 stories are used in this investigation. This set includes two steel moment-resisting frame structures and four reinforced concrete moment-resisting frame structures. Four were previously studied by Haselton (2006), and two were previously studied by

Vamvatsikos and Cornell (2005b). Results are described in the sections that follow.

G.1 Four-Story Code-Compliant Reinforced Concrete Building

The subject building is a four-story reinforced concrete special perimeter moment frame designed in accordance with modern building code provisions (ICC 2003, ASCE 2002, ACI 2002). The building has a story height of 15 ft in the first story, and 13 ft in the remaining stories. The design base shear coefficient was 0.092. The building was modeled in OpenSEES and analyzed using incremental dynamic analysis using 80 recorded time histories which were scaled at twenty-two different ground motion intensities. The pushover analysis was conducted using a lateral force distribution in accordance with ASCE/SEI 7-05 *Minimum Design Loads for Buildings and Other Structures* (ASCE, 2006). Ground motions were scaled to increasing values of the pseudo-acceleration spectral ordinate at the fundamental period of vibration of the building (T_1=1.12s). For a more detailed description of the building and its modeling, the reader is referred to Haselton (2006).

Figure G-1 shows the results from the nonlinear static (pushover) analysis. The figure on the left shows the force-deformation curve while the figure on the right shows the distribution of story drift ratios at a roof drift ratio of 6%. It can be seen that story drifts primarily concentrate in the lower two stories. The force-deformation pushover curve is characterized by a gradual loss in lateral strength for roof drift ratios between 1% and 3.5%, followed by a more pronounced loss in lateral strength for roof drift ratios greater than 3.5%.

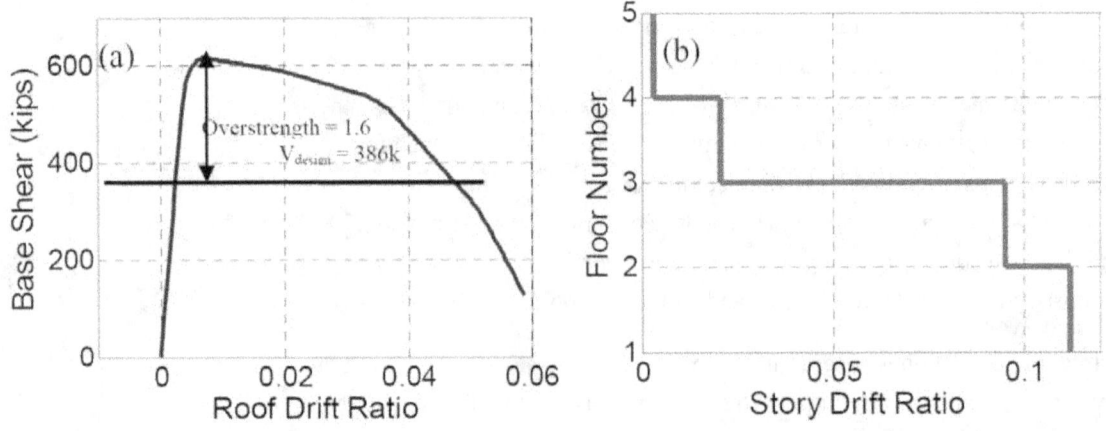

Figure G-1   (a) Monotonic pushover force-deformation curve and (b) story drifts at a roof drift ratio of the 0.06 in a four-story concrete frame building (Haselton 2006).

Figure G-2 shows one of three simplified tri-linear force-displacement capacity boundaries selected to estimate the seismic response of the structure. Alternates are shown in Figure G-4 and Figure G-5. Although a sloping intermediate segment might have been somewhat more appropriate for this structure, a horizontal intermediate segment was selected in order to evaluate the proposed equation for R_{di} and results using SPO2IDA.

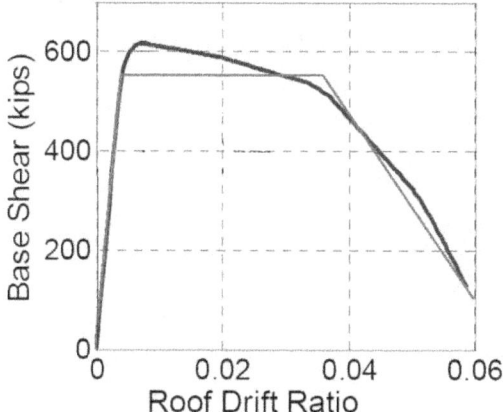

Figure G-2   Tri-linear capacity boundary selected for approximate analysis.

Figure G-3 shows the median seismic behavior computed from incremental dynamic analyses conducted by Haselton (2006). These results are indicated as MDOF IDA in the figure. Also shown are results computed using the proposed equation for R_{di} and approximate results from SPO2IDA. In the figure, R_{di} and SPO2IDA both provide a good approximation of the collapse capacity of the building.

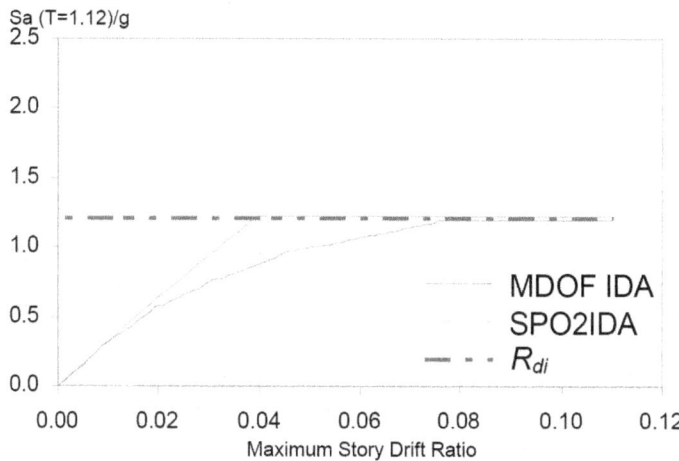

Figure G-3   Comparison of median collapse capacity for a four-story code-compliant concrete frame building computed using incremental dynamic analysis and approximate procedures.

To explore sensitivity to the idealization of the force-displacement capacity boundary, two alternate idealizations, along with corresponding results, are shown in Figure G-4 and Figure G-5. Although median collapse capacities change with the selection of the force-displacement capacity boundary, the observed changes are relatively small.

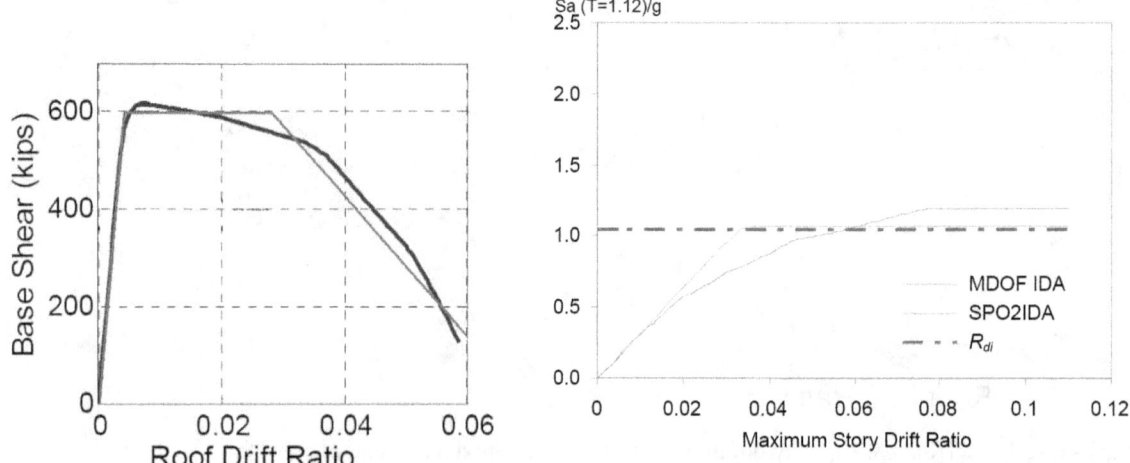

Figure G-4   Effect of selecting an alternate force-displacement capacity boundary on estimates of median collapse capacity for a four-story code-compliant concrete frame building.

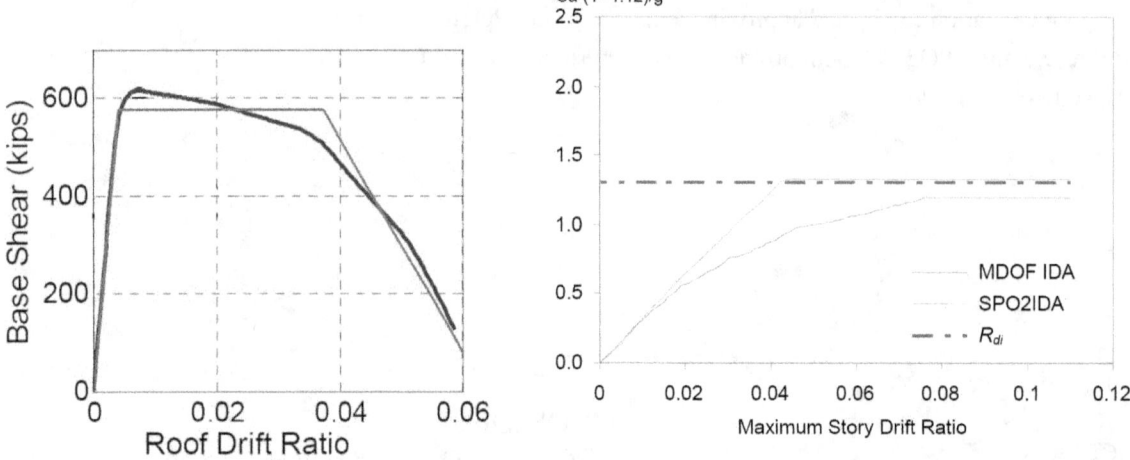

Figure G-5   Effect of selecting an alternate force-displacement capacity boundary on estimates of median collapse capacity for a four-story code-compliant concrete frame building.

The median results shown above represent a measure of the central tendency of the response of the system; however, considerable dispersion exists around the median. To illustrate record-to-record variability, Figure G-6 shows incremental dynamic analysis results for all 80 ground motions. It can be seen that there are ground motions that produce the collapse of the structure at intensities equal to one third of the median intensity. Similarly, there are ground motions that require an intensity that is twice as large as the median intensity in order to produce the collapse of the structure.

Also shown in Figure G-6 are the 16th and 84th percentiles of the results. Approximately 70% of the ground motions fall between these two dashed lines. When estimating the collapse probability of a structure, it is important to consider this variability. For more information, the reader is referred to Haselton (2006).

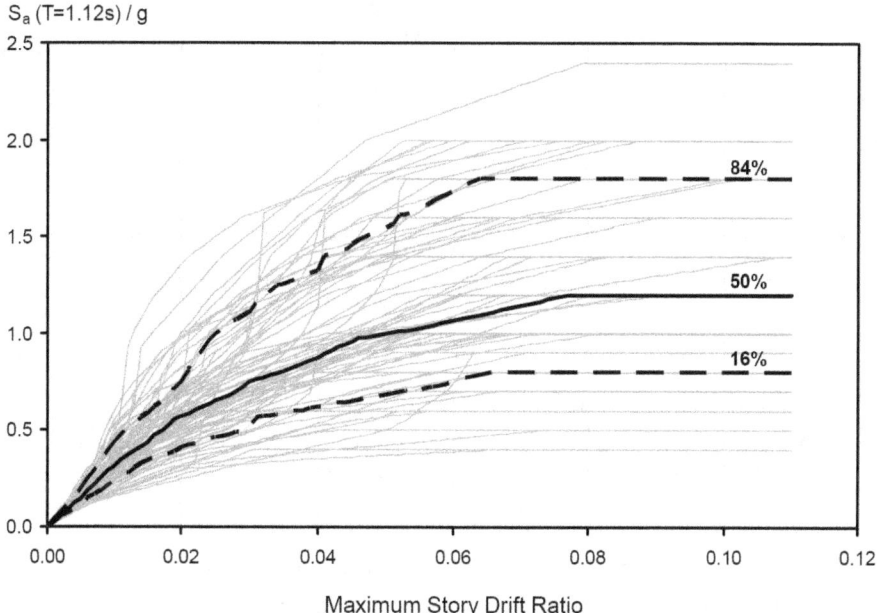

S_a (T=1.12s) / g

Maximum Story Drift Ratio

Figure G-6   Incremental dynamic analysis results for a four-story code-compliant concrete frame building subjected to 80 ground motions (adapted from Haselton, 2006).

G.2   Eight-Story Code-Compliant Reinforced Concrete Building

The subject building is an eight-story reinforced concrete special perimeter moment frame designed in accordance with modern building code provisions (ICC 2003, ASCE 2002, ACI 2002). The building has a story height of 15 ft in the first story, and 13 ft in the remaining stories. The design base shear coefficient was 0.05. The building was modeled in OpenSEES and analyzed using incremental dynamic analysis with the same 80 recorded ground motions that were used to analyze the four-story building. The pushover analysis was performed using a lateral force distribution in accordance with

ASCE/SEI 7-05. The fundamental period of vibration of the building is $T_1=1.71s$. For a more detailed description of the building and its modeling, the reader is referred to Haselton (2006).

Figure G-7 shows the results from the nonlinear static (pushover) analysis of the building. The figure on the left shows the force-deformation curve while the figure on the right shows the distribution of story drift ratios at a roof drift ratio of 2.6%. It can be seen that story drifts primarily concentrate in the lower four stories. The force-deformation pushover curve is characterized by a hardening segment for roof drift ratios between 0.3% and 0.8%, followed by softening segment for roof drift ratios greater than 0.8%.

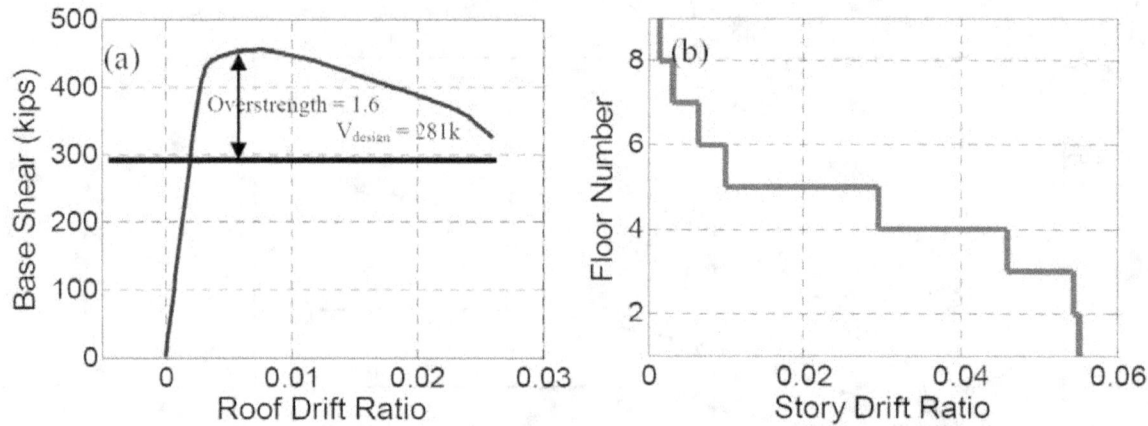

Figure G-7   (a) Monotonic pushover force-deformation curve and (b) distribution of story drift demands at a roof drift ratio of 2.6% in an eight-story concrete frame building (Haselton 2006).

Figure G-8 shows the simplified tri-linear force-displacement capacity boundary selected to evaluate the proposed equation for R_{di} and results using SPO2IDA.

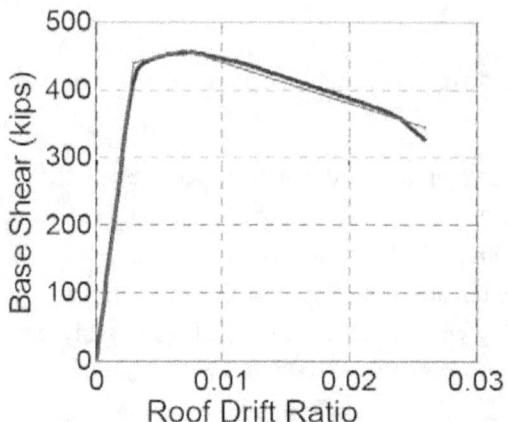

Figure G-8   Tri-linear capacity boundary selected for approximate analyses using SPO2IDA.

Figure G-9 shows the median seismic behavior computed from incremental dynamic analyses conducted by Haselton (2006). These results are indicated as MDOF IDA in the figure. Also shown are results computed using the proposed equation for R_{di} and approximate results from SPO2IDA. In the figure, R_{di} provides a good estimate of the median collapse capacity, while SPO2IDA overestimates the collapse capacity somewhat. Figure G-10 shows incremental dynamic analysis results for all ground motion records.

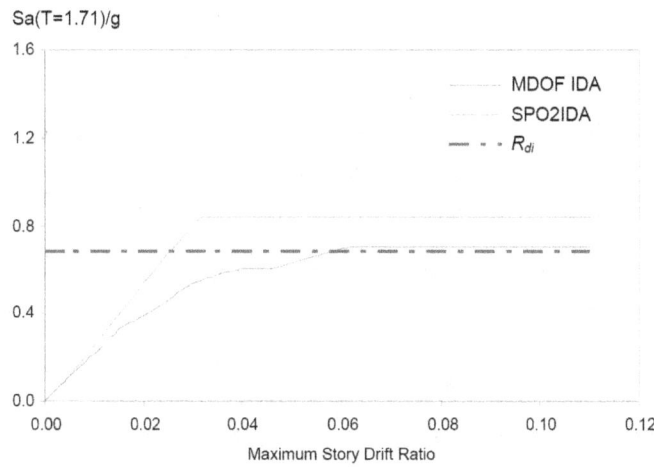

Figure G-9   Comparison of median collapse capacity for an eight-story code-compliant concrete frame building computed using incremental dynamic analysis and approximate procedures.

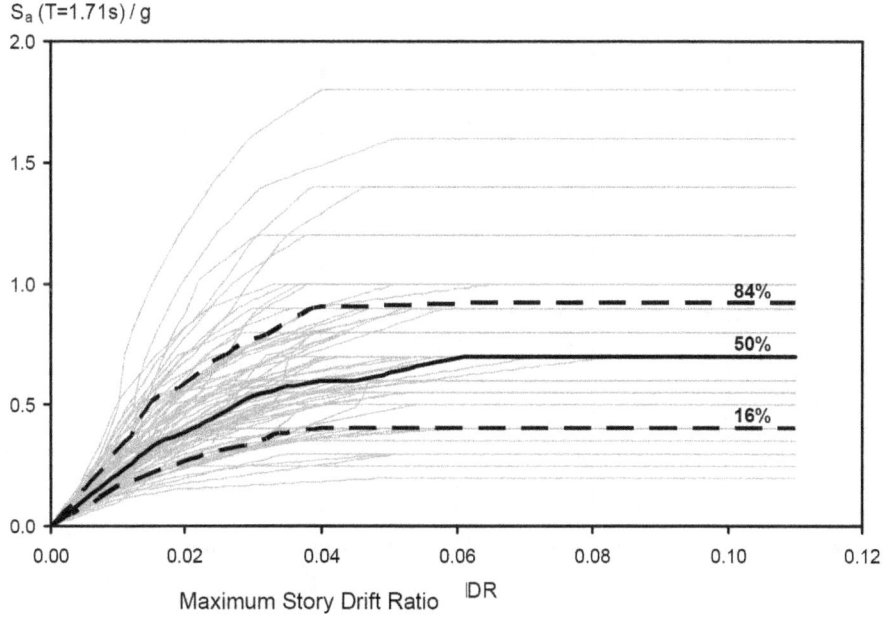

Figure G-10   Incremental dynamic analysis results for an eight-story code-compliant concrete frame building subjected to 80 ground motions (adapted from Haselton, 2006).

G.3 Twelve-Story Code-Compliant Reinforced Concrete Building

The subject building is a twelve-story reinforced concrete special perimeter moment frame designed in accordance with modern building code provisions (ICC 2003, ASCE 2002, ACI 2002). Similarly to the two previous buildings, the story height is 15 ft in the first story and 13 ft in the remaining stories. The design base shear coefficient was 0.044. The building was modeled in OpenSEES and analyzed using an incremental dynamic analysis using the same 80 recorded ground motions that were used to analyze the four-story building. The pushover analysis was again done using a lateral force distribution in accordance with ASCE/SEI 7-05. The fundamental period of vibration of the building is T_1=2.01s. For a more detailed description of the building and its modeling, the reader is referred to Haselton (2006).

Figure G-11 shows the results from the nonlinear static (pushover) analysis of the building. The figure on the left shows the force-deformation curve while the figure on the right shows the distribution of story drift ratios at a roof drift ratio of 2.7%. It can be seen that story drifts decrease approximately linearly with increasing height with the largest story drifts occurring in the two lower stories. The force-deformation pushover curve is characterized by a hardening segment for roof drift ratios between 0.3% and 0.8%, followed by softening segment for roof drift ratios greater than 0.8%.

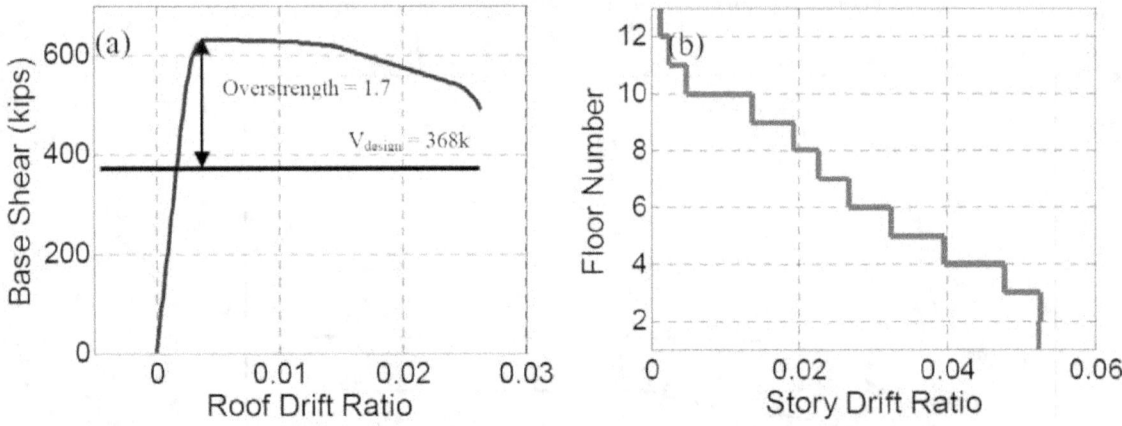

Figure G-11   (a) Monotonic pushover force-deformation curve and (b) distribution of story drift demands at a roof drift ratio of 2.7% in a twelve-story concrete frame building (Haselton 2006).

Figure G-12 shows the simplified tri-linear force-displacement capacity boundary selected to evaluate the proposed equation for R_{di} and results using SPO2IDA. It is assumed that at a roof drift ratio of 2.6% the structure reaches its maximum deformation capacity and a total loss in strength occurs.

Figure G-13 compares the median seismic behavior computed from incremental dynamic analyses conducted by Haselton (2006), indicated in the figure as MDOF IDA, with results computed using the proposed equation for R_{di} and approximate results from SPO2IDA. In the figure, both approximate methods somewhat overestimate the collapse capacity of the structure. Figure G-14 shows incremental dynamic analysis results for all ground motion records.

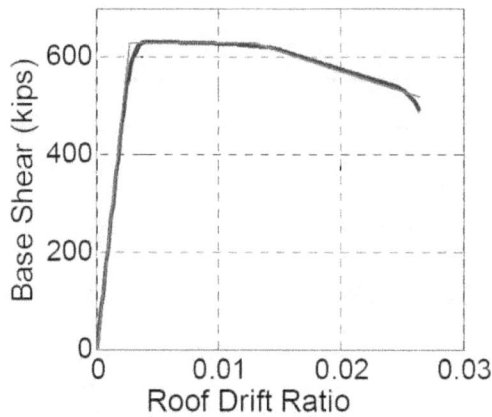

Figure G-12   Tri-linear capacity boundary selected for approximate analyses using *SPO2IDA*.

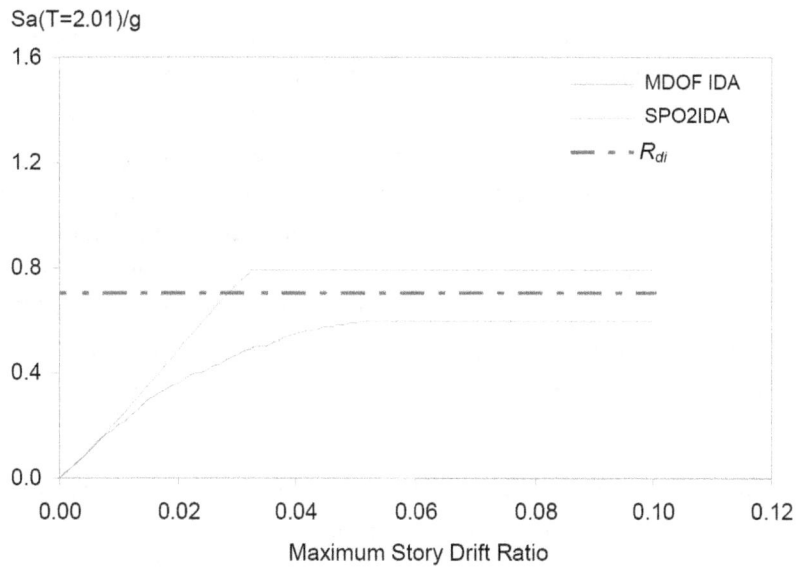

Figure G-13   Comparison of median collapse capacity for a twelve-story code-compliant concrete frame building computed using incremental dynamic analysis and approximate procedures.

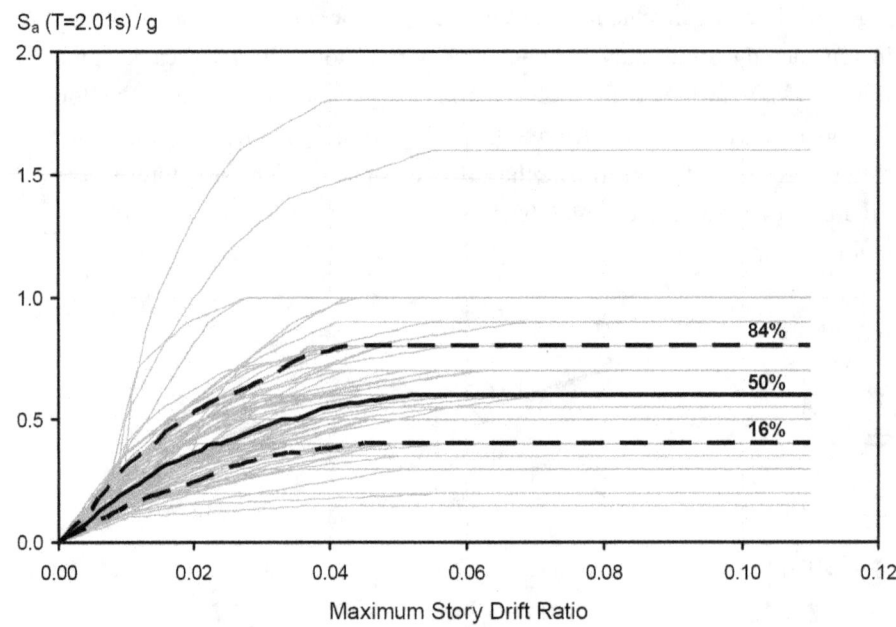

Figure G-14   Incremental dynamic analysis results for a twelve-story code-compliant concrete frame building subjected to 80 ground motions (adapted from Haselton, 2006).

G.4   Twenty-Story Code-Compliant Reinforced Concrete Building

The subject building is a twenty-story reinforced concrete special perimeter moment frame designed in accordance with modern building code provisions (ICC 2003, ASCE 2002, ACI 2002). The story height is 15 ft in the first story and 13 ft in the remaining stories. The design base shear coefficient was 0.044. The building was modeled in OpenSEES and analyzed using an incremental dynamic analysis using the same 80 recorded ground motions that were used to analyze the four-story building. The pushover analysis was again done using a lateral force distribution in accordance with ASCE/SEI 7-05. The fundamental period of vibration of the building is T_1=2.63s. For a more detailed description of the building and its modeling, the reader is referred to Haselton (2006).

Figure G-15 shows the results from the nonlinear static (pushover) analysis of the building. The figure on the left shows the force-deformation curve while the figure on the right shows the distribution of story drift ratios at a roof drift ratio of 1.8%. It can be seen that story drifts decrease approximately linearly with increasing height, with the largest story drifts occurring in the lower two stories. The force-deformation pushover curve is characterized by a slight softening segment for roof drift ratios between 0.3%

and 0.9%, followed by steeper softening segment for roof drift ratios greater than 0.9%.

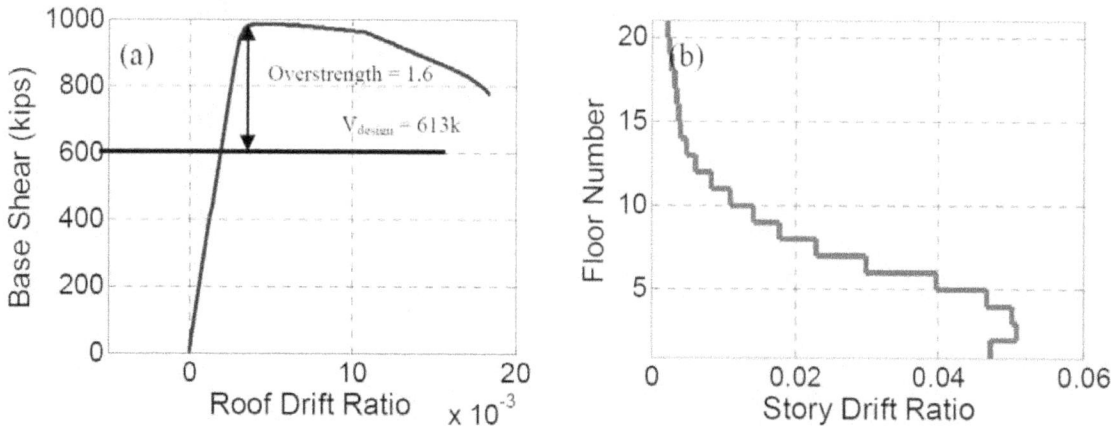

Figure G-15   (a) Monotonic pushover force-deformation curve and (b) distribution of story drift demands at a roof drift ratio of 1.8% in a twenty-story concrete frame building (Haselton 2006).

Figure G-16 shows the simplified tri-linear force-displacement capacity boundary selected to evaluate the proposed equation for R_{di} and results using SPO2IDA. It is assumed that at a roof drift ratio of 1.85% the structure reaches its maximum deformation capacity and a total loss in strength occurs.

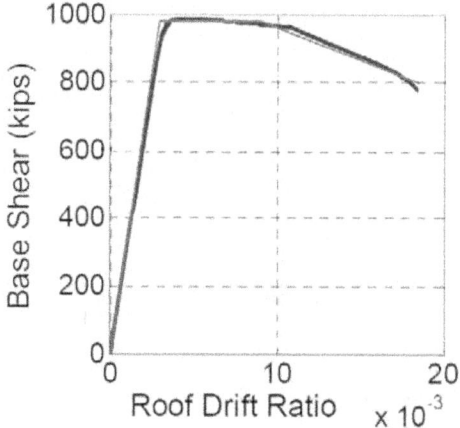

Figure G-16   Tri-linear capacity boundary selected for approximate analyses using *SPO2IDA*.

Figure G-17 compares the median seismic behavior computed from incremental dynamic analyses conducted by Haselton (2006), indicated in the figure as MDOF IDA, with results computed using the proposed equation for R_{di} and approximate results from SPO2IDA. In the figure, proposed equation for R_{di} provides a good estimate of the median collapse capacity, while SPO2IDA somewhat overestimates the collapse capacity. Figure G-18 shows incremental dynamic analysis results for all ground motion records.

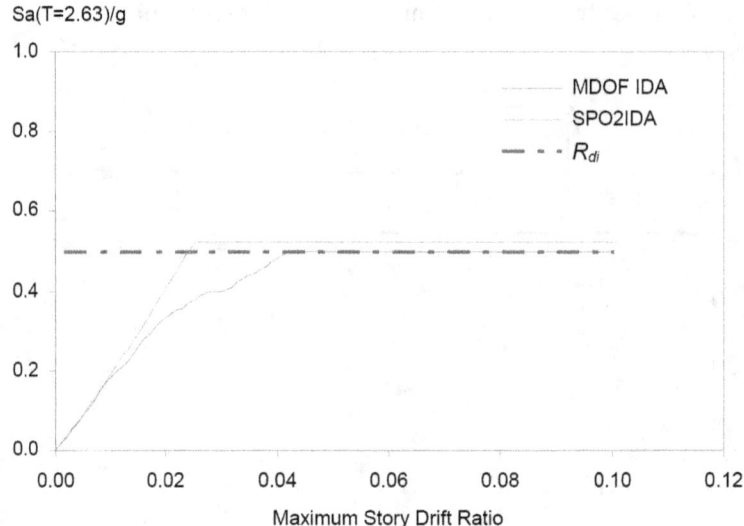

Figure G-17   Comparison of median collapse capacity for a twenty-story code-compliant concrete frame building computed using incremental dynamic analysis and approximate procedures.

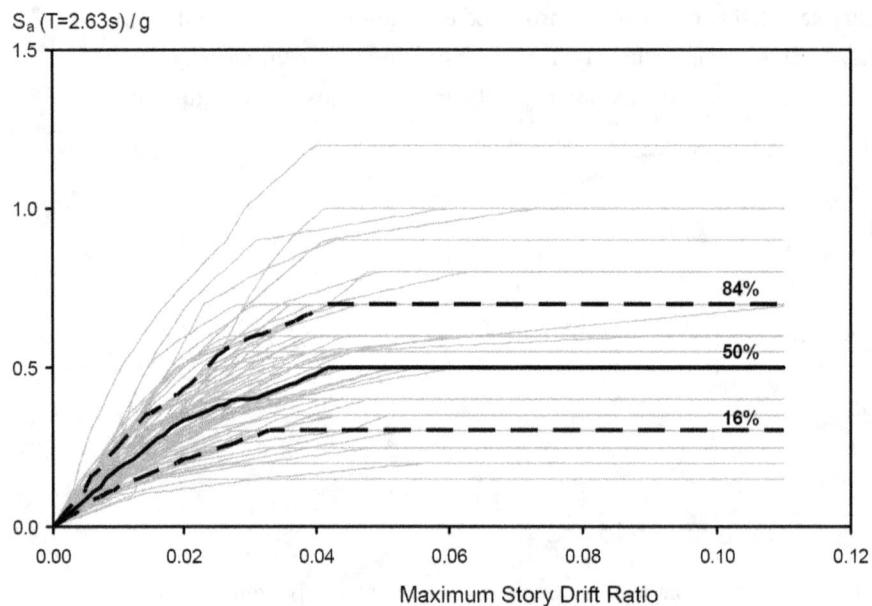

Figure G-18   Incremental dynamic analysis results for a twenty-story code-compliant concrete frame building subjected to 80 ground motions (adapted from Haselton, 2006).

G.5 Nine-Story Pre-Northridge Steel Moment-Resisting Frame Building

The subject building is a nine-story steel moment-resisting frame designed for the FEMA-funded SAC project in accordance with pre-Northridge code requirements for Los Angeles (ICBO, 1994). The building has a single-story basement that is 12 ft in height. The first story height is 18 ft and the remaining stories are 13 ft uniformly. The building is symmetric in plan with six bays of 30 ft in each direction. There is a perimeter moment-resisting frame designed for lateral-force-resistance, while internal gravity columns carry most of the vertical load. The building was modeled in OpenSEES and analyzed using incremental dynamic analysis with 30 "ordinary" ground motions. The pushover analysis was done using a triangular lateral force distribution. The fundamental period of vibration of the building is T_1=2.3s. For a more detailed description of the building and its modeling, the reader is referred to Gupta and Krawinkler (1999).

The results from a nonlinear static (pushover) analysis of the building are shown in Figure G-19. The force-deformation pushover curve is characterized by a hardening segment for roof drift ratios between 1% and 2.5%, followed by a softening segment that terminates when the building reaches zero strength at 5% roof drift. The simplified tri-linear force-displacement capacity boundary, also shown in Figure G-19, was selected to evaluate the proposed equation for R_{di} and results using SPO2IDA. In both cases the hardening segment has 13% of the elastic stiffness while the negative stiffness is -74% of elastic.

Figure G-20 shows the median seismic behavior computed from incremental dynamic analyses conducted by Vamvatsikos and Fragiadakis (2006). These results are indicated as MDOF IDA in the figure. Also shown are results computed using the proposed equation for R_{di} and approximate results from SPO2IDA. In the figure, both R_{di} and SPO2IDA provide a good approximation of the collapse capacity of the building.

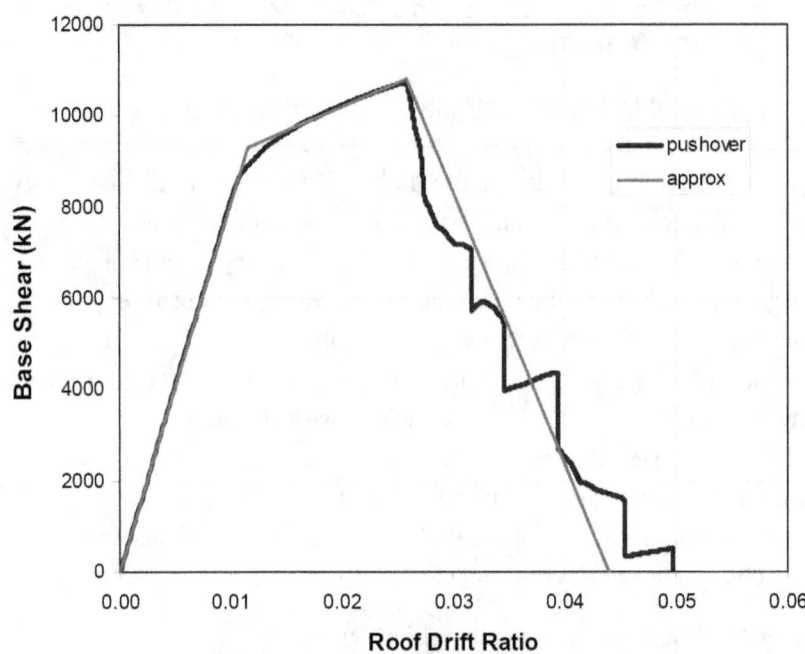

Figure G-19   Monotonic pushover force-deformation curve, and tri-linear approximation, for a nine-story pre-Northridge steel moment frame building (adapted from Gupta and Krawinkler, 1999).

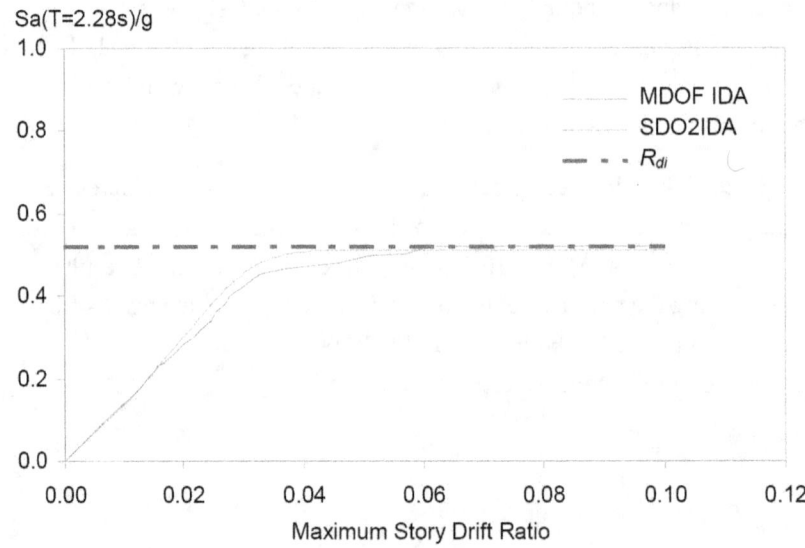

Figure G-20   Comparison of median collapse capacity for a nine-story pre-Northridge steel moment frame building computed using incremental dynamic analysis and approximate procedures.

G.6   Twenty-Story Pre-Northridge Steel Moment-Resisting Frame Building

The subject building is a twenty-story steel moment resisting frame designed for the FEMA-funded SAC project in accordance with pre-Northridge code requirements for Los Angeles (ICBO, 1994). The building has a basement consisting of two stories that are 12 ft in height. The first story height is 18 ft and the remaining stories are 13 ft uniformly. The building is slightly asymmetric in plan, with five bays of 20 ft in one direction and six bays of 20 ft in the other direction. There is a perimeter moment-resisting frame designed for lateral-force-resistance. Four internal gravity columns carry the vertical loads. The building was modeled in Drain-2DX and analyzed using incremental dynamic analysis with 30 "ordinary" ground motions. The pushover analysis was done using a parabolic ($k = 2$) lateral force distribution. The fundamental period of vibration of the building is $T_1=4.0s$. For a more detailed description of the building and its modeling, the reader is referred to Gupta and Krawinkler (1999).

The results from the nonlinear static (pushover) analysis of the building are shown in Figure G-21. The force-deformation pushover curve is characterized by a short hardening segment (5% stiffness ratio) from 0.7% to 1.2% roof drift ratio that then turns negative (-24% stiffness ratio) and terminates when the building reaches zero strength at 4% roof drift. The simplified tri-linear force-displacement capacity boundary, also shown in Figure G-21, was selected to evaluate the proposed equation for R_{di} and results using SPO2IDA.

Figure G-22 shows the median seismic behavior computed from incremental dynamic analyses conducted by Vamvatsikos and Cornell (2006). These results are indicated as MDOF IDA in the figure. Also shown are results computed using the proposed equation for R_{di} and approximate results from SPO2IDA. In the figure, R_{di} overestimates the collapse capacity of the building by about 25%, while SPO2IDA provides a good approximation.

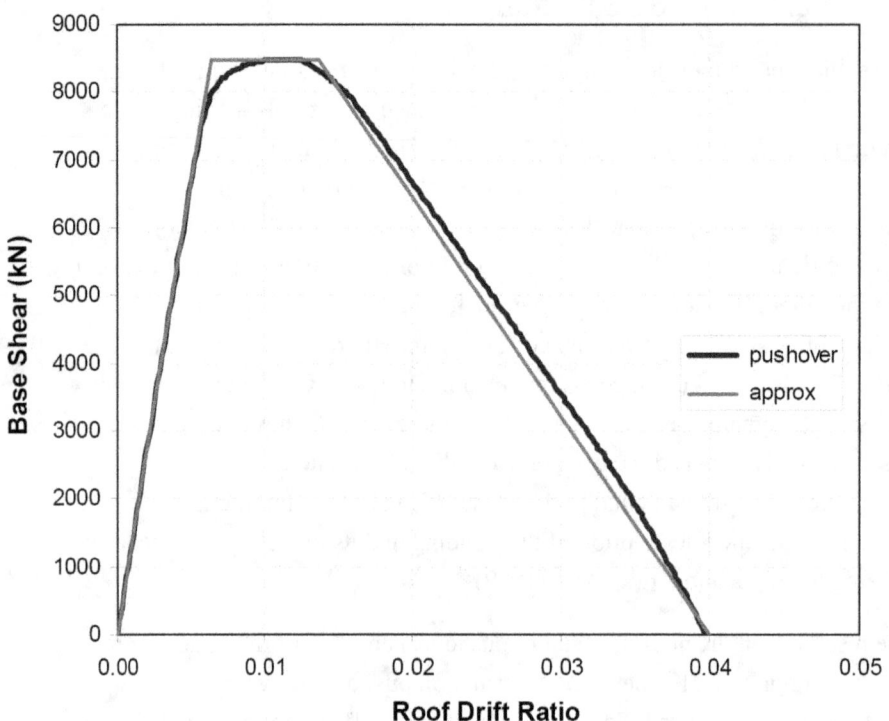

Figure G-21   Monotonic pushover force-deformation curve, and tri-linear approximation, for a twenty-story pre-Northridge steel moment frame building (adapted from Gupta and Krawinkler, 1999).

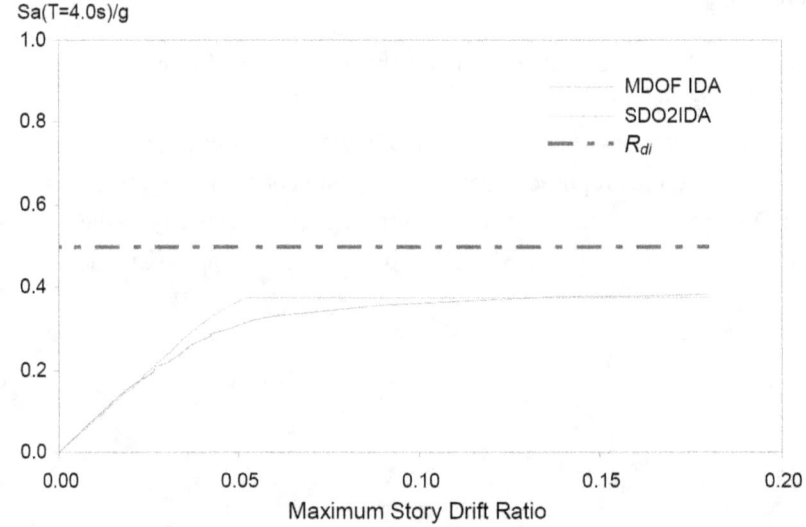

Figure G-22   Comparison of median collapse capacity for a twenty-story pre-Northridge steel moment frame building computed using incremental dynamic analysis and approximate procedures.

G.7 Summary and Recommendations

The studies documented above indicate that the application of procedures developed for SDOF systems to several representative MDOF moment frame systems produces reasonable approximations of the median intensity causing lateral dynamic instability. This was true in the case of both the proposed equation for R_{di} and simplified nonlinear dynamic analysis using SPO2IDA. These results lead to a recommendation for more thorough investigation of MDOF systems to modify, or further refine, the procedures presented here.

References and Bibliography

ACI, 2002, *Building Code Requirements for Structural Concrete (ACI 318-02)* and *Commentary (ACI 318R-02)*, American Concrete Institute, Farmington Hills, MI.

Allahabadi, R. and Powell, G. H., 1988, *DRAIN-2DX: User Guide*, Report No. UCB/EERC-88/06, Earthquake Engineering Research Center, University of California, Berkeley, California.

Anagnostopoulos, S.A., 1972, *Nonlinear Dynamic Response and Ductility Requirements of Building Structures Subjected to Earthquakes*, Report No. R72-54, Civil Engineering Dept., Massachusetts Institute of Technology, Cambridge, Massachusetts.

Archambault, M.H., Tremblay, R., and Filiatrault, A., 1995, *Etude du Comportement Sismique des Contreventements Ductiles en X avec Profiles Tubulaires en Acier.* Rapport no. EPM/GCS-1995-09, *in* Departement de Genie Civil, Ecole Polytechnique, Montreal, Canada, (in French).

ASCE, 2002, *Minimum Design Loads for Buildings and Other Structures*, ASCE 7-02, American Society of Civil Engineers, Reston, VA.

ASCE, 2003, *Seismic Evaluation of Existing Buildings*, ASCE Standard SEI/ASCE 31-03, American Society of Civil Engineers, Reston, Virginia.

ASCE, 2006a, *Minimum Design Loads for Buildings and Other Structures*, ASCE Standard ASCE/SEI 7-05, including Supplement No. 1, American Society of Civil Engineers, Reston, Virginia.

ASCE, 2006b, *Seismic Rehabilitation of Existing Buildings*, ASCE Standard ASCE/SEI 41-06, American Society of Civil Engineers, Reston, Virginia.

Astaneh-Asl, A., Goel, S.C., and Hanson, R.D., 1982, *Cyclic Behavior of Double Angle Bracing Members with End Gusset Plates*, Rep. No. UMEE 82R7, Dept. of Civil Engineering, University of Michigan, Ann Arbor, Michigan.

Astaneh-Asl, A., Goel, S.C., 1984, "Cyclic in-plane buckling of double angle braces," *Journal of Structural Engineering*, ASCE, 110(9), pp. 2036-2055.

ATC, 1992, *Guidelines for Seismic Testing of Components of Steel Structures*, ATC-24 Report, Applied Technology Council, Redwood City, California.

ATC, 1996, *Seismic Evaluation and Retrofit of Concrete Buildings*, ATC-40 Report, Volumes 1 and 2, Applied Technology Council, Redwood City, California.

ATC, 2007, *Guidelines for Seismic Performance Assessment of Buildings*, ATC-58 Report, 35% draft, prepared by the Applied Technology Council for the Federal Emergency Management Agency, Redwood City, California.

ATC, 2008b, *Interim Guidelines on Modeling and Acceptance Criteria for Seismic Design and Analysis of Tall Buildings*, ATC-72-1 Report, 90% Draft, prepared by the Applied Technology Council for the Pacific Earthquake Engineering Research Center, Redwood City, California.

Baber, T.T. and Wen, Y., 1981, "Random vibration hysteretic, degrading systems," *J. Eng. Mech. Div.,* 107, 6, 1069-1087

Baber, T.T., and Noori, M.N., 1985, "Random vibration of degrading, pinching systems." *J. Engrg. Mech.,* ASCE, 111(8), 1010–1026.

Baber, T.T. and Noori, M.N., 1986, "Modeling general hysteresis behavior and random vibration application," *Trans. ASME,* 108, 411-420

Berg, G.V. and Da Deppo, D.A., 1960, "Dynamic analysis of elastoplastic structures," *Proc. American Society of Civil Engineers*, Vol. 86, No. 2, pp 35-58.

Bernal, D., 1998, "Instability of buildings during seismic response," *Engineering Structures*, Vol. 20, No. 4-6, pp. 496-502.

Bernal, D., 1992, "Instability of buildings subjected to earthquakes," *Journal of Structural Engineering* , ASCE, Vol. 118, No. 8, pp. 2239-2260.

Bernal, D., 1987, "Amplification factors for inelastic dynamic P-Δ effects in earthquake analysis," *Earthquake Eng. Struct. Dyn.,* 15(5), pp. 117-144.

Black, R.G., Wenger, W.A.B. and Popov, E.P., 1980, *Inelastic Buckling of Steel Struts under Cyclic Load Reversals,* Report No. UCB/EERC-

80/40, Earthquake Engineering Research Center, University of California, Berkeley, California.

Bouc, R., 1967, "Modele mathematique d'hysteresis." *Acustica*, France, 24(1), 16–25 (in French).

Bouc, R., 1967, "Forced vibration of mechanical systems with hysteresis," *Proc., 4th Conf. on Nonlinear Oscillations,* Prague, Czechoslovakia

Bruneau, M. and Vian, D., 2002, "Tests to collapse of simple structures and comparison with existing codified procedures," *Proc. 7th U.S. National Conference on Earthquake Engineering*, Boston, MA.

Bruneau, M. and Vian, D., 2002, "Experimental investigation of P-Δ effects to collapse during earthquakes," *Proc. 12th European Conference on Earthquake Engineering*, London, UK.

Casciati, F., 1989, "Stochastic dynamics of hysteretic media." *Struct. Safety*, Amsterdam, 6, 259–269.

Casciati, F., and Faravelli, L., 1985, "Nonlinear stochastic dynamics by equivalent linearization, *Proc. 2nd Int. Workshop on Stochastic Methods in Structural Mechanics*, Pavia, Italy, 571-586

Caughey, T.K., 1960a, "Sinusoidal excitation of a system with bilinear hysteresis," *Journal of Applied Mechanics,* American Society of Mechanical Engineers 27(4), pp. 640–648.

Caughey, T.K., 1960b, "Random excitation of a system with bilinear hysteresis," *Journal of Applied Mechanics,* American Society of Mechanical Engineers, 27(4), pp. 649–652.

Chenouda, M. and Ayoub, A., 2007, "Inelastic displacement ratios of degrading systems." *Journal of Structural Engineering*, 134(6): 1030-1045.

Chopra, A.K., Chintanapakdee, C., 2004, "Inelastic deformation ratios for design and evaluation of structures: single-degree-of-freedom bilinear systems." *Journal of Structural Engineering*, 130(9): 1309-1319.

Clough, R.W., 1966, *Effects of Stiffness Degradation on Earthquake Ductility Requirement*, Rep. No. 6614, Struct. and Mat. Res., University of California, Berkeley, California.

Clough, R.W., and Johnston, S.B., 1966, "Effects of stiffness degradation on earthquake ductility requirements." *Proc. 2nd Japan Earthquake Engrg. Symp.*, 227–232.

Dolsek, M., 2002, *Seismic Response of Infilled Reinforced Concrete Frames*, Ph.D. thesis, University of Ljubljana, Faculty of Civil and Geodetic Engineering, Ljubljana, Slovenia [in Slovenian].

Dolsek, M. and Fajfar, P., 2004, "Inelastic spectra for infilled reinforced concrete frames," *Earthquake Engineering and Structural Dynamics*, Vol. 33, 1395–1416.

Dolsek ,M. and Fajfar, P., 2005, "Simplified nonlinear seismic analysis of infilled reinforced concrete frames," *Earthquake Engineering and Structural Dynamics*, Vol. 34, 49–66.

Dotiwala, F.S., 1996, *A Nonlinear Flexural-Shear Model for RC Columns Subjected to Earthquake Loads*, MS thesis, Department of Civil and Environmental Engineering, University of Wisconsin-Madison.

EERI, 2006, *New Information on the Seismic Performance of Existing Concrete Buildings*, EERI Technical Seminar developed by PEER and funded by FEMA, video download @ www.eeri.org.

EERI, 2008, *World Housing Encyclopedia*, www.world-housing.net,

El-Bahy, A., Kunnath, S.K., Stone, W.C. and Taylor, A.W., 1999, "Cumulative seismic damage of circular bridge columns: variable amplitude tests," *ACI Structural Journal*, Vol. 96, No.5, 711-719.

El-Tawil, S., 1996, *Inelastic Dynamic Analysis of Mixed Steel-Concrete Space Frames*, Ph.D. Dissertation, Cornell University, Ithaca, NY.

Elwood, K.J., 2002, *Shake Table Tests and Analytical Studies on the Gravity Load Collapse of Reinforced Concrete Frames*, Ph.D. Dissertation, University of California, Berkeley, California.

Elwood, K.J., and Moehle, J.P., 2003, *Shake-Table Tests and Analytical Studies on the Gravity Load Collapse of Reinforced Concrete Frames*, PEER Report 2003/01, Pacific Earthquake Engineering Research Center, University of California, Berkeley, California.

Elwood, K.J., 2004, "Modeling failures in existing reinforced concrete columns," *Can. J. Civ. Eng.* 31: 846–859.

Elwood, K.J., and Moehle, J.P., 2005, "Drift capacity of reinforced concrete columns with light transverse reinforcement," *Earthquake Spectra*, Volume 21, No. 1, pp. 71–89.

Elwood, K.J. and Moehle, J.P., 2006, "Idealized backbone model for existing reinforced concrete columns and comparisons with FEMA 356 criteria," *The Struct. Design of Tall Spec. Buildings*, 15, 553–569

Elwood, K.J., Matamoros, A.B., Wallace, J.W., Lehman, D.E., Heintz, J.A., Mitchell, A.D., Moore, M.A., Valley, M.T., Lowes, L.N., Comartin, C.D., Moehle, J.P., 2007, "Update to ASCE/SEI 41 concrete provisions," *Earthquake Spectra*, Earthquake Engineering Research Institute, Oakland, California.

FEMA, 1997, *NEHRP Guidelines for the Seismic Rehabilitation of Buildings*, FEMA 273 Report, prepared by the Building Seismic Safety Council and the Applied Technology Council for the Federal Emergency Management Agency, Washington, D.C.

FEMA, 1998, *Handbook for the Seismic Evaluation of Buildings – A Prestandard*, FEMA 310 Report, prepared by the American Society of Civil Engineers for the Federal Emergency Management Agency, Washington, D.C.

FEMA, 2000, *Prestandard and Commentary for the Seismic Rehabilitation of Buildings*, FEMA 356 Report, prepared by the American Society of Civil Engineers for the Federal Emergency Management Agency, Washington, D.C.

FEMA, 2004a, *NEHRP Recommended Provisions for Seismic Regulations for New Buildings and Other Structures, Part 1: Provisions*, FEMA 450-1, 2003 Edition, Federal Emergency Management Agency, Washington, D.C.

FEMA, 2004b, *NEHRP Recommended Provisions for Seismic Regulations for New Buildings and Other Structures, Part 2: Commentary*, FEMA 450-2, 2003 Edition, Federal Emergency Management Agency, Washington, D.C.

FEMA, 2005, *Improvement of Nonlinear Static Seismic Analysis Procedures*, FEMA 440 Report, prepared by the Applied Technology Council for the Federal Emergency Management Agency, Washington, D.C.

FEMA, 2009, *Quantification of Building Seismic Performance Factors*, FEMA P695 Report, prepared by the Applied Technology Council for the Federal Emergency Management Agency, Washington, D.C.

Foutch, D.A. and Shi, S., 1998, "Effects of hysteresis type on the seismic response of buildings," *Proc. 6th U.S. National Conference on Earthquake Engineering*, Seattle, Washington; Earthquake Engineering Research Institute, Oakland, California.

Goel, SC., El-Tayem A., 1986, "Cyclic load behavior of angle X-bracing," *Journal of Structural Engineering*, ASCE, 112(11), pp. 2528-2539.

Goel, S.C., and Stojadinovic, B., 1999, University of Michigan, Ann Arbor, Michigan.

Gugerli, H., and Goel, S.C., 1982, *Inelastic Cyclic Behavior of Steel Bracing Members*, Rep. No. UMEE 82R1, Dept. of Civil Engineering, University of Michigan, Ann Arbor, Michigan.

Gupta, A., and Krawinkler, H., 1998, "Effect of stiffness degradation on deformation demands for SDOF and MDOF structures," *Proc. Sixth U.S. National Conference on Earthquake Engineering*, Seattle, Washington.

Gupta, A., and Krawinkler, H., 1999, *Seismic Demands for Performance Evaluation of Steel Moment-Resisting Frame Structures*. Rep. No. 132, John A. Blume Earthquake Engineering Center, Dept. of Civil and Environmental Engineering, Stanford University, Stanford, California.

Gupta, B., and Kunnath, S., 1998, "Effect of hysteretic model parameters on inelastic seismic demands," *Proc., 6th Nat. Conf. Earthq. Engrg.*, Seattle, Washington.

Haselton, C.B., 2006, *Assessing Seismic Collapse Safety of Modern Reinforced Concrete Moment Frame Buildings*, Ph.D. Dissertation, Department of Civil and Environmental Engineering, Stanford University.

Haselton, C., Liel, A., Taylor Lange, S., and Deierlein, G.G., 2007, *Beam-Column Element Model Calibrated for Predicting Flexural Response Leading to Global Collapse of RC Frame Buildings* (In Preparation). Pacific Earthquake Engineering Research Center 2007/03, University of California, Berkeley, California.

Hassan, O.F. and Goel, S.C., 1991, *Modeling of Bracing Members and Seismic Behavior of Concentrically Braced Steel Structures*, Research Report UMCE 91-1, Department of Civil Engineering, University of Michigan, Ann Arbor, Michigan,

Higginbotham, A.B. and Hanson, R.D., 1976, "Axial hysteretic behavior of steel members," *J. Struct. Div.*, ASCE, 102, pp. 1365-81.

Hisada, T., Nakagawa, K., and Izumi, M., 1962, "Earthquake response of structures having various restoring force characteristics," *Proc. Japan National Conference on Earthquake Engineering*, pp. 63-68.

Ibarra, L., Medina, R., Krawinkler, H., 2002, "Collapse assessment of deteriorating SDOF systems," *Proc. 12th European Conference on*

Earthquake Engineering, London, UK, Paper 665, Elsevier Science Ltd.

Ibarra, L.F., and Krawinkler, H., 2005, *Global Collapse of Frame Structures under Seismic Excitations*, Report No. 152, John A. Blume Earthquake Engineering Research Center, Department of Civil and Environmental Engineering, Stanford University, Stanford, California.

Ibarra, L., Medina, R., and Krawinkler, H., 2005, "Hysteretic models that incorporate strength and stiffness deterioration, *Earthquake Engineering and Structural Dynamics*, Vol. 34, no. 12, pp. 1489-1511

ICBO, 1994, *Uniform Building Code*, International Conference of Building Officials, Whittier, California.

ICC, 2003, *International Building Code*, International Code Council, Washington, D.C.

Ikeda, K., Mahin, S.A., and Dermitzakis, S.N., 1984, *Phenomenological Modeling of Steel Braces under Cyclic Loading*, Rep. No. UCB/EERC-84/09, Earthquake Engineering Research Center, University of California, Berkeley, California.

Ikeda, K., and Mahin, S.A., 1984, *A Refined Physical Theory Model for Predicting the Behavior of Braced Steel Frames*, Rep. No. UCB/EERC-84/12, Earthquake Engineering Research Center, University of California, Berkeley, California.

Ikeda, K., and Mahin, S.A., 1986, "Cyclic response of steel braces." *J. Struct. Eng.*, 112(2), 342–361.

Ingham J.M., Liddell D. and Davidson B., 2001, Influence of loading history on the response of a reinforced concrete beam, *Bulletin of the New Zealand Society of Earthquake Engineering*, 34(2), 107-124

Iwan, W.D., 1961, *The Dynamic Response of Bilinear Hysteretic Systems*, Ph.D. Thesis, California Institute of Technology, Pasadena, California.

Iwan, W.D., 1966, "A distributed-element model for hysteresis and its steady-state dynamic response." *J. Appl. Mech.*, 33(42), 893–900.

Iwan, W.D., 1967, "On a class of models for the yielding behavior of continuous and composite systems," *J. Appl. Mech.*, 34, 612–617.

Iwan, W.D., 1973, "A model for the dynamic analysis of deteriorating structures," *Proc. Fifth World Conference on Earthquake Engineering,* Rome, Italy, pp. 1782-1791.

Iwan, W.D., 1977, "The response of simple stiffness degrading structures," *Proc. Sixth World Conference on Earthquake Engineering,* pp. 1094-1099.

Iwan, W.D., 1978, "The earthquake response of strongly deteriorating systems including gravity effects", *Proc. Sixth European Conference on Earthquake Engineering,* pp. 23-30, 1978.

Iwan, W.D., and Gates, N.C., 1979a, "Estimating earthquake response of simple hysteretic structures." *J. Eng. Mech. Div.,* ASCE, 105(3), 391–405.

Iwan, W.D., and Gates, N.C., 1979b, "The effective period and damping of a class of hysteretic structures," *Earthquake Engineering and Structural Dynamics.* Vol. 7, no. 3, pp. 199-211.

Iwan, W.D., 1980, "Estimating inelastic response spectra from elastic spectra," *Earthquake Engineering and Structural Dynamics.* Vol. 8, no. 4, pp. 375-388.

Jacobsen, L.S., 1958, "Behavioristic models representing hysteresis in structural joints," *Proc. Festskrift in Honor of Prof. Anker Engelund's seventieth birthday,* Technical University of Denmark

Jain, A.K., Goel, S.C., and Hanson, R.D., 1978a, *Hysteresis Behavior of Bracing Members and Seismic Response of Braced Frames with Different Proportions,* Rep. No. UMEE 78R3, Dept. of Civil Engineering, University of Michigan, Ann Arbor, Michigan.

Jain, A.K., Goel, S.C., and Hanson, R.D., 1978b, "Inelastic response of restrained steel tubes." *Journal of Structural Division,* ASCE, 104(6), pp. 897-910.

Jain, A.K. and Goel, S.C., 1978, *Hysteresis Models for Steel Members Subjected to Cyclic Buckling or Cyclic End Moments and Buckling,* (User's Guide for DRAIN-2D: EL9 and EL10), Report UMEE 78R6, Department of Civil Engineering, University of Michigan, Ann Arbor, Michigan.

Jain, A.K., Goel, S.C., and Hanson, R.D., 1980, "Hysteretic cycles of axially loaded steel members." *Journal of Structural Division,* ASCE, 106(ST8), pp. 1777-1795.

Jennings, P.C., 1963, *Response of Simple Yielding Structures to Earthquake Excitation*, Ph.D. Thesis, California Institute of Technology, Pasadena, California.

Jun Jin and Sherif El-Tawil, 2003, "Inelastic cyclic model for steel braces," *J. Engrg. Mech.*, 129(5), 548-557.

Kanvinde, A.M., 2003, "Methods to evaluate the dynamic stability of structures – shake table tests and nonlinear dynamic analyses," EERI Annual Student Paper Competition, *Proceedings of 2003 EERI Annual Meeting*, Portland.

Kaul, R., 2004, *Object-Oriented Development of Strength and Stiffness Degrading Models for Reinforced Concrete Structures*, Ph.D. Thesis, Department of Civil and Environmental Engineering, Stanford University, Stanford, California.

Kawashima, K., MacRae, G.A., Hoshikuma, J., and Nagaya K., 1998, "Residual displacement response spectrum." *Journal of Structural Engineering*, 124(5), pp. 523-530.

Khatib, I. and Mahin, S., 1987, "Dynamic inelastic behavior of chevron-braced steel frames", *Fifth Canadian Conference on Earthquake Engineering*, Balkema, Rotterdam.

Khatib, I.F., Mahin, S.A., and Pister, K.S., 1988, *Seismic Behavior of Concentrically Braced Steel Frames*, Report UCB/EERC-88/01, Earthquake Engineering Research Center, University of California. Berkeley, California.

Kunnath, S.K., Reinhorn, A.M. and Park, Y.J., 1990, "Analytical modeling of inelastic seismic response of RC structures," *Journal of Structural Engineering*, ASCE, 116, 996–1017.

Kunnath S.K., Mander J.B. and Fang L., 1997, "Parameter identification for degrading and pinched hysteretic structural-concrete systems," *Engineering Structures*, 19(3), 224-232.

Lee, L.H, Han, S.W., and Oh, Y.H., 1999, "Determination of ductility factor considering different hysteretic models," *Earthquake Engineering and Structural Dynamics*, Vol. 28, 957–977.

Lignos, D.G., (2008), *Sidesway Collapse of Deteriorating Structural Systems under Seismic Excitations*, Ph.D. Thesis, Department of Civil and Environmental Engineering, Stanford University, Stanford, California.

Lin, M.L., Weng, Y.T., Tsai, K.C., Hsiao, P.C., Chen, C.H. and Lai, J.W., 2004, Pseudo-dynamic test of a full-scale CFTBRB frame: Part 3 - Analysis and performance evaluation. *Proc. 13th World Conference on Earthquake Engineering*, Vancouver, B.C., Canada, Paper No. 2173, August 1-6.

Liu, J. and Astaneh-Asl, A., 2004, "Moment-rotation parameters for composite shear-tab connections," *Journal of Structural Engineering*, ASCE, Vol. 130, No. 9, 1371-1380.

MacRae, G.A., and Kawashima, K., 1997, "Post-earthquake residual displacements of bilinear oscillators," *Earthquake Engineering and Structural Dynamics* 26, pp. 701-716.

Mahin, S.A. and Bertero, V.V., 1972, *Rate of Loading Effect on Uncracked and Repaired Reinforced Concrete Members*, Report EERC 73/6, Earthquake Engineering Research Center, University of California, Berkeley, California

Mahin, S.A. and Lin, J., 1983, *Construction of inelastic response spectra for single-degree-of-freedom systems: computer program and applications*, Report No. UCB/EERC-84/09, Earthquake Engineering Research Center, University of California, Berkeley, California.

Maison, B.F., and Popov, E.P., 1980, "Cyclic response prediction for braced steel frames," *J. Struct. Div.*, ASCE, 106(7), 1401–1416.

Masing, G., 1926, "Eigenspannungen und verfestigung beim messing." *Proc., 2nd Int. Cong. Appl. Mech.*, 332–335.

McKenna, F.T., 1997, *Object-Oriented Finite Element Programming: Framework for Analysis, Algorithms, and Parallel Computing*, Ph.D. Dissertation, University of California, Berkeley, California.

Medina, R., 2002, *Seismic Demands for Nondeteriorating Frame Structures and their Dependence on Ground Motions*, Ph.D. Dissertation, Department of Civil and Environmental Engineering, Stanford University.

Medina, R. and Krawinkler, H., 2004, "Influence of hysteretic behavior on the nonlinear response of frame structures," *Proc. 13th World Conference on Earthquake Engineering*, Vancouver, Canada.

Mehanny, S.S.F., and Deierlein, G.G., 2001, "Seismic collapse assessment of composite RCS moment frames," *Journal of Structural Engineering*, ASCE, Vol. 127(9).

Menegotto, M., and Pinto, P., 1973, "Method of analysis for cyclically loaded RC plane frames including changes in geometry and non-elastic behavior of elements under combined normal force and bending." *Proc., Symp. Resistance and Ultimate Deformability of Struct. Acted on by Well-Defined Repeated Loads*, IABSE Reports, Vol. 13. 2557–2573.

Miranda, E., 1993, "Evaluation of site-dependent inelastic seismic design spectra," *Journal of Structural Engineering*, ASCE, 119(5), pp. 1319-1338.

Miranda, E., 2000, "Inelastic displacement ratios for structures on firm sites," *Journal of Structural Engineering*, ASCE, 126(10), pp. 1150-1159.

Miranda, E. and Akkar, S.D., 2003, "Dynamic instability of simple structural systems," *Journal of Structural Engineering*, ASCE, 129(12), pp 1722-1727.

Miranda, E. and Ruiz-Garcia, J., 2002, "Influence of stiffness degradation on strength demands of structures built on soft soil sites," *Engineering Structures*, Vol. 24, No. 10, pp. 1271-1281.

Mostaghel, N., 1998, *Analytical description of pinching, degrading hysteretic systems*, Rep. No. UT-CE/ST-98-102, Dept. of Civ. Engrg., University of Toledo, Toledo, Ohio.

Mostaghel, N., 1999, "Analytical description of pinching, degrading, hysteretic systems," *J. Engrg. Mech.*, ASCE, 125(2), 216–224.

Mostaghel, N. and Byrd, R.A., 2002, "Inversion of Ramberg-Osgood equation and description of hysteresis loops," *International Journal of Non-Linear Mechanics*. Vol. 37, no. 8, pp. 1319-1335.

Nakashima, M. and Wakabayashi, M., 1992, "Analysis and design of steel braces and braced frames in building structures," *in* Fukumoto Y, Lee GC, editors, *Stability and Ductility of Steel Structures under Cyclic Loading*, Boca Raton, FL: CRC Press; p. 309-21.

Nassar, A.A. and Krawinkler, H., 1991, *Seismic Demands for SDOF and MDOF Systems*, Report No. 95, John A. Blume Earthquake Engineering Center, Stanford University, Stanford, California, 204 pages.

Nielsen, N.N. and Imbeault, F.A., 1970, "Validity of various hysteretic systems," *Proc. Third Japan Earthquake Engineering Symposium*, Tokyo, Japan, pp. 707-714.

Nonaka, T., 1973, "An elastic-plastic analysis of a bar under repeated axial loading," *Int. J. Solids Struct.*, 9, 569–580.

Nonaka, T., 1977, "Approximation of yield condition for the hysteretic behavior of a bar under repeated axial loading," *Int. J. Solids Struct.*,13, 637–643.

Otani, S. and Sozen, M.A., 1972, *Behavior of Multistory Reinforced Concrete Frames during Earthquakes*, Civil Engineering Studies, SRS No. 392, University of Illinois at Urbana, 551 pages.

Otani, S., 1981, "Hysteresis model of reinforced concrete for earthquake response analysis," *Journal, Fac. of Eng.*, University of Tokyo, Series B, Vol. XXXVI-11, 2, 407-441.

Ozdemir, H., 1976, *Nonlinear Transient Dynamic Analysis of Yielding Structures*, Ph.D. Dissertation, University of California, Berkeley, California.

Pampanin, S, Christopoulos, C, and Priestley, M.J.N., 2002, *Residual Deformations in the Performance-Seismic Assessment of Frame Structures*, Research Report No. ROSE-2002/02, European School for Advanced Studies in Reduction of Seismic Risk, Pavia, Italy.

Park, Y.J., and Ang, A., 1985, "Mechanistic seismic damage model for reinforced concrete," *Journal of Structural Engineering*, ASCE, 111-4, pp. 722-739.

Park, Y. J., Reinhorn, A.M., and Kunnath, S.K., 1987, *IDARC: inelastic damage analysis of reinforced concrete frame, shear-wall, structures*, Tech. Rep. NCEER-87-0008, State University of New York at Buffalo, Buffalo, New York.

PEER, 2005, *Van Nuys Hotel Building Testbed Report: Exercising Seismic Performance Assessment*, PEER Report 2005/11, Pacific Earthquake Engineering Research Center, College of Engineering, University of California, Berkeley.

PEER, 2006, *PEER NGA Database*, Pacific Earthquake Engineering Research Center, University of California, Berkeley, California, http://peer.berkeley.edu/nga/

Penzien, J., 1960a, "Elastoplastic response of idealized multistory structures subjected to a strong-motion earthquake, *Proc. Second World Conf. on Earthquake Engineering*, Vol. II, Tokyo and Kyoto, Japan.

Penzien, J., 1960b, "Dynamic response of elastoplastic frames," *Proc. American Society of Civil Engineers*, Vol. 86, No. 7, pp 81-94.

Pincheira, J.A., and Dotiwala, F.S., 1996, "Modeling of nonductile RC columns subjected to earthquake loading," *Proc. 11th World Conf. on Earthquake Engineering*, Paper No. 316, Acapulco, Mexico.

Pincheira, J.A, Dotiwala, F.S., and D'Souza, J.T., 1999, "Spectral displacement demands of stiffness- and strength-degrading systems," *Earthquake Spectra*, 15(2), 245–272.

Pinto, P. E. and Giuffre, A., 1970, "Behavior of reinforced concrete sections under cyclic loading of high intensity," *Giornale del Genio Civile*, No. 5, (in Italian).

Popov, E.P. and Black, G.R., 1981, "Steel struts under severe cyclic loadings," *Journal of Structural Engineering*, ASCE, 107(9), pp. 1857-1881

Popov, E.P., Zayas, V.A., and Mahin, S.A., 1979, "Cyclic inelastic buckling of thin tubular columns." *J. Struct. Div.*, ASCE, 105(11), pp. 2261–2277.

Powell, G.H., and Row, D.G., 1976, *Influence of Analysis and Design Assumptions on Computed Inelastic Response of Moderately Tall Frames*, Report No. UCB/EERC-76/11, Earthquake Engineering Research Center, University of California at Berkeley, Berkeley, California.

Prathuangsit, D., Goel, S.C. and Hanson, R.D., 1978, "Axial hysteresis behavior with end restraints," *J. Struct. Div.*, ASCE, 104, 883-95.

Rahnama, M, and Krawinkler, H., 1993, *Effects of Soft Soil and Hysteretic Models on Seismic Demands*, Report No. 108, John A. Blume Earthquake Engineering Center, Department of Civil and Environmental Engineering, Stanford University, Stanford, California, 258 pages.

Rai, D.C., Goel, S.C., and Firmansjah, J., 1996, *User's Guide Structural Nonlinear Analysis Program (SNAP)*, Department of Civil and Environmental Engineering; University of Michigan; College of Engineering, Ann Arbor, Michigan.

Ramberg, W., and Osgood, W.R., 1943, *Description of Stress-Strain Curves by Three Parameters*, Tech. Note 902, National Advisory Committee on Aeronautics.

Remennikov, A.M. and Walpole, W.R., 1997a, "Analytical prediction of seismic behavior for concentrically-braces steel systems," *Earthquake Eng. Struct. Dyn.*, 26, 859–874.

Remennikov, A.M. and Walpole, W.R., 1997b, "Modeling the inelastic cyclic behavior of a bracing member for work-hardening material," *Int. J. Solids Struct.*, 34, 3491–3515.

Riddell, R. and Newmark, N.M., 1979, *Statistical Analysis of the Response of Nonlinear Systems Subjected to Earthquakes*, Rep. No. UILI-ENG79-2016, Structural Research Series, Dept. of Civil Engineering, University of Illinois, Urbana.

Riddell, R. and Newmark, N.M., 1979, "Force-deformation models for nonlinear analyses," *J. Struct. Div.*, ASCE, 105, pp. 2773–2778.

Ruiz-Garcia, J. and Miranda, E., 2003, "Inelastic displacement ratio for evaluation of existing structures," *Earthquake Engineering and Structural Dynamics*, 32(8), 1237-1258, 2003.

Ruiz-Garcia, J. and Miranda, E., 2004, "Inelastic displacement ratios for structures built on soft soil sites", *Journal of Structural Engineering*, Vol. 130, No. 12, December 2004, pp. 2051-2061

Ruiz-Garcia, J. and Miranda, E., 2005, *Performance-Based Assessment of Existing Structures Accounting for Residual Displacements*, Report No. 153, John A. Blume Earthquake Engineering Center, Department of Civil and Environmental Engineering, Stanford University, Stanford, California, 444 pages.

Ruiz-Garcia, J. and Miranda, E., 2006a, "Residual displacement ratios for the evaluation of existing structures," *Earthquake Engineering and Structural Dynamics*, Vol. 35, pp. 315-336, 2006.

Ruiz-Garcia, J. and Miranda, E., 2006b, "Inelastic displacement ratios for evaluation of structures built on soft soil sites," *Earthquake Engineering and Structural Dynamics*, Vol. 35(6), pp.679–94.

Sezen, H., 2002, *Seismic response and modeling of lightly reinforced concrete building columns*, Ph.D. Dissertation, U.C. Berkeley, CA

Shi, S. and Foutch, D.A., 1997, *Evaluation of Connection Fracture and Hysteresis Type on the Seismic Response of Steel Buildings*, Report No. 617, Civil Engineering Studies, Structural Research Series, University of Illinois at Urbana-Champaign, Urbana, Illinois.

Shibata, M., 1982, "Analysis of elastic-plastic behavior of a steel brace subjected to repeated axial force," *Int. J. Solids Struct.*, 18, 217–228.

Shing, P.B., Noland, J.L., Spaeh, H.P., Klamerus, E.W., and Schuller. M.P., 1991, "Response of Single-Story Reinforced Masonry Shear Walls to In-Plane Lateral Loads," *TCCMaR Report 3.1(a)-2*, U.S.-Japan

Coordinated Program for Masonry Building Research, University of Colorado, Boulder, Colorado.

Sivaselvan, M.V. and Reinhorn, A.M., 1999, *Hysteretic Models for Cyclic Behavior of Deteriorating Inelastic Structures*, Tech. Rep. MCEER-99-0018, Multidisciplinary Ctr. for Earthquake Engrg. Res., State University of New York at Buffalo, Buffalo, New York

Sivaselvan, M. and Reinhorn, A., 2000, "Hysteretic models for deteriorating inelastic structures," *Journal of Engineering Mechanics*, Vol. 126. No. 6., pp. 633-640.

Song J.-K., and Pincheira, J.A., 2000, "Seismic analysis of older reinforced concrete columns," *Earthquake Spectra*, 16(4), 817–851.

Takeda, T., Sozen, M.A. and Nielsen, N.N., 1970, "Reinforced concrete response to simulated earthquakes," *Journal of the Structure Division*, ASCE, ST12, pp. 2557-2573.

Takemura, H. and Kawashima, K., 1997, "Effect of loading hysteresis on ductility capacity of reinforced concrete bridge piers," *Journal of Structural Engineering*, Japan, Vol. 43A, pp. 849-858, (in Japanese).

Tang, X. and Goel, S.C., 1987, *Seismic Analysis and Design Considerations of Braced Steel Structures*, Research Report UMCE 87-4, Department of Civil Engineering, University of Michigan, Ann Arbor, Michigan.

Thyagarajan, R.S., 1989, *Modeling and Analysis of Hysteretic Structural Behavior*, Rep. No. EERL 89-03, California Institute of Technology, Pasadena, California.

Thyagarajan, R.S., and Iwan, W.D., 1990, "Performance characteristics of a widely used hysteretic model in structural dynamics." *Proc., 4th U.S. Nat. Conf. on Earthquake Engrg.*, Vol. 2, 177–186.

Tremblay, R., Tchebotarev, N. and Filiatrault, A., 1997, "Seismic performance of RBS connections for steel moment-resisting frames: influence of loading rate and floor slab", *Behavior of Steel Structures in Seismic Areas, Proceedings of the Second International Conference*, STESSA, Kyoto, August 3-8, 1997, p. 664-671.

Tremblay, R., 2002, "Inelastic seismic response of steel bracing members", *Journal of Constructional Steel Research*, 58, pp. 665-701.

Tremblay, R., Archambault, M.-H., and Filiatrault, A., 2003, "Seismic response of concentrically braced steel frames made with rectangular hollow bracing members," *Journal of Structural Engineering,* ASCE, 129(12), pp. 1626-1636.

Uang, C.M., Kent, Yu, K., and Gilton, C., 2000, *Cyclic Response of RBS Moment Connections: Loading Sequence and Lateral Bracing Effects,* Report No. SSRP 99/13, University of California, San Diego, California.

Uang, C.M. and Gatto, K., 2003, "Effects of finish materials and dynamic loading on the cyclic response of woodframe shearwalls," *Journal of Structural Engineering,* 129(10), 1394-1402

Uriz, P. and Mahin, S., 2004, "Seismic performance of concentrically braced steel frame buildings," *Proc. 13th World Conf. on Earthquake Engineering,* Vancouver, Canada.

Uriz, P., 2005, *Towards Earthquake Resistant Design of Concentrically Braced Steel Structures,* Ph.D. Thesis, University of California at Berkeley, Berkeley, California, 467 p.

Valles, R.E., Reinhorn, A.M., Kunnath, S.K., Li, C., and Madan, A., 1996, *IDARC 2D version 4.0: A program for the inelastic damage analysis of buildings,* Tech. Rep. NCEER-96-0010, Nat. Ctr. For Earthquake Engrg. Res., State University of New York at Buffalo, Buffalo, New York.

Vamvatsikos, D. and Cornell, C.A., 2002, "Incremental Dynamic Analysis," *Earthquake Engineering and Structural Dynamics,* Vol. 31, Issue 3, pp. 491-514.

Vamvatsikos, D., and Cornell, C.A., 2005, *Seismic Performance, Capacity and Reliability of Structures as seen through Incremental Dynamic Analysis,* Report No. 151, John A. Blume Earthquake Engineering Research Center, Department of Civil and Environmental Engineering, Stanford University, Stanford, California.

Vamvatsikos, D. and Cornell, C.A., 2006, "Direct estimation of the seismic demand and capacity of oscillators with multi-linear static pushovers through Incremental Dynamic Analysis", *Earthquake Engineering and Structural Dynamics,* 35(9): 1097-1117.

Veletsos, A.S. and Newmark, N.M., 1960, "Effects of inelastic behavior on the response of simple systems to earthquake ground motions," *Proc. Second World Conf. on Earthquake Engineering,* Vol. II, Tokyo and Kyoto, Japan, pp 895-912.

Venti, M. and Engelhardt, M.D., 1999, *Brief Report of Steel Moment Connection Test, Specimen DBBW (Dog Bone, Bolted Web),* SAC

Joint Venture - Steel Project, Internal SAC Phase 2 Background Report.

Vian, D. and Bruneau, M., 2001, *Experimental Investigation of P-Δ Effects to Collapse During Earthquakes*, Report MCEER-01-0001, Multidisciplinary Research for Earthquake Engineering Research Center, University at Buffalo, New York.

Vian, D. and Bruneau, M., 2003, "Tests to structural collapse of single degree of freedom frames subjected to earthquake excitations." *Journal of Structural Engineering*, ASCE, 129(12), 1676-1685.

Wen, Y.K., 1976, "Method for random vibration of hysteretic systems," *J. Engrg. Mech. Div.*, ASCE, 102(2), 249–263.

Wen, Y.K., 1989, "Methods of random vibration for inelastic structures," *Appl. Mech. Rev.*, 42(2), 39–52.

Zayas, V.A., Popov, E.P., and Mahin, S.A., 1980, *Cyclic Inelastic Buckling of Tubular Steel Braces*, Rep. No. UCB/EERC-80/16, Earthquake Engineering Research Center, University of California, Berkeley, California.

Project Participants

ATC Management and Oversight

Christopher Rojahn
Project Executive Director
Applied Technology Council
201 Redwood Shores Parkway, Suite 240
Redwood City, California 94065

Jon A. Heintz
Project Quality Control Monitor
Applied Technology Council
201 Redwood Shores Parkway, Suite 240
Redwood City, California 94065

William T. Holmes
Project Technical Monitor
Rutherford & Chekene
55 Second Street, Suite 600
San Francisco, California 94105

Federal Emergency Management Agency

Michael Mahoney
Project Officer
Federal Emergency Management Agency
500 C Street, SW
Washington, DC 20472

Mai (Mike) Tong
Project Monitor
Federal Emergency Management Agency
500 C Street, SW
Washington, DC 20472

Robert D. Hanson
FEMA Technical Monitor
Federal Emergency Management Agency
2926 Saklan Indian Drive
Walnut Creek, California 94595

Project Management Committee

Craig D. Comartin
Project Technical Director
Comartin Engineers
7683 Andrea Avenue
Stockton, California 95207

Eduardo Miranda
Senior Advisor on Strength Degradation
Stanford University
Civil & Environmental Engineering
Terman Room 293
Stanford, California 94305

Michael Valley
Senior Advisor for Structural Engineering
Magnusson Klemencic Associates
1301 Fifth Avenue, Suite 3200
Seattle, Washington 98101

Working Group

Dimitrios Vamvatsikos
University of Cyprus
75 Kallipoleos Street
P.O. Box 20537
Nicosia, 1678, Cyprus

Project Review Panel

Kenneth Elwood
University of British Columbia
Dept. of Civil Engineering
6250 Applied Science Lane, Room 2010
Vancouver, British Columbia V6T 1Z4 Canada

Subhash C. Goel
University of Michigan
Dept. of Civil and Envir. Engineering
2350 Hayward, 2340 G.G. Brown Building
Ann Arbor, Michigan 48109-2125

Farzad Naeim
John A. Martin & Associates, Inc.
1212 S. Flower Street, 4th Floor
Los Angeles, California 90015

Workshop Participants

Mark Aschheim
Santa Clara University
500 El Camino Real
Dept. of Civil Engineering
Santa Clara, California 95053

Michael Cochran
Weidlinger Associates
4551 Glencoe Avenue, Suite 350
Marina del Rey, California 90292-7927

Craig D. Comartin
Comartin Engineers
7683 Andrea Avenue
Stockton, California 95207

Anthony Court
A. B. Court & Associates
4340 Hawk Street
San Diego, California 92103

Kenneth Elwood
University of British Columbia
Department of Civil Engineering
6250 Applied Science Lane, Room 2010
Vancouver, British Columbia V6T 1Z4 Canada

Subhash C. Goel
University of Michigan
Department of Civil and Envir. Engineering
2350 Hayward, 2340 G.G. Brown Building
Ann Arbor, Michigan 48109

Robert D. Hanson
Federal Emergency Management Agency
2926 Saklan Indian Drive
Walnut Creek, California 94595

Curt Haselton
California State University, Chico
Department of Civil Engineering
Langdon 209F
Chico, California 95929

Jon A. Heintz
Applied Technology Council
201 Redwood Shores Pkwy., Suite 240
Redwood City, California 94065

YeongAe Heo
University of California at Davis
Dept. of Civil Engineering
Davis, California 95616

William T. Holmes
Rutherford & Chekene
55 Second Street, Suite 600
San Francisco, California 94105

Sashi Kunnath
University of California, Davis
Dept. of Civil & Env. Engineering
One Shields Ave., 2001 Engr III
Davis, California 95616

Joseph Maffei
Rutherford & Chekene
55 Second Street, Suite 600
San Francisco, California 94105

Stephen Mahin
University of California, Berkeley
777 Davis Hall, Dept. of Civil Engineering
Berkeley, California 94720

Michael Mahoney
Federal Emergency Management Agency
500 C Street, SW
Washington, DC 20472

Michael Mehrain
URS Corporation
915 Wilshire Blvd., Suite 700
Los Angeles, California 90017

Eduardo Miranda
Stanford University
Civil & Environmental Engineering
Terman Room 293
Stanford, California 94305

Mark Moore
ZFA Consulting
55 Second Street, Suite 600
San Francisco, California 94105

Charles Roeder
University of Washington
Department of Civil Engineering
233-B More Hall Box 2700
Seattle, Washington 98195

Mark Sinclair
Degenkolb Engineers
225 Bush Street, Suite 1000
San Francisco, California 94104

Peter Somers
Magnusson Klemencic Associates
1301 Fifth Avenue, Suite 3200
Seattle, Washington 98101

Mai (Mike) Tong
Federal Emergency Management Agency
500 C Street, SW
Washington, DC 20472

Luis Toranzo
KPFF Consulting Engineers
6080 Center Drive, Suite 300
Los Angeles, California 90045

Michael Willford
ARUP
901 Market Street, Suite 260
San Francisco, California 94103